U. Karrenberg

An Interactive Multimedia Introduction to Signal Processing

With 240 Figures

Springer

Dipl.-Ing. Ulrich Karrenberg
Mintarder Weg 90

40472 Düsseldorf
Germany

E-mail: ulrich.karrenberg@t-online.de

ISBN 3-540-43509-3 Springer-Verlag Berlin Heidelberg New York

CIP data applied for

Die Deutsche Bibliothek - CIP-Einheitsaufnahme
Karrenberg, Ulrich: An interactive multimedia introduction to signal processing / Ulrich Karrenberg. - Berlin ; Heidelberg ;
New York ; Barcelona ; Hong Kong ; London ; Milan ; Paris ; Tokyo : Springer, 2002
ISBN 3-540-43509-3

This work is subject to copyright. All rights are reserved, whether the whole or part of the material is concerned, specifically the rights of translation, reprinting, reuse of illustrations, recitation, broadcasting, reproduction on microfilm or in other ways, and storage in data banks. Duplication of this publication or parts thereof is permitted only under the provisions of the German Copyright Law of September 9, 1965, in its current version, and permission for use must always be obtained from Springer-Verlag. Violations are liable for prosecution act under German Copyright Law.

Springer-Verlag Berlin Heidelberg New York
a member of BertelsmannSpringer Science+Business Media GmbH

http://www.springer.de

© Springer-Verlag Berlin Heidelberg 2002
Printed in Germany

DASY*Lab*® is a Trademark of NATIONAL INSTRUMENTS SERVICES GmbH & Co. KG
Copyright © 1993 - 2002 by National Instruments Services GmbH. All rights reserved.
National Instruments Services GmbH & Co. KG, Postfach 401264, D-41182 Mönchengladbach, Germany

The use of general descriptive names, registered names, trademarks, etc. in this publication does not imply, even in the absence of a specific statement, that such names are exempt from the relevant protective laws and regulations and therefore free for general use.

Multiple licenses for schools, university faculties, and institutions of further and higher education of industry and the public authorities can be acquired via the author. The author acts in the name and on behalf of National Instruments Services with regard to the DASYLab S program.

Typesetting: Dataconversion by author
Cover-design: design & production, Heidelberg
Printed on acid-free paper   SPIN: 10876144   62 / 3020 hu - 5 4 3 2 1 0 -

An Interactive Multimedia Introduction to Signal Processing

PLEASE ASK AT ISSUE DESK
FOR ACCOMPANYING CD ROM

Springer

*Berlin
Heidelberg
New York
Barcelona
Hongkong
London
Mailand
Paris
Tokio*

This book is dedicated to Claude E. Shannon, a pioneer of modern communications technology. He died on 25 February 2001. Only few people know in depth the fundamental content of his 55-page book "Mathematical Theory of Communication" published in 1948. This, however, does not diminish his genius and the uniqueness of his findings. They have changed the world more than any other discovery because communication is now the keyword in our society and in life itself.

His work will be accomplished once his theory has been integrated into modern physics thus leading to a greater understanding of the central principles of nature. This is a task which has not yet been fulfilled.

- *If you want to build a ship, don't round up people to procure wood, don't allocate the different tasks and jobs, but arouse their longing for the open sea which stretches to infinity!*
  (Antoine de Saint-Exupery)

- *The success of language in conveying information ist vastly overrated, especially in learned circles. Not only is language highly elleptical, but also nothing can supply the defect of first-hand experience of types cognate to the things explicitly mentioned. ..... First-hand knowledge is the ultimate basis of intellectual life. ..... The second-handedness of learned world is the secret of its mediocrity. It is tame because it never has been scared by facts.*
  (Alfred North Whitehead)

- *People ought to be ashamed who take the miracles of science and technology for granted without understanding more about them than a cow does about the botanical principles behind the plants it happily munches.*
  (Albert Einstein at the Berlin Funkausstellung - telecommunications exhibition - in 1930)

- *Real problems ignore the fact that education has been arbitrarily divided into different school subjects.*
  (the author)

- *The purpose of computing is insight, not numbers!*
  (R. W. Hamming)

- *Information and uncertainty find themselves to be partners.*
  (Warren Weaver)

Table of contents

Introduction 1
    A science that people can understand 1
    Target groups 2
    Graphic programming 3
    The electronic document 4
    The Camtasia video player 10

Chapter 1
The concept: methods - content - objectives 11
    Everything under one roof 12
    Hardware: Systems on a chip 12
    The software is the instrument 13
    A case for up-to-date education 14
    On the unity of theory and practice 14
    Multimedia and interactive learning 14
    Science and Mathematics 15
    In search of other "tools" 17
    Physics as the point of departure 22
    Clarification of Objectives 24
    Preliminary conclusions: the concept takes on clearer contours 28
    Exercises on Chapter 1 : 30

Chapter 2
Signals in the time and frequency domain 33
    The FOURIER Principle 33
    Periodic oscillations 34
    Our ear as a FOURIER-analyzer 35
    FOURIER -Transformation: from the time domain to the frequency domain and back 42
    Important periodic oscillations/signals 47
    Comparison of signals in the time and frequency domain 48
    The confusing phase spectrum 50
    Interference: nothing to be seen although everything is there 50
    Opposites which have a great deal in common: sine and d-pulse ... 52
    Non-periodic and one-off signals 56
    Pure randomness: stochastic noise 57
    Noise and information 58
    Exercises for Chapter 2: 61

Chapter 3
The Uncertainty Principle 65
    A strange relationship between frequency and time
    and its practical consequences 65
    Sinusoidal signal and d-pulse as a limiting case of the Uncertainty Principle 69
    Why ideal filters cannot exist 70
    Frequency measurements in the case of non-periodic signals 74
    Near-periodic signals 80

    Tones, sounds and music 81
    Exercises on Chapter 3 86

Chapter 4
Language as a carrier of information 89
    How speech, tones and sounds are generated and perceived 97
    Case study: a simple system for voice recognition..................................... 105
    Refinement and optimisation phase.. 110
    Pattern recognition.. 113
    Exercises on Chapter 4 115

Chapter 5
The Symmetry Principle 117
    For reasons of symmetry: negative frequencies 117
    Proof of the physical existence of negative frequencies............................. 117
    Periodic spectra... 125
    Inverse FOURIER transformation and GAUSSian plane 128
    Exercises on Chapter 5 140

Chapter 6
System analysis 141
    Sweep... 143
    Modern test signals... 148
    The d-pulse 149
    The step function 153
    The GAUSSian pulse.. 159
    The GAUSSian oscillation pulse 160
    The Burst signal 161
    The Si-function and the Si-oscillation pulse... 162
    Noise 164
    Transients in systems... 168
    Exercises on Chapter 6 173

Chapter 7
Linear and non-linear processes 175
    System analysis and system synthesis... 175
    Measuring a process to reveal whether it is linear or non-linear............... 175
    Line and space 176
    Inter-disciplinary significance 176
    Mirroring and projection 177
    A complex component: the transistor... 179
    There are only few linear processes 179
    Multiplication of a signal by a constant.. 180
    Addition of two or more signals 181
    Delay.. 181
    Differentiation 183
    Integration 190
    Malicious functions or signal curves 197
    Filters 199
    Non-linear processes.. 204

    Multiplication of two signals 205
    Formation of the absolute value 209
    Quantization 211
    Windowing 214
    Exercises on Chapter 7 215

Chapter 8
Classical modulation procedures 217
    Transmission media 217
    Modulation with sinusoidal carriers 217
    Modulation and demodulation in the traditional sense 218
    Amplitude modulation and demodulation AM 219
    Wasting energy: double sideband AM with carrier 226
    Single sideband modulation without a carrier 227
    Frequency multiplex 235
    Mixing 238
    Frequency modulation FM 240
    Demodulation of FM-signals 250
    The phase locked loop PLL 250
    Phase modulation 256
    Immunity to interference of modulation processes 258
    Practical information theory 261
    Exercises on Chapter 8 262

Chapter 9
Digitalisation 265
    Digital technology does not always mean the same thing 265
    Digital processing of analog signals 265
    The gateway to the digital world: the A/D converter 267
    Principle of a D/A converter 269
    Analog pulse modulation processes 272
    DASYLab and digital signal processing 274
    Digital signals in the time and frequency domain 276
    The period length of digital signals 277
    The periodic spectrum of digital signals 286
    The Sampling Principle 288
    Retrieval of the analog signal 293
    Non-synchronicity 295
    Signal distortion as a result of signal windowing 298
    Check list 299
    Exercises on Chapter 9 301

Chapter 10
Digital filters 303
    Hardware versus software 303
    How analog filters work 303
    FFT filters 306
    Digital filtering in the time domain 311
    Convolution 315
    Case study: Design and application of digital filters 317

Avoiding ripple content in the conducting state region 322
Exercises on Chapter 10 327

Chapter 11
Digital transmission technology I: source encoding 329

Encoding and decoding of digital signals and data 331
Compression 331
Low-loss and lossy compression 333
RLE encoding 334
Huffman encoding 334
LZW encoding 335
Source encoding of audio signals 338
Delta encoding or delta modulation 338
Sigma-delta modulation or encoding (S-D-M) 343
Noise shaping and decimation filter 345
Exploiting psycho-acoustic effects (MPEG) 345
Encoding and physics 352
Exercises on Chapter 11 353

Chapter 12
Digital transmission technology II: channel encoding 355

Error protection encoding for the reduction of bit error probability 355
Distance 356
Hamming codes and Hamming distance 358
Convolutional encoding 360
Viterbi decoding 364
Hard and soft decision 366
Channel capacity 368
Exercises on Chapter 12 370

Chapter 13
Digital Transmission Techniques III: Modulation 371

Keying of discrete states 374
Amplitude Shift Keying (2-ASK) 374
Phase Shift Keying (2-PSK) 374
Frequency Shift Keying 2-FSK 376
Signal space 377
Quadrature Phase Shift Keying – QPSK 380
Digital Quadrature Amplitude Modulation (QAM) 383
Multiple Access 387
Discrete Multitone 390
Orthogonal Frequency division Multiplex (OFDM) 395
Coded OFDM (COFDM) and Digital Audio Broadcasting (DAB) 400
Global System for Mobile Communications (GSM) 402
Asymmetric Digital Subscriber Line (ADSL) 402
Spread Spectrum 405
Exercises on Chapter 13 409

Bibliography 411

Index 419

Introduction

Education, further education and training in the field of microelectronics/computer and communications technology are currently at the focus of public interest. Highly qualified experts are desperately needed in these areas which have an excellent future market potential. Future-oriented approaches for university and school courses and for home study and exploratory learning are, however, non-existent.

A science that people can understand

When students chose their subjects they generally shun those which involve the theory and the technical aspects of the field of signals - processes - systems. These subjects have a bad reputation because they are regarded as "difficult" and because there are many hurdles to overcome, both with regard to the preparation for a degree course and to the course itself.

Universities and industry have done very little so far to remove these obstacles, although the above-mentioned field is connected with the most important, high-turnover industry and service sector.

As a instructor for teachers, the author was shocked when he found that worldwide there was apparently no convincing basic didactic concept in the field of microelectronics, computer-, communications- and automation technology for those preparing to study these subjects or already attending a university course. He asked himself why university courses overemphasised the theoretical aspects, whereas job training was practice-orientated. He arrived at the conclusion that, particularly in the above- mentioned field, theory and practice must form an integrated whole.

The following anecdote highlights the theory/practice dilemma: 14 trainee teachers are taking part in a seminar on telecommunications/technical informatics. All of them are engineers with university diplomas, some of them even have practical job experience. One of them is faced with the problem of having to conduct a sequence of lessons on control engineering at a vocational school. Consequently, the agenda of the seminar is changed to "didactic reduction and elementarisation". All the trainee teachers attending took part in lectures, seminars and practicals on control engineering as a part of their degree course. When the trainer asks them if they remember a basic term or concept in control engineering they hesitate and eventually come up with *Laplace transformation*. When they are asked what the term is about they have to admit, rather sheepishly, that they can do calculations using the relevant formula, but do not really have any idea about the fundamental background.

It cannot be denied that topics that are taught at universities are largely taken in as mere facts and then applied without reflection and a deeper understanding of the subject matter. But there are alternative ways of teaching topics to improve the efficiency of the learning process and use the time involved economically.

In short: the learning system presented here uses various different methods aimed at

- providing access to the fascinating discipline of Signals - Processes - Systems even for those who have not had any previous scientific experience in this academic field.

- improving the symbiosis of theory and practice while taking a degree course

- facilitating the transition from university to job life for graduates to avoid a "practice-shock".

This book has an interdisciplinary approach and involves sciences which deal with communication in the widest sense if they are of didactic relevance in the achievement of an objective.

Teaching and learning are *communicative phenomena*. This book considers the findings of recent brain research on visual learning and the development of consciousness (by interaction with the outside world) and the findings of the psychology of learning. Over two hundred high-quality illustrations and designs for transparencies, simulations and experiments form the core of this learning system.

The subject matter is illustrated and backed up mainly with reference to physics. Electromagnetic oscillations, waves and quantums are information carriers; there is a physical exchange between transmitter and receiver. The technology involved here is defined quite simply as the sensible and responsible application of the laws of nature because in technology - including the field of Signals - Processes - Systems - nothing works unless it is in accord with these laws.

There is a wealth of specialist literature in all languages available using mathematical models to explain phenomena in signalling systems. This approach is also very popular with university lecturers. But instead of adding yet another publication based on this approach, the most important methodological measure in this book is to avoid using mathematical models. This methodological approach aims at removing obstacles for learners and facilitating access to this discipline.

This learning system thus complements in an ideal way the materials used by university lecturers. In addition, it caters for the vast number of people for whom access to this discipline has been very difficult.

Target groups

The above exposition has already given you an idea as to who the target groups of this book are:

- University lecturers

    - who want to use high-quality visual material, interactive simulations and graphic explanations of signalling processes for their lectures and seminars.
    - who want to visualise the role mathematics plays in signalling systems when they talk about mathematical models in their lectures.
    - who appreciate being able to design and undertake laboratory experiments and exercises almost free of charge, or have them designed by students at their PCs.

- Students of engineering sciences at technical colleges and universities such as microelectronics, technical informatics, control-, measuring- and automation technology, information- and communications technology etc. who feel that they have lost track of the basic content of their course in the "mathematical jungle" of lectures on systems theory.

- Students of other technical or scientific disciplines who have to deal with computer-assisted processing, analysis and representation of measuring data (signals) but who wish to avoid mathematical and programming barriers.

- Student teachers in the above disciplines whose problem consists in translating the "theory of signals - processes - systems which is mainly formulated in mathematical terms into language consonant with the imaginative potential of school children. (simplifying and presenting in an elementary way in accordance with educational method).

- Teachers of the above-mentioned disciplines at vocational schools and colleges who are looking for up-to-date approaches and teaching materials which they wish to use in their teaching.

- Engineers in a given profession, whose university training took place some time ago and who as a result of deficits in maths and information science (programming languages, algorithms) have not been able up to now to deal with modern aspects of computer-based signal processing.

- Skilled workers and technicians in the above disciplines and professions who would like to qualify further in their profession by home study.

- Physics teachers at the secondary level who would like to demonstrate the importance of their subject for understanding modern technologies using the example of the complex "signals - processes - systems" for example, in the framework of an advanced course in "Oscillations and Waves".

- Students in professions related to information technology or in the microelectronics - computer technology-communications technology profession who are undergoing training at vocational schools, vocational colleges and engineering polytechnics.

- Those who are interested in a popular presentation of science in order to obtain a lively overview of this highly topical field.

- Students who have not yet decided on a profession or course of study at a university and who would like to inform themselves about this discipline but who up to now had no access to this field as a result of the mathematical bias.

- Firms which work in the field of measurement, control and automation technology and who are interested in-house training and further education.

Graphic programming

The central idea of this system of learning is the implementation of a professional development environment for the graphic programming of signal processing systems. In accordance with this further obstacles have been removed along with algorithms and programming languages whereby it becomes possible to focus on signal processing itself.

DASY*Lab* working in the background makes real signal processing and simulation possible. The software provides an almost ideal and complete experimental laboratory with all imaginable "equipment" and measuring instruments. DASY*Lab* is distributed worldwide by National Instruments Services GmbH & Co KG (Moenchengladbach) - a subsidiary of National Instruments in Austin, Texas - in many countries and languages with great success and is used in the field of measurement and automatic control technology. Whereas the individual industrial licence is by no means affordable in the context of training or education programmes this educational system comes with a study version which has the same performance and which is supplied practically free of charge. It is very easy to operate and offers all the possibilities of developing, modifying, optimising, discarding and redesigning one's own systems or applications.

The electronic document

The CD contains the complete electronic document - plus multimedia and interactive features - including all the programs, videos, handbooks etc. The learning system SiProSys.pdf is identical with the book.

How to install the system

Boot your PC. After loading Windows insert the CD into the drive. After a brief moment you will see the following display on your screen:

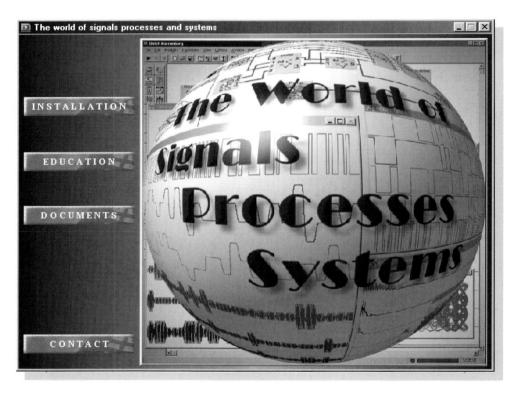

Illustration 1:   *The display after starting the CD*

In principle you simply need not to install any program on your PC! You only have to link *any one* of the DASY*Lab* files (*.dsb in the folder dasylab) on the CD with this program. Finally, click on to the button "Education". Everything else then runs directly off the CD.

However, you are recommended to install the complete system on your hard disk. Everything is then much faster and the CD-ROM-drive does not need to be booted each time.

Installation of DASY*Lab*

Press the button "Installation" and select DASY*Lab*. This starts the installation.

Installation of Acrobat Reader

This is a special version of Acrobat Reader with extended search function. If you have already installed a different Acrobat Reader you should first uninstall it. Installation now proceeds as usual.

Activating the *.dsb files

There are almost 200 files in the dasylab folder with the ending *.dsb. These files contain the signalling systems with which you will work interactively in the "Education".

You must now „tell" your PC with which program this file type is to be opened. Try opening any *.dsb file with a double click. Then the menu in Illustration 2 appears (in this case "abb013.dsb" was selected).

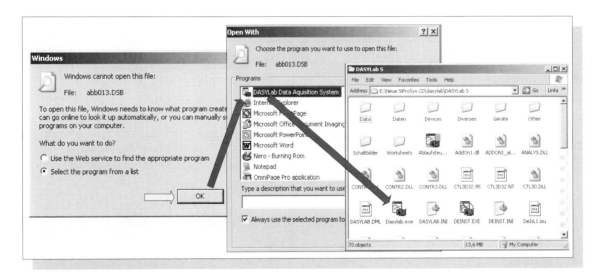

Illustration 2:   ***Linking the *.dsb files with Dasylab.exe***

You will not find at once the file **Dasylab.exe** *in the window. Therefore press the button "Select...." and find the file* **Dasylab.exe** *in the explorer via the folders „programs" and „dasylab". Click on once and then confirm in the window "open with". Then DASYLab will start and load this file.*

Installation of the "learning system" on the hard disc

Create a new folder on partition C (or a different partition) and call it for example SiProSys. Mark (as shown in Illustration 3) the four folders "dasylab", "documents", "index" and "video" and the two files SiProSys.pdf and index.pdx in the CD contents by keeping the control key (Ctrl) pressed and clicking onto the folders and files mentioned with the left mouse button. Copy the folders and files using Ctrl C (hold Ctrl and press the C button). Now open this empty folder SiProSys and insert all the folders and files using Ctrl V (hold the Ctrl and press the V button).

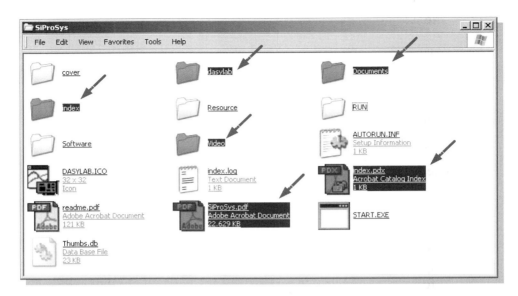

Illustration 3: **Installing the „learning sytem" on harddisk**

After installing DASYLab and the Acrobat Reader you should copy these 6 files in a new folder „SiProSys" on one partition of the Harddisk. So the the „Learning system" will run faster.

In order to be able to start the learning system easily, finally install a link of the file SiProSys.pdf on the desktop. You can now start the learning system directly from desktop with a double click.

Keyword search

The installed version of the Acrobat Reader has a very convenient search function for the electronic book. Illustration 4 shows how it works. The location where an entry was found - even in the illustrations - is displayed successively.

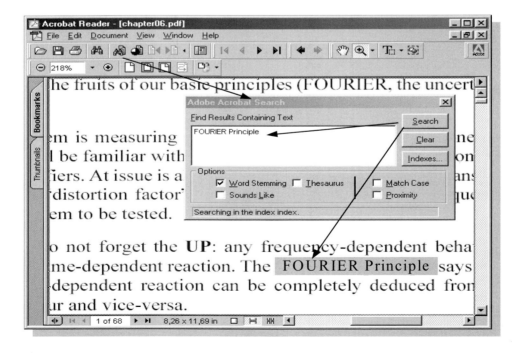

Illustration 4: **Keyword search with Acrobat Reader**

Installation of the software via My Computer or Explorer

Should for any reason the installation of the software (Acrobat Reader and DASY*Lab*) not be possible via the operating surface of the CD, it is always possible - just like with any other CD - to install the software via My Computer or Explorer. Open the folder Software (see Illustration 3). A double click on rp505enu.exe will start the installation of Acrobat Reader. In the folder DASYLab S you will find the Setup.exe file for DASY*Lab*. The installation will also start after a double click.

The interactive learning system

The file SiProSys.pdf is designed as an interactive medium. There are therefore active links which lead to other programmes or pages of the learning system.

Users are recommended to familiarise themselves with the use of Acrobat Reader. Many things can be grasped and carried out intuitively by experienced users. You will find the (official) manual for Acrobat Reader in the folder Documents should any problems arise.

The links in the pdf-documents are designed as junction surfaces. If they are not specially designated the cursor will inform you. The normal cursor symbol "hand" changes to "index finger" (see Illustration 4).

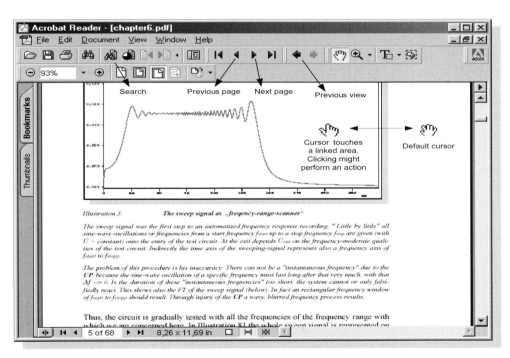

Illustration 5: ***Handling the interactive document***

There are three different links in the learning system:

- Each capital heading on the first page of the chapter is enclosed in a yellow frame. Click on to this surface to start a "screencam" video with an introduction to the relevant chapter. Switch on the loudspeaker of the PC.

- The major part of the roughly 250 Illustrations in the book are linked to the relevant DASY*Lab* applications which carry out the experiment corresponding to the Illustration. This can be modified or changed at will. The change can be stored and filed under a different name. To return to the pdf-document simply terminate DASY*Lab*. If DASY*Lab* remains active in the background no further DASY*Lab* link can be started.

- Pointers to other parts of the document - for instance from the table of contents or other Illustrations - are also linked. To return to the previous part of the document select the relevant arrow key above right in the menu (see Illustration 4).

Technical features and prerequisites

The learning system runs on all Windows operating systems from Windows 95. A PC is required which is suitable for multimedia applications with at least 64 MB main memory. A full-duplex sound board for the input and output of analog signals with the driver suitable for the operating system (for instance, Windows NT?!) is also essential. The stereo input and output of the sound board gives two analog input and output channels.

Digital signals can be inputted and outputted in multichannel mode via the parallel port. In the folder "documents" you will find the relevant manuals for operating DASY*Lab* and for the input and output of real signals via sound board and parallel port.

In networked PC systems the learning system should on principle be installed locally on each PC. So far there is no typical network version available. However, expert users say that it is possible to install the learning system - for instance, in a fast NOVELL network - purely on the server.

System development with block diagrams

DASY*Lab* has a range of modules (components) each of which embodies a signalling process. By synthesizing these modules to form a block diagram a signalling system is created.

These modules are to be found in the menu under generic terms (such as "data reduction"). You should only place the modules you are using in a module bar on the left of the display in order to have quicker access to them. The best thing is to create your own new module bar.

The video on Chapter 1 shows you how to work with DASY*Lab*. Help in DASY*Lab* and the manuals in the folder "Documents" give you detailed information.

The simplest way to learn how the individual modules interact is by using the many examples in the learning system and in DASY*Lab* (open Worksheet in the menue and than the folder "Examples". It's much better to try things out than to study the manuals.

All the examples in the learning system are optimised for the XGA-screen resolution (1024 * 768). Of course a higher resolution can be used as pictures cannot in principle be big enough. However, you will hardly find a beamer with a higher resolution for teaching purposes.

The possibilities of DASY*Lab* of measuring precisely visualised data and measurement results with the cursor are fascinating -

- Enlarging signal segments (zooming),

- Using "3D-representations" for the visualisation of large amounts of data ("waterfall" and "frequency-time landscapes", including colour sonograms),

- integrating these vector graphics directly into your documentation (likewise the block diagrams) by just pressing a button.

The simplest possibility of inputting and outputting real (analog) signals is provided by an inexpensive headset (headphones plus microphone). This "hardware" is above all practical when students are working together in groups.

Please remember that in this S-Version of DASY*Lab* the processing of real signals by means of the technical features of the sound card, the parallel port and - this is completely new - the serial interface is limited. For instance, slow fluctuating signals cannot be inputted via the sound card because it has coupling capacitors at its inputs. The manual, however, gives a circuit for the conversion of the instantaneous value of a (slowly changing) measurement voltage into an (instantaneous) frequency by using a simple voltage- frequency converter (VCO). This is similar to processes in telemetrics.

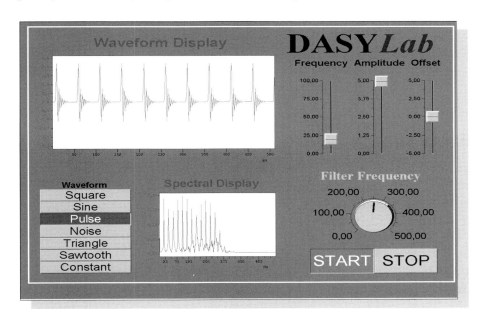

Illustration 6: **Layout Presentation**

For the presentation of visualised data and the easy handling of the created system the so-called **layout presentation** *is available. It is equivalent to what is usually to be found on the front panel of a measuring system which conceals the technology in the casing.*

As the aim of this learning system is precisely to convey these systems and the technical background it is not used in this book.

For professional applications - for instance in automation technology - you will require the industrial version of DASY*Lab* and special multifunction and interface boards together with DASY*Lab* drivers.

The S-Version has the same range of functions as the industrial version; for obvious reasons only the data format was changed and the possibilities of communicating with hardware peripherals reduced.

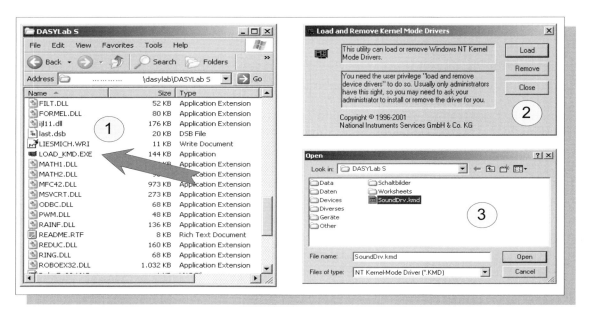

Illustration 7: **Using Windows NT, 2000, XP**

You want to use the "Learning system" from CD-ROM only without installing DASYLab on your harddisk? First open the folder **DASYLab S** *(inside the folder* **dasylab***). At second start* **LOAD_KMD.exe** *and at third open the file* **SoundDrv.kmd***. This is the precondition for sending or receiving real signals via the soundcard.*

The Camtasia video player

You can get an overview about the content of every chapter by starting the chapter-video. Each capital heading on the first page of the chapter is enclosed in a yellow frame. Now waite some seconds (!) and the Camtasia Player will start the video automatically.

Camtasia Player is a simple, standalone movie player, optimized for playing screen recordings. Camtasia Player always plays back movies at 100% of their original size so that they remain readable. It runs on any 32-bit Windows platform. The Player is a single and very small EXE file that does not need any setup, nor does it use the registry or any configuration files.

The videos have an XGA resolution, i.e. 1024 * 768 pixel size. If you have an XGA screen resolution, please choose „full screen" in the menue (see Illustration 8 at the bottom). Otherwise you will not see the whole picture.

Illustration 8: **Starting the Camtasia video**

Chapter 1

The concept: methods - content - objectives

Microelectronics is already the key industry (see Illustration 1). In the view of experts in this field it has changed our lives more than any other technology and will continue to do so. It is probably impossible for us to imagine the social, political and economic impact of this technology.

Professional mobility will probably be virtually synonymous with the qualified and responsible application of microelectronics in the widest sense. It is likely to influence education and science more profoundly and change them more rapidly than ever before.

Microelectronics has long been characterised by confusing variety and increasingly rapid innovation. The question of a concept for university degree courses, training, further education and advanced training in the field of microelectronics is becoming more and more urgent. In view of the tendency arising from this apparent complexity to regard the "engineer as the skilled worker of tomorrow" (VDI-Nachrichten - news bulletin of the German Association of Engineers) the question inevitably arises - What is to become of today's army of skilled workers?

Automobil and transport • Air traffic control • Traffic control systems • Automotive diagnosis systems • Anti-blocking system • Radar distance control • On-board computers • Traffic light control system • Motor control • Global positioning system	**Energy and environment** • Solar technology • Heat pumps • Lighting control • Heating control • Air conditioning • Air/water surveillance • Optimisation of combustion processes • Recycling • Weather forecast	**Offices and trade** • Word processing • Voice output • Voice recognition • Bar-code scanners • Character recognition • Copiers • Office computers • Printers • Optical recognition of patterns
Entertainment/leisure • Musical instruments • Games • Radio/hifi • Cameras • Camcorders • Personal computers • Digital video	**Applications** **of** **microelectronics**	**Industry** • Machine-tool control • Measuring instruments • Process controls • CAD/CAM/CIM • Robots • Safety appliances • Transport installations
Household/ private consumption • Cookers/stoves • Clocks and watches • Dishwashers • Washing machines • Pocket calculator • Alarm system • Systems for apportioning heating costs	**Communications** • Telephone systems • Data networks • Satellite communications • Broadband communications • Telemonitoring/locating • Encoding and decoding • Mobile telephony • Storage technology • Notebooks/PDA	**Medical applications** • Patient monitoring • Optical aids • Pace-makers • Laboratory instruments • Anaesthetic apparatus • Hearing aids • Prostheses • Sonography • NMR tomography

Illustration 9:   ***Key industry microelectronics***

(Source: ZVEI -Association of the German electronics industry)

At the present time there are already entire systems integrated on one chip which consists of millions of transistors. The question is what should be taught and how should it be presented to give both academic and non-academic young people access to this fascinating and irreversible technology which is vital for us all.

In the first place this is a question for the microelectronics speciality itself to answer. After all, a skilled field is only elevated to the rank of a specialist academic discipline if it is possible to demonstrate clarity and transparency (a structure) underlying the infinite variety by means of appropriate intellectual strategies - i.e. a well-thought out *concept*.

Everything under one roof

There are, therefore, good reasons for making this subject accessible to the greatest possible number of people. And there appears to be a master key which makes access easier. All the examples in Illustration 9 show that what is involved is *signal processing* - e.g. measuring, controlling, regulating. This is true even of (modern) washing machines which work through an electronically controlled program monitoring water-level and temperature. All that happens in a computer is the processing of signals, though in this context it is usually referred to as data processing.

> *The entire field of microelectronics does nothing but signal processing*

This key statement makes it possible to subsume practically all the fields of microelectronics under one roof. On the basis of this idea microelectronics can perhaps most easily be represented as a triad consisting of the three columns *Hardware, Software* and the *"theory of signals, processes and systems"*. Whereas present-day hardware and software will be obsolete in a relatively short space of time, this does not apply to the third column, the "*Theory of signals, processes and systems*". It is practically timeless as it is based on natural laws!

The three columns of microelectronics must be examined more closely under the aspect of future development.

Hardware: Systems on a chip

Today it is possible by means of the computer to carry out practically any kind of signal-processing, -analyzing or -visualising. The computer is a kind of universal hardware for signals or data. The hardware of a PC can already be fitted on to an area the size of a telephone chip card. As a result of the progress made in the field of large-scale integrated circuits - at present up to 600 million transistors can be integrated on a single chip - it is not unreasonable to predict that in the foreseeable future the complete PC hardware including memory, graphics, sound, interfaces, video codec etc will be integrated on a *single* chip.

For all the problems that cannot be solved by means of this universal PC-chip there will soon also be highly complex *freely programmable chips* both for analogue and digital circuits. They will be programmed on the computer; at the touch of a button the circuit design will be "burnt" into the chip. The competence required for this belongs, however, without doubt to the field of *signals - processes - systems*!

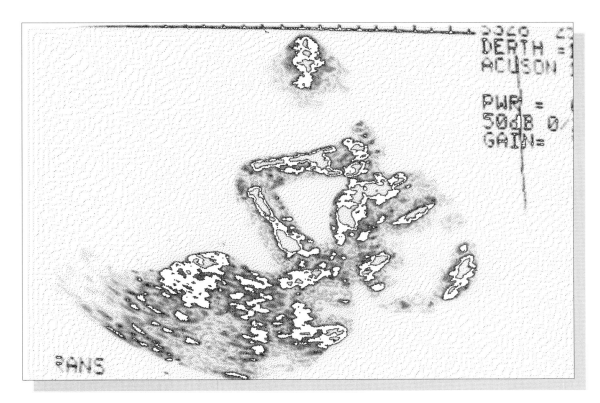

Illustration 10: ***Sonography***

This is also applied signal processing - a sonogram by means of ultra-sound sonography in the 22nd week of pregnancy. The unborn child is sucking its thumb (Source: Geo Wissen No., 2/91, p107). Are first-rate programming and hardware skills and mathematical competence sufficient to develop a scanning machine of this kind? In other words, what information will be essential to be able to develop such complex instruments in microelectronics and computer technology in the future?

Instead of the confusing variety of more that 100,000 different ICs (Integrated Circuits) the trend is clearly towards a "PC Standard IC" or freely programmable ICs. Application-specific integrated circuits (ASICs) optimised for function and cost will probably be used at best in mass production.

What are the consequences? The focus of micro-electronics will shift more and more towards software in future. *Algorithms instead of hardware*!

The software is the instrument

This motto of the USA-based company National Instruments thus seems to be justified. But what kind of software and software programming will prevail? This question is of particular importance in view of the *high degree of learning effectiveness and economic use of time* in the learning process at which we are aiming. It will therefore be dealt with in more detail at a later stage. Wouldn´t it be marvellous to create programmes in a fraction of the time previously needed without having to use cryptic progamming languages and algorithms involving higher mathematics? We are thinking of programs whose content and structure can be designed and understood by young students, skilled workers, technicians and engineers without taking a degree course in mathematics.

The PC is a programmable „calculator". The actual characteristics of the system result from the software. To a certain extent they are *virtual systems* as they are "intangible".

Where will the user get his software from in future? From the internet, of course. What does he need to select the appropriate programs for each specific form of signal or data processing? Above all competence in signal technology. At this point the third column, the *theory of Signal Processing Systems* becomes involved.

A case for up-to-date education

The signals - processes - systems triad is synonymous with *communication*, the key concept in our society, indeed of our very existence.

Communication is almost the definition of being alive. If the brain cells no longer communicate with each other the individual is clinically dead. At present something of a change in paradigm is taking place in medicine. Whereas it was hitherto regarded as an empirical science based on experience, molecular biology and genetics now offer powerful instruments with which to unravel and understand the causal chain leading to illness. Illness is seen more and more as as a disturbance in communication between cells! The resulting possibilities of organisation, prevention and treatment are not only challenging the limits of ethics and morality but also our ability to finance the social system.

Without communication there would be no peoples, states, industry, economies, schools and families. And what is necessary for a teacher's effectiveness - communication!

In short, in all the fields of everyday, professional and academic life innumerable questions arise which in the final analysis are inextricably linked with communication or can be answered with reference to it. There is not sufficient correlation of this in education and in many areas of science. At the beginning of the new millennium there is a discussion about which subjects are necessary to update the curriculum of secondary schools, that is to adapt it more effectively to the requirements of the real world. Possible subjects such as Economics and Technology are discussed and at the same time vehemently rejected. "Communications" is hardly ever even mentioned!

On the unity of theory and practice

What is needed are universal explanatory modes which are relevant for the intellectual understanding of numerous communicative phenomena from scientific theory and practice and which make it possible to create order, structure and transparency within a specialism. This is a first indication of what we mean by theory.

Teaching and learning denote one particular group of communicative phenomena, communications technology and the theory of signals-processes-systems another. The physiology of our bodies is a further group of communicative phenomena. Is it possible to see them as belonging under *one* roof?

Multimedia and interactive learning

Multimedia and interactive communication in the learning process simply means *communicating with all the senses*!

Up to now language has dominated educational and cultural communication. Apart from the fact that language is highly fragmented and redundant a physical bandwidth of 3 kHz is sufficient for the acoustic flow of information as in the case of the telephone.

In order to transmit a flow of pictures of the same quality which corresponds to the stereoscopic image of our eyes roughly 300 MHz are required. We thus absorb up to 100,000 times as much information with the eyes per unit of time as compared with language!

In addition, the sense of hearing was the last sense to be acquired in the course of evolution. We may therefore suspect that our brain is mainly structured for the processing of information in images. *Anyone reading a novel creates his own film at the same time.*

All the findings of modern brain research - see BURDA ACADEMY OF THE THIRD MILLENNIUM („Envisioning Knowledge") - point to a paradigm shift in thinking - from the teaching of knowledge dominated at present by writing and language to an image-based processing of knowledge ("pictorial turn").

Imagine someone wants to learn how to fly a plane. He buys the appropriate book. After studying it carefully he welcomes passengers on the gangway of the plane with the words: "Do not be afraid, I have studied the book very carefully". Would you get on to the plane? This describes roughly the traditional situation at schools and universities (lectures). In contrast, the flight simulator represents an educational quantum leap in that it addresses all the senses and makes independent exploratory learning possible without students having to bear the consequences of their mistakes. Roughly 70% of the training of pilots now takes place in the flight simulator.

Many teachers of all kinds still forget that computer-based technologies like these for creating models, simulation and visualisation for instance in the development and testing of new technical systems - chips, cars, planes, ships - save billions of dollars. They don´t accept that it is possible to learn, plan and develop far more effectively and with a far more economical use of time (say, by the factor 5). If nothing is done the new media may be deprived of this recognition as a result of the traditional attitudes of the universities and schools.

The *system of learning* presented here attempts as far as possible to implement these insights .

Science and Mathematics

These remarks are intended to achieve the following:

- To explain to the university lecturer why he will not find mathematics in this book apart from the four basic arithmetic operations.

- To explain clearly to students both at school and university the value and power of mathematics so that it may be seen more as an opportunity and inspiration than as a source of torture.

- To explain briefly for everybody else what exactly mathematics does and why it is indispensable for the (natural) sciences.

Nothing seems possible in the exact sciences without mathematics. Any student of electrical engineering, information technology or physics knows this to his cost. They spend a considerable part of their degree course on "pure and applied mathematics". For the practically oriented skilled worker and technician mathematics represents an insuperable barrier. Only too often the result of this over-insistence on the scientific nature of the subject is:

> *The important essentials are lost in the tangled undergrowth of a mathematical jungle.*

On the other hand for many people - including the present writer - mathematics represents the greatest intellectual achievement ever attained by man. But what role does it play in concrete terms? A brief answer to this complex of questions can only be incomplete. However, the following propositions may help to clarify the issue in general terms:

- In the first instance, mathematics is simply a powerful intellectual tool. On closer examination it represents the only method - in contrast to language - with which to guarantee *precision*, *lack of ambiguity*, *simplification, communicability, verifiability, predictability* and *lack of redundancy* in the description of relationships.

- It also represents the most brilliant method of presenting information in a concentrated form. Thus in physics a single mathematical equation - the famous *Schroedinger equation* - to a large extent describes the cosmos, i.e. the world of atoms, chemical bonds, solid state physics etc.

- All the conclusions arrived at by mathematics are also correct on a common sense level as all the premises and steps have been proven correct.

Speaking somewhat metaphorically, mathematics offers the possibility of moving from the furrowed plain of inexact linguistic concepts, the confusing number of external influences in real processes, contradictions, redundancies, and rising to a virtual plane in which all possible coherent and logical mental strategies can be tried out with any number of variables and initial and peripheral values. The reference to reality can be created by allocating meaningful and experimentally verifiable quantities to the mathematical variables and constants which in the case of information technology are of a physical nature.

It is obvious that scientists prefer this "paradise" for the application of new strategies of thinking. However, this "paradise" is only accessible by means of a science degree. On the other hand, mathematics pervades our everyday life. We calculate the paths of the stars and satellites, the stability of buildings and airplanes, simulate the behaviour of dynamic processes and use mathematics to make predictions. Even the worlds of computer games consist in the final analysis of *nothing but mathematics*.

For these reasons mathematics is the most common method of creating models for the theory of signals, processes and systems at university. Hundreds of books provide well-substantiated accounts of this kind and every university lecturer uses them to design his own course of lectures. There is not much point in writing yet another book of this kind. The study system presented here makes it possible to demonstrate directly what lies behind the mathematics and how it reacts to real signals - an ideal addition to a traditional course of lectures. Let us try to convince you!

> ## Mathematical model of the electrical conductor resistance R
>
> $$R = \rho \times \left(\frac{l}{A} \right)$$
>
> l: = length of conductor (in m); A := cross section (in mm²); (l: = length of conductor (in m);
> (ρ := specific resistance of the material (in $\Omega mm^2/m$).
>
> Note: the model is limited to three variables (ρ, l, A). Thus the minimal influence of the temperature of the conductor on electrical resistance R is left out of account. The model is valid for 20°.
>
> The model guarantees ...
>
> - **Precision**: any combination of specific values for (ρ, l and A produces an exact result which only depends on the accuracy of measurement of the quantities given.
>
> - **Lack of ambiguity**: any combination of numbers of (ρ, l and A describes a discrete physical situation and produces a single result for R.
>
> - **Simplification**: the mathematical model describes the method for calculation of the electrical resistance of the conductor for an infinite variety of different materials, lengths and cross-sections.
>
> - **Communicability**: the mathematical model is valid irrespective of language or other barriers, i.e. worldwide
>
> - **Verifiability**: the mathematical model is experimentally verifiable. Innumerable measurements with the most varied of materials, lengths and cross-sections confirm without exception the validity of the model.
>
> - **Predictability**: the conductor resistance for a given material, length and cross-section can be predicted. In the case of a short-circuit it is possible by measuring the conductor resistance in a cable to predict where the defect is located provided the material and the cross-section are known and the conductor has a homogeneous structure.
>
> - **Lack of redundancy**: the mathematical model does not contain any "padding", it produces pure information.
>
> Note: mathematical models in physics are not provable in the sense of strict mathematical logic. Here the tenet -"the experiment (meaning experimental verification) - is the sole arbiter of scientific truth" - is applicable.

*Illustration 11:   **Features of mathematical models demonstrated by means of a simple example.***

In search of other "tools"

The "theory of signals, processes and systems" is understood as the creation of mathematical models of signal technology processes on the basis of physical oscillation and wave phenomena and quantum physics (still not completely understood). This applies against the background that *nothing* works in technology that contravenes natural laws. All the framework conditions and explanatory models in technology must therefore inevitably derive from natural laws, and more specifically from physics.

> **Wiener-Khintchine Theorem**
> For a well behaved stationary random process the power spectrum is equal to the Fourier transform of the autocorrelation function.
>
> $$S_x(e^{j\omega}) = \sum_{k=-\infty}^{\infty} R_x(k)e^{-j\omega k}$$
>
> Sloppy proof:
>
> $$\begin{aligned}
S_x(e^{j\omega}) &= \lim_{N\to\infty} \frac{1}{2N+1} E[|X_N(e^{j\omega})|^2] \\
&= \lim_{N\to\infty} \frac{1}{2N+1} E\left[\left(\sum_{n=-N}^{N} x(n)e^{-j\omega n}\right)\left(\sum_{k=-N}^{N} x(k)e^{-j\omega k}\right)^*\right] \\
&= \lim_{N\to\infty} \frac{1}{2N+1} E\left[\sum_{n=-N}^{N}\sum_{k=-N}^{N} x(n)x(k)e^{-j\omega(n-k)}\right] \\
&= \lim_{N\to\infty} \frac{1}{2N+1} \sum_{n=-N}^{N}\sum_{k=-N}^{N} E[x(n)x(k)]e^{-j\omega(n-k)} \\
&= \lim_{N\to\infty} \frac{1}{2N+1} \sum_{n=-N}^{N}\sum_{k=-N}^{N} R_x(n-k)e^{-j\omega(n-k)} \\
&= \lim_{N\to\infty} \lim_{M\to\infty} \frac{1}{2N+1} \sum_{n=-N}^{N}\sum_{k=-M}^{M} R_x(n-k)e^{-j\omega(n-k)} \\
&= \lim_{N\to\infty} \frac{1}{2N+1} \sum_{n=-N}^{N} \lim_{M\to\infty} \sum_{k=-M}^{M} R_x(n-k)e^{-j\omega(n-k)} \\
&= \lim_{N\to\infty} \frac{1}{2N+1} \sum_{n=-N}^{N} \left(\sum_{k=-\infty}^{\infty} R_x(k)e^{-j\omega k}\right) \\
&= \left(\sum_{k=-\infty}^{\infty} R_x(k)e^{-j\omega k}\right) \lim_{N\to\infty} \frac{1}{2N+1} \sum_{n=-N}^{N} 1 \\
&= \sum_{k=-\infty}^{\infty} R_x(k)e^{-j\omega k}
\end{aligned}$$

Illustration 12:   ***Mathematics as a barrier***

A "sample" from a source on the signal processing. It contains an important statement on the relationship between the time and frequency domain (Wiener-Khintchine-theorem). Up to now attempts to demonstrate clearly such relationships without mathematics have not been successful. The formula shows the dominance of mathematics; physics have been almost totally submerged. Only terms such as FOURIER transform and power spectrum point to the physical basis. (Source: http://ece.www.purdue.edu/).

This approach is not recognised by all scientists who are working with the theory of signals, processing and systems. There are well-founded specialist books in which the words "physics" or "natural laws" do not even occur. Claude Shannon's Information theory - of fundamental importance for all modern Communications systems - is in addition presented as a purely mathematical theory based only on statistics and the calculus of probability and appears to have nothing to do with physics. This approach appears to derive from the fact that the concept of information has not yet been properly anchored in physics.

As for the reasons described above a "theory of Signal Processing Systems" eschewing paradisiacal (mathematical) possibilities should nevertheless be accessible and comprehensible, inevitably the question of other "tools" arises. Are words and texts adequate, for example? Does language have the possibilities for conveying information ascribed to it in literature?

In the present context there seems to be a better alternative. If clear and graphic mean that "image expresses more than a thousand words" this appears to mean for communications technology that images similar to mathematics are capable of expressing information in a condensed and relatively distinct way. Our brain appears generally - consciously or unconsciously - to transform texts into images. As already mentioned anybody reading a book makes his own "film" of it.

In order to avoid as much as possible this complicated "transformation process" the present writer sees it as an exciting challenge to base the whole complex of signals, processes and systems on images.

This is done in two ways:

(1) The DASY*Lab* professional development environment working in the background of the learning system permits the virtually unrestricted graphic programming of applications that can be run in measuring, automatic control engineering and automation technology.

(2) At every point in the system it is possible to visualise the signals. By comparing the imput signal and output signal of each module or system it is possible to analyse and understand the technical behaviour of a signal.

> *The graphic programming of virtual and real systems will be the trump card of the future both in science and scientific based teaching methods.*

It makes it possible, without a great deal of outlay in terms of time and money and in an environmentally friendly way, to generate, parameterise, simulate, optimise any communications system as a block diagram on the screen and bring about contact with the outside world in real terms via special hardware (such as a sound card). By linking up the building blocks (processes) on the screen to form a block diagram a virtual system is created in the "background" in which the appropriate communication algorithms for the entire process are linked together (with the addition of important sub-programs, for example for mathematics and graphics).

As a result, traditional programming languages and the entire mathematical instruments become less important, the barrier between the problem as such and the solution is drastically reduced. Programming languages and mathematics thus no longer distract attention from the real communication content and problems.

The intellectual gap between the "theoretical" engineer and the "practical" technician could be largely removed in this way. Factual knowledge which soon becomes obsolete - brought about at the present time by the enormous diversity of microelectronics - as an essential part of education and training programmes could be reduced to the bare minimum by use of progressive technology and methods. As a result the costs of education and training programmes could be reduced and job mobility in the sector could be dramatically increased.

> *The graphic programming of signal systems implies that theory and practice form a single unit.*

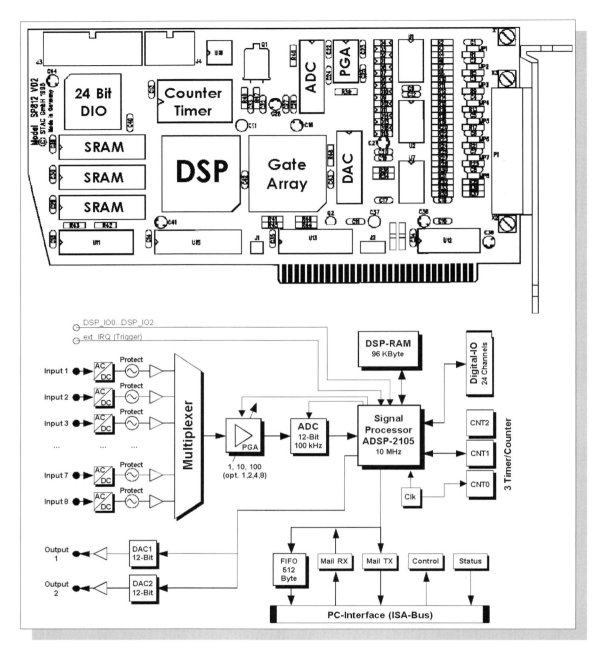

Illustration 13: **_Real picture of a signal system and H block diagram_**

This picture of a PC multi-function board with a signal processor (DSP) illustrates the dominance of digital technology. The only analog ICs are the programmable amplifier (PGA) and the multiplexer. But even here the setting is purely digital. This multi-function board can read and output both digital and analog signals, as it were linking the PC with the outside world. The signal can be processed extremely rapidly on the board by means of the signal processor. This reduces the workload of the actual PC. This real picture of the multi-function board only gives an impression of the structure of the hardware.

The hardware block diagram (H block diagram) below provides more information. At least it clearly shows the structure of the hardware components and the way they work together to receive or send out signals. Multi-function boards of this kind are structured in such a way that they have universal application within the framework of measuring and automatic control engineering. The disadvantage of H block diagrams is that the hardware structure is visible but not the signal processes which are carried out by the signal processor. In actual fact, these are all-important since the program decides the signal processing.

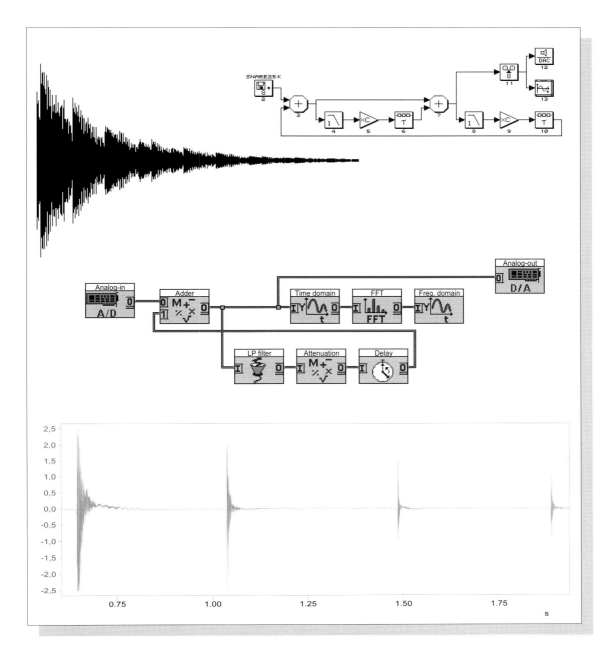

Illustration 14: ***S frame diagram and signal block in the time domain***

The above illustrates a combined echo-hall-system for the low frequency range. This block diagram was also compiled on the screen from standard components - called up from a library - and linked together. Certain parameters must be set by mouse-click for some of the components, for example the lowpass filter (4 and/or 8), the cut-off frequency and the filter order. The input signal goes directly to "oscilloscope" (13) (see screen below left) and to the PC speaker (12). Parallel to this the input signal is lowpass filtered (4), weakened (5) and finally with a time delay (6) is added to the input signal (7) ("echo"). The sum of both signals is fed back to the inlet, after the operations described above have been carried out (8), (9), (10), (3) ("reverb"). The program with the linked algorithms is running in the background and carries out the echo-hall effect on a real level.

An echo device produced by DASYLab is described below. The production of echoes is also a branch of the feedback phenomenon which contains the three "components" lowpass filtering, delay, damping (multiplication by a number smaller than 1).

Turn on the microphone and loudspeaker of your PC. Click on to the picture and the DASYLab experimental laboratory opens immediately. Start the experiment and go on experimenting to your heart's content.

Physics as the point of departure

The science-oriented description of the overall complex of "signal processing systems" implies compliance with the following criterion which is valid for the (exact) sciences:

> *A structure can be created on the basis of a small number of basic phenomena or axioms which enables an infinite number of "individual cases" to be categorised and explained. Put less abstractly this means that all the individual cases can be attributed to the same basic phenomena according to a unified scheme. Only in this way is it possible to retain an overview of a specialism in view of the present explosion of knowledge - the doubling of knowledge in intervals that are getting shorter and shorter.*

These basic phenomena must in this context be of a physical nature as physics, i.e. nature uniquely dictates the framework of what is technologically feasible. Physics is thus the point of departure, the framework and marginal conditions of the theory of signals, processes and systems. The primary explanatory models are to be found not in mathematics but in oscillation, wave and quantum physics. For that reason it *must* be possible to convey, explain and form a model of communication technology processes with reference to this field.

It is essential that there should be an alternative concept, largely free from mathematics, to describe the complex "signals - processes - systems" in a scientific way, i.e. on the basis of a small number of basic physical phenomena. Thus, for example, the "sonography" illustration shows a close proximity to oscillation and wave physics. It is above all important to understand in terms of content on the basis of a number of physical phenomena what effects the various signal processes have.

The point of departure are at this moment only three physical phenomena, whereby the second is strictly speaking only a consequence of the first:

- *FOURIER Principle (**FP**)*
- *Uncertainty Principle (**UP**)* and
- *Symmetry Principle (**SP**)*.

All the signals, processes and systems are to be explained by reference to these phenomena! This concept has never been used and comes into its own here as a central didactic principle for the first time.

Perhaps you will notice that the information contained in the signal is not included as a structuring element here. The reason - there are many theories in physics but up to now no "information theory". Scientists such as Dennis Gabor, R.V.L. Hartley, K. Küpfmüller etc. have tried in vain to put communications theory on a physical basis. The clear "winner" so far is the American mathematician Claude Shannon.

Simply imagine the scientific discipline and specialist field communications technology as a tree with roots, a trunk, branches and leaves. Each leaf on the tree is equivalent to one communication technology problem. The tree has three main roots - the three principles **FP**, **UP** and **SP**. Every single leaf, i.e. practically any communications technology problem can be reached via these roots, the trunk and the branches.

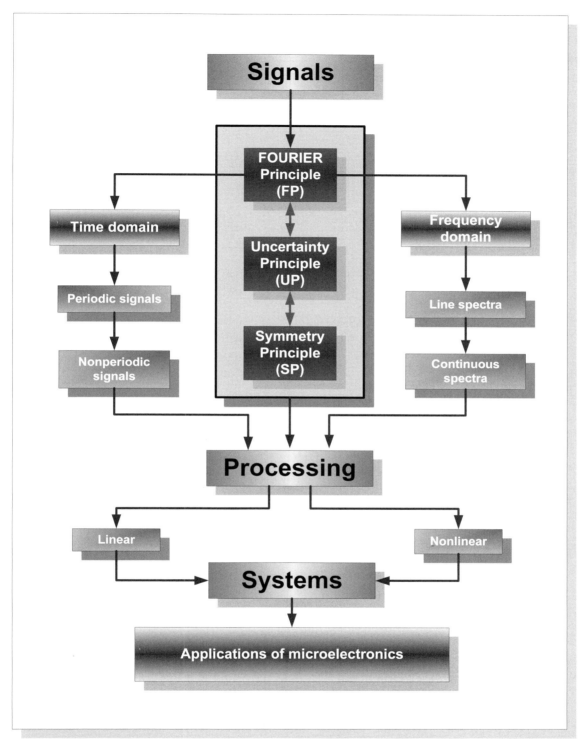

Illustration 15: **Structure for this concept of "Signals - Processing -Systems"**

In order to make the confusing variety of signal processes manageable and transparent we shall constantly refer back to the structure described here. It does not, however, make any claim to be complete and will be supplemented and extended where necessary.

*Perhaps you will notice that the **information** contained in the signal is not included as a structuring element here. The reason - there are many theories in physics but up to now no (complete) "information theory".*

Clarification of Objectives

Up until now, we have described the *methods* to be used in this introductory course. It is now time to focus on content. Here too, a kind of preview will show the reader what awaits him and enable him to get his bearings.

To use another image: imagine the content as a kind of platform which is supported and fenced in. The railing is described by the technical framework conditions that exist at present.

The basic phenomena and basic concepts and their definitions form the supports of the platform. Beyond the railing the horizon can be seen more or less clearly. It extends to the boundaries defined by natural laws - i.e. by physics.

The platform is roughly as follows:

- Signals can be defined as (physical) oscillations or waves which are carriers of information.

- Information exists in the form of agreed patterns which constitute meaning which the recipient must know and recognise. Data are informative patterns. The essential task of communications technology or signal technology is thus the recognition of patterns.

- All the measurable signals of the real physical world - even if they are bit patterns - are present in analog continuous form. They are usually transformed into electric signals by sensors (converters) and are present at the measuring points usually in the form of AC voltage.

- (Storage) oscilloscopes are instruments which can display the chronological course of AC voltages in the form of a graph on the screen.

- Every signal technology process (e.g. filtering, modulation and demodulation) can in general be described in mathematical terms (theory!) and subsequently be carried out by means of the appropriate algorithms or programs on a purely arithmetic level. Thus, from the theoretical point of view an analog system - created by the combination of several individual circuits/ processes/ algorithms - represents an analog computer which works through a sequence of algorithms in real time. This aspect is often overlooked. For example, this is also true of conventional radio receivers.

- *Real time processing* of (frequency band limited) signals is understood as the capacity to record the flow of desired information without gaps or to retrieve it from the signal.

- Analog components are above all resistors, spools, capacitors and, of course, diodes, transistors etc. Their fundamental disadvantages are inaccuracy (tolerance), noise features, lack of long-term consistency (aging), temperature dependence, non-linearity (where it is desired) and above all their combined behaviour. Thus, a real spool behaves like a combination of (ideal) inductance L and resistance R. A real resistance has the same equivalent circuit, when current flows through a magnetic field forms around it and therefore inductance exists in addition to the resistance. Every diode is not only a rectifier, it also distorts in a nonlinear fashion in an unwanted way. This combined behaviour is the reason why it is not possible to ascribe only one signal operation to a component; it always embodies several signal operations. As a circuit usually consists of many components, the real behaviour of the circuit may diverge considerably from the behaviour planned. Every conductor track of a motherboard also has an ohmic resistance R and an inductance L as a result of the magnetic field which arises when electricity flows through it. An electrical field exists between two parallel conductor tracks; they thus form a capacitance. These characteristics of the motherboard are rarely taken into account in the design of the circuit (except in the high and maximum frequency area).

- As a result of the influences or inteference effects listed here analogue technology has its limitations especially where the greatest precision in the execution of a desired behaviour (e.g. filtering) is required. This means that analog circuits cannot be produced in practice with the quality or characteristics which are theoretically possible.

- There is, however, a way to go beyond what is possible with analog technology and enter a field whose absolute limits are defined by physics: signal processing by means of digital computers. As their computational precision - unlike the analog computer and the circuits represented by it - can be increased ad infinitum, there is the possibility of carrying out signal operations with the desired degree of precision which previously could not be carried out.

- A communications system can also be represented by a program which links up a number of algorithms of signal operations. In conjunction with a computer or microprocessor system (hardware) this produces a system for the digital processing of signals. The program decides what the system can do.

- The limits of this technology with digital computers (microprocessors) in the area defined by physics are at present defined by the speed of computing and the problems that arise in the conversion and reconversion of signals (A/D and D/A conversion) which is necessary for computation. They are constantly displaced to the outside. The aim of this development is the real-time processing of signals, i.e the arithmetical processing of signals at such a high speed that no unwanted loss of information occurs.

- It is already clear: the measurement and control technology processes which are executed by means of microprocessor circuits are essentially superior to traditional analogue processes. Thus for example it is not possible with traditional technology to store data precisely for a longer period of time before they are re-used, even the precise short-term time delaying of analog signals causes considerable difficulties.

- Analog technology is being pushed more and more to the periphery of microelectronics, to the source and drain of the signal, and the actual transmission path which in retrospect can be seen as a "crutch" with which it was possible to carry out signal processes in a very inadequate and faulty fashion which was prone to interference. The trend is therefore to reduce the analogue part of a system to a minimum by means of an A/D converter to arrive at numbers as quickly as possible with which it is possible to compute (Sampel&Hold, quantization and encoding). The precision of A/D and D/A converters now depends practically only on factors such as the height and constancy of the quartz frequency reference and/or the constancy of a constant current source or reference voltage. Any modern digital multimeter is based on this technology.

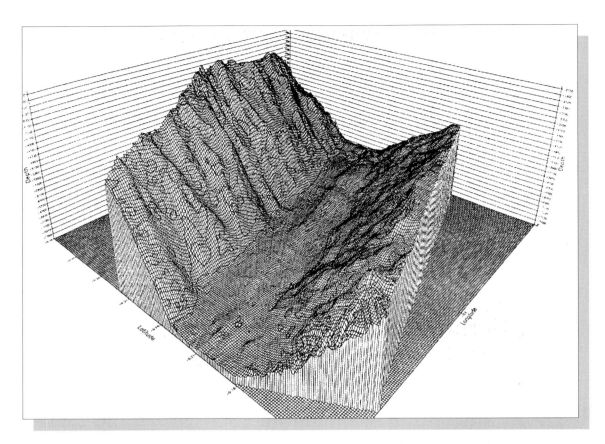

Illustration 16: **Data analysis for the graphic representation of measurement data**

Deep-sea echo sounding, for example, provides a very large amount of the most various measurement data, which after being interpreted physically (i.e. subjected to various mathematical and physical processes) must finally be available in an ordered presentation, here as the relief map of a deep-sea rift. The interpretation of measurement data of the most various kinds, their processing and presentation in a structured form is increasingly becoming an essential vocational qualification. (Source: Krupp-Atlas Elektronik)

- Communications technology is more and more synonymous with computer-aided signal processing. Digital signal processors which are optimised for such operations are used more and more and today make possible the real time processing of two- and multi-dimensional signals - i.e. pictures.

- As at present and even more so in the future a communications system can be represented by a program of linked algorithms. There is now the possibility of placing the individual components of a system as a "virtual" component on the screen by a mouse click and by joining up these components to form a block diagram and thus generating a virtual system (VS). At the same time the appropriate algorithms are linked in the background, i.e. by adding important sub-programs (for mathematics and graphics) they are linked together for the overall process.

- Virtual systems provide real results by means of hardware (computer, peripherals) and the programs which forms the virtual system. It is impossible to tell from the outside whether the signal processing was carried out on a purely hardware basis or by using a virtual system -i.e. a program.

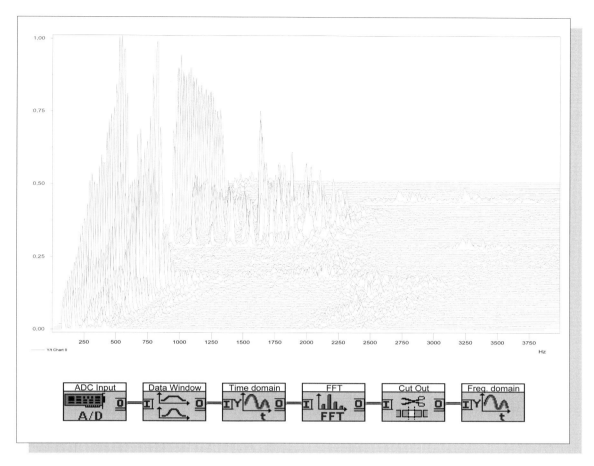

Illustration 17:   ***Frequency-time landscape of a sequence of tones***

Frequency-time-landscapes are used among other things for the analysis of voice or in the investigation of certain transient processes. The way in which a signal changes over time is represented. The frequency axis is horizontal and the time axis goes back obliquely. The vertical axis gives the level.
Only a few years ago measuring instruments of this kind were very expensive (up to $ 25.000). In this case this measuring instrument was put together by means of DASYLab using only 6 modules and of these two are superfluous (modules "time domain" and "cut out") in this application.
It takes weeks or months to program a complex measuring instrument of this kind in the traditional fashion. Using DASYLab it takes a few minutes including all settings and trial attempts. Is it necessary to say any more about the benefits and advantages of the graphic programming of technical signal systems? You only need to plug a microphone into the sound card and click on to the picture in the electronic PDF-document). Have fun experimenting.

- An important special field of computer-aided signal analysis is data analysis. Here the essential thing is to present a large amount of (stored) data - for instance, a large number of measurements - in an ordered and structured fashion which makes it possible to interpret them. In Illustration 16, for example, this is the graphic, three-dimensional representation of the contours of a deep-sea rift which results from millions of echolocation measurements. These "measurements" can, for example, also be share prices at the stock exchange and the aim of the analysis of data an improved estimate or prediction of stock exchange trends. In this context completely new technologies are used which can "learn" or which can be optimised by training : fuzzy logic and neuronal networks or a combination of the two - neuro-fuzzy.

Preliminary conclusions: the concept takes on clearer contours

All the facts, propositions and arguments advanced so far would be useless if it were not possible to extrapolate a clear and up-to-date concept which will be valid in the future - a concept which will be valid many years from now and which is convincing in its simplicity. Like the thesis formulated on the first page: *microelectronics does nothing else but signal processing!*

- The huge number of discrete analog circuits will in future no longer be state of the art and is therefore not dealt with here. Thus the multi-function board represented in Illustration 13 shows that analog technology will at most remain in existence at the beginning (source) and end (drain) of a communications system. The "core" of the system is purely digital. There are exceptions only in the field of high and maximum frequency, for instance, on the actual path of transmission.

- The entire (digital) hardware - as the example of the multi-function board in Illustration 13 again shows - consists of only a small number of chips (A/D, D/A conversion, multiplexer, timer, memory etc and above all a processor). In future more and more of these components will be integrated *on a single chip*. This is already the case with many microcontrollers, indeed with entire systems. It cannot therefore be the aim of this manuscript to discuss in detail an infinite or even a large number of different IC chips. In future they will no longer exist. Hardware will, therefore, in the following always be presented as a block diagram (see Illustration 13). This block diagram consists of standard components/circuits which are linked with each other. We shall refer to this kind of block diagram as a *hardware block diagram (H block diagram)*.

- The (digital) hardware has the task of providing the processor (computer) with the measurement data or signals in a suitable form. The program contains the signal processes in algorithmic form. What the processor does with the data is determined by the program. The "intelligence" of the overall system lies in the software. As recent development shows, software can largely replace hardware. *Algorithms instead of circuits!* Thus only a few standard components remain even for *digital* hardware.

- Programs for the processing of signals will presumably no longer be represented as a "cryptic code" but also as a block diagram. This shows the order and linking of the processes to be carried out. The block diagram can be programmed graphically on the screen and produces in the background the source code in a certain program language (e.g. C++).

- We shall refer to block diagrams of this kind as *signal block diagrams (S block diagram)*. Almost all the signal systems illustrated in this book are S block diagrams, behind which *virtual* systems are concealed. They were above all generated by means of DASY*Lab*.

- The actual signal processes are to be understood by means of the pictorial comparison of the input and output signal (in the time and frequency domain). In this way it is possible to see how the process has *changed* the signal.

The essential forms of representation used in this book have now been described. They are of a visual nature and are in keeping with the human capacity for thinking in pictures.

They are as follows

- H block diagrams
- S block diagrams (as a visual diagrammatic representation of the signal processes) and
- progression of signal (in the time and frequency domain)

Thus, many intellectual and psychological barriers have been removed at the outset. In order to understand the "theory of signal processing systems" you do not need

- to have mastered one or several program languages
- to have studied mathematics
- to have a detailed knowledge of hundreds of very different IC chips

The starting point are simply three physical phenomena:

- *FOURIER Principle (**FP**)*
- *Uncertainty Principle (**UP**)* and
- *Symmetry Principle (**SP**)*

The first important thing is to have a sound understanding of these fundamental principles. Precisely this is the purpose of the following chapters.

Exercises on Chapter 1 :

The DASY*Lab* program will accompany us from now on. It is a superb work platform, a fully equipped laboratory for measurement and development by means of which practically all systems of measurement and control can be constructed.

The school version is completely functional and can input and output real analogue and digital signals (via a sound card or the parallel interface).

General skills in the use of Microsoft Windows are important for handling the program.

Exercise 1:

Familiarise yourself carefully with the basic DASY*Lab* functions. Any questions will be thoroughly explained under the *Help* menu option and all the components (modules) are described in detail.

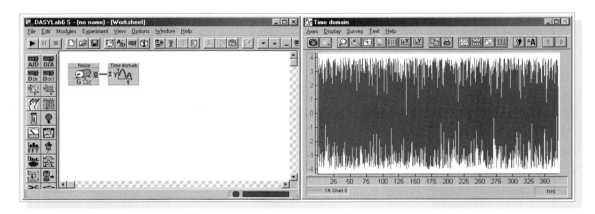

Illustration 18:   ***First exercise with DASYLab***

(a) Restrict yourself in the first instance to the above two modules (laboratory equipment) generator and "oscilloscope" (screen). Try as above to produce a noise signal and make it visible.

(b) Put other signals on the screen by setting the signal generator appropriately (form of signal, amplitude, frequency, phase) by a double click on the component. Do a bit of experimenting to familiarise yourself with the possible settings.

(c) Try via the screen menu to "magnify" an excerpt by means of the magnifying lens (zoom). Then reverse this representation.

(d) Switch the cursor on. You will see two vertical lines on the screen. At the same time a further display window opens in which the chronological position of the two lines is given in figures. Now move the cursor lines, measure instantaneous values, the time interval between them, etc.

(e) Put the screen with the time interval of the signal into the Windows clipboard and print out the picture as a document.

Exercise 2

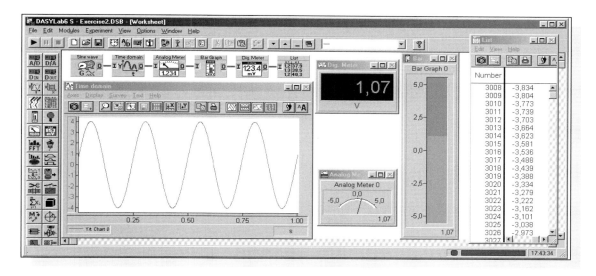

Illustration 19: ***Visualisation of measurement data***

The *visualisation of measurement data and signal processes* is the most important aid in this manuscript for understanding signal processes. DASY*Lab* provides many different visualisation methods for measurement data and signal processes. First create the circuit illustrated by means of various visualisation components (see above). Try to design the size and position of the displays as on the screen. Select a sinusoidal signal with the frequency f = 2 Hz.

(a) Now start the system above left and watch all the displays for some length of time. Try to find out which measurement data refer to the analogue instrument, the digital instrument or the bar chart.

(b) Try to see the correlation between the development of the signal on the screen of the plotter and the measurement data on the list (f = 0,2 Hz). At what intervals are the spot measurements of the signal ascertained or stored. How high is the so-called sample rate with which "samples" of the course of the signal are taken?

(c) For what kind of measurements are analog, digital instrument and bar graph suitable? What measurement from a whole block sequence of measurements do they reproduce?

(d) Which of the "display instruments" most clearly provides the readings which the computer could then process?

(e) Find out how to set the block format and the sample frequency in the menu (A/D). What exactly do these two quantities indicate?

(f) Set a block format of 1024 and a sample frequency of 1024 for all further experiments. How long does the recording of a measurement sequence (of a block) take and how many readings does it consist of?

Exercise 3:

You will find your "components" (processes) either in the "cabinet" on the left hand side (simply click on to the symbol and place it on the screen) or in the menu under *module*.

Work closely with the simplest of the processes mapped there. Design very simple circuits using the *Help* option in the DASY*Lab* menu.

(a) Begin with a simple circuit which at intervals of 1s switches the "lamp" on and off.

(b) Link up two different signals by means of the mathematics component - for instance, addition or multiplication - and look at all three signals one below the other on the same screen.

(c) Examine the examples for "action" and "message" in the DASY*Lab* S-version. Try designing such circuits yourself.

(d) Examine the "black box component" and consider when and for what purpose it might be used.

Exercise 4:

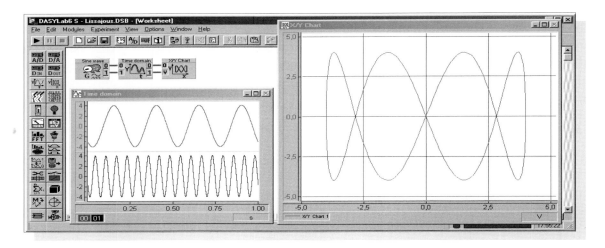

Illustration 20:   ***LISSAJOUS figures***

Try to create the circuit illustrated representing the so-called LISSAJOUS figures. Use sinusoidal signals in each case.

(a) At what frequencies do you get a stationary picture and at what frequencies a picture which "rotates" more slowly or more quickly?

(b) Try by means of specific experiments to find out what this "piece of equipment" or measuring instrument could be used for.

Chapter 2

Signals in the time and frequency domain

From a physical point of view signals are oscillations or waves. They are imprinted with certain information by changing according to a certain pattern.

Only electrical or electromagnetic signals are used in information technology. They have incomparable advantages compared with other forms of signals - e.g. acoustic signals.

Electric signals ....

- spread at (almost) the speed of light,
- can be directed by means of cables to where they are needed,
- can be transmitted around the world and even into space by means of aerials through the atmosphere and vacuum without cables,
- are unrivalled in the way they can be received, processed and transmitted accurately and interference-proof,
- use hardly any energy compared with other electrical and mechanical systems,
- are processed by the tiniest of chips which can all be manufactured very cheaply (fully automated production in large series),
- when used properly they do not pollute the environment and are not a health hazard.

If a signal contains information then there must be an infinite number of different signals as there is an infinite variety of information.

If one wanted to know everything about all signals and how they react to processes or systems, a course of study would inevitably tend to be infinitely long too. Since this is not possible it is necessary to look for a way of describing all signals according to a unified pattern.

The FOURIER Principle

The FOURIER Principle makes it possible to regard all signals as composed of the same unified "components". Simple experiments with DASY*Lab* or with a signal generator ("function generator"), an oscilloscope, a loudspeaker with a built-in amplifier and - most important ! - your sense of hearing, lead to the insight which the French mathematician, natural scientist and advisor to Napoleon discovered mathematically almost two hundred years ago.

Illustration 21:   **Jean Baptiste FOURIER (1768-1830)**

Fourier is regarded as one of the founders of mathematical physics. He developed the foundations of the mathematical theory of heat conduction and made important contributions to the theory of partial differential equations. He could not have dreamt of the importance that "his" FOURIER transformation would have in natural sciences and technology.

Periodic oscillations

These experiments are to be carried out with various periodic oscillations.

> *Periodic oscillations* are oscillations which are repeated over and over again in the same way after a specific period length T. Theoretically - i.e. seen in an idealised way - they last for an infinite period of time in the past, the present and the future. In practical terms this is never the case but it simplifies the approach.

In the case of many practical applications - for instance, in quartz clocks and other clock pulse generators ("timers") or in the case of network AC voltage the length of signal is so great that it almost corresponds to the ideal "infinitely long". The precision of measurement of time depends largely on how precisely periodic the reference voltage was and is and how periodic it stays.

Although it is very important for many applications, periodic oscillations are not typical signals. They hardly provide new information as their future course can be precisely predicted. The greater the uncertainty about the development of the signal at the next moment, the greater the information *may be* that is contained in it. The more we know what message will be conveyed by a source the less the uncertainty and therefore the information value. Information often seems associated more with knowledge than with the idea of uncertainty.

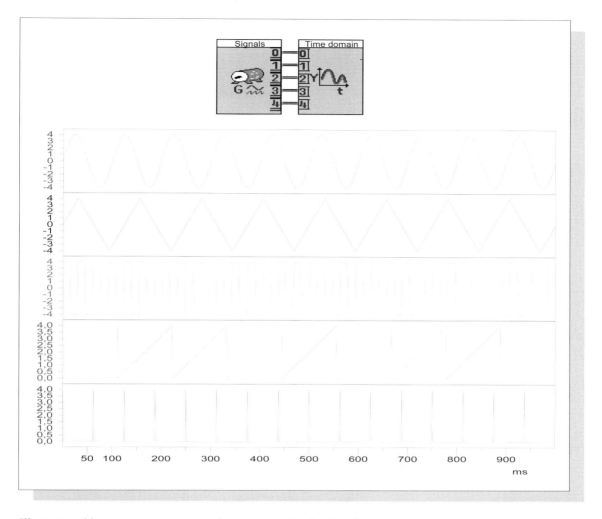

Illustration 22: ***Important periodic signals***

Here you see five important forms of periodic signals, from the top to the bottom: sine, triangle, rectangle, saw tooth and (extremely short!) pulses. From a theoretical point of view periodic signals are of infinite duration, that is, they extend far into the past and future beyond the illustrated segment. Try to determine the period length and frequency of the individual signals.

Surprisingly however, we must say that language and music are not conceivable without "near periodic" oscillations inspite of what has just been said. Periodic oscillations are easier to describe in their behaviour and that is why we are dealing with them at the beginning of this book.

Our ear as a FOURIER-analyzer

By means of very simple experiments it is possible to establish fundamental common features of different oscillation and signal forms. Simple instruments to be found in almost any collection of teaching aids are adequate for this purpose.

A function generator is able to produce different periodic AC voltages. It represents the source of the signal. The signal can be heard over the loudspeaker and can be seen on the screen of the oscilloscope or computer.

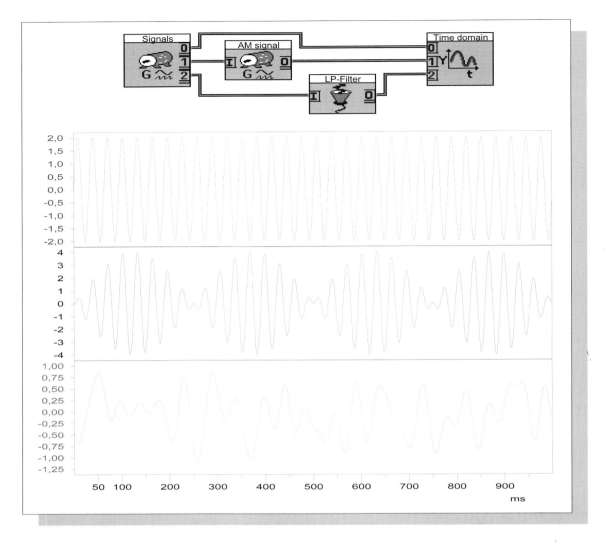

Illustration 23: **_Signal and information_**

A generator module produces in the first instance three different signals the lower two of which are subsequently "manipulated". The information value of the above signals increases from the top to the bottom. The signal above is a sine whose course can be predicted exactly. After a time there is therefore no new information. The middle signal is a modulated sine signal, the amplitude follows a certain sinusoidal pattern. Finally the signal bottom right has a rather "random" course (it is filtered noise). It can be least well predicted but contains, for example, all the information about the special characteristics of the filter.

As an example first choose a periodic sawtooth voltage with the period length T = 10 ms (Frequency f = 200Hz). If one listens carefully several tones of different frequency can be heard. The higher the tone the weaker they seem in this case. If one listens longer one finds that the second lowest tone is exactly one octave higher than the lowest, i.e. twice as high as the base tone.

In the case of all the other periodic forms of signal there are several tones to be heard simultaneously. The triangle signal in Illustration 22 sounds soft and round, very similar to a recorder note. The "saw tooth" sounds much sharper, more like the tone of a violin. In this signal there are more stronger high tones (overtones) than in the "triangle". Apparently the overtones contribute to the sharpness of the tone.

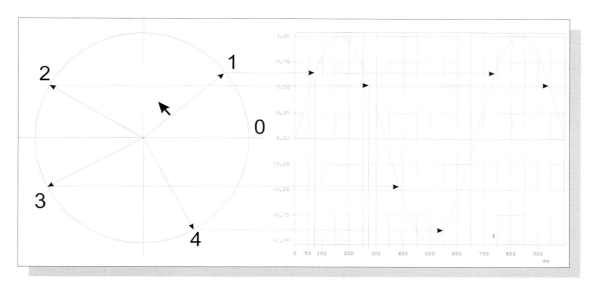

Illustration 24: ***Geometric model for the way in which a sinusoidal signal arises***

Let a pointer rotate uniformly in anti-clockwise direction, beginning in the diagram at 0. When for example the numbers express time values in ms the pointer is in position 1 after 70 ms, after 550ms in position 4 etc. The period length (of 0 to 6.28) is T = 666ms, i.e. the pointer turns 1.5 times per second. Only the projection of the pointer on to the vertical axis can be measured physically. The visible/ measurable sine course results from the pointer projections at any given moment. It should be noted that the (periodic) sinusoidal signal existed before 0 and continues to exist after 1000 ms as it lasts for an infinite length of time in theory! Only a tiny time segment can be represented, here slightly more than the period length T.

There is a single form of AC voltage which only has one audible tone: the sinusoidal signal! In these experiments it is only a question of time before we begin to feel suspicious. Thus in the "sawtooth" of 100Hz there is an audible sine of 200Hz, 300Hz etc. This means that if we could not see that a periodic sawtooth signal had been made audible our ear would make us think that we were simultaneously hearing a sinusoidal signal of 100 Hz, 200Hz, 300Hz etc.

Preliminary conclusions:

(1) There is only one single oscillation which contains only one tone: the (periodic) sinusoidal signal

(2) All the other (periodic) signals or oscillations - for instance tones and vowels contain several tones.

(3) Our ear tells us

- one tone = one sinusoidal signal
- this means: several tones = several sinusoidal signals
- All periodic signals/oscillations apart from the sine contain several tones

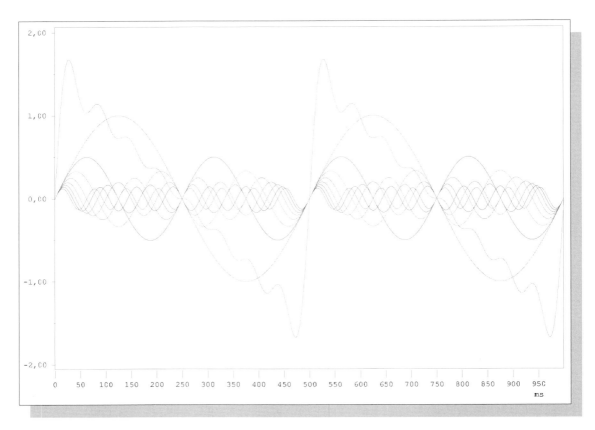

Illustration 25: **Addition of oscillations/signals from uniform components**

This is the first illustration of the FOURIER synthesis. Using the example of a periodic sawtooth signal it is shown that sawtooth-like signals arise by adding appropriate sinusoidal signals. Here are the first six of the (theoretically) infinite number of sinusoidal signals which are required to obtain a perfect linear sawtooth signal with a sudden change. This example will be further investigated in the next few illustrations. The following can be clearly seen: (a) in some places (there are five visible here) all the sinusoidal functions have the value zero: at those points the "sawtooth" or the sum has the value zero. (b) near the "jump zero position" all the sinusoidal signals on the left and the right point in the same direction, the sum must therefore be greatest here. By contrast, all the sinusoidal signals almost completely eliminate each other near the „flank zero position", so that the sum is very small.

From this the FOURIER Principle results which is fundamental for our purposes.

> *All oscillations/signals can be understood as consisting of nothing but sinusoidal signals of differing frequency and amplitude.*

This has far-reaching consequences for the natural sciences - oscillation and wave physics - , technology and mathematics. As will be shown, the FOURIER Principle holds good for all signals, i.e. also for non-periodic and one-off signals.

The importance of this principle for signal and communications technology is based on its reversal.

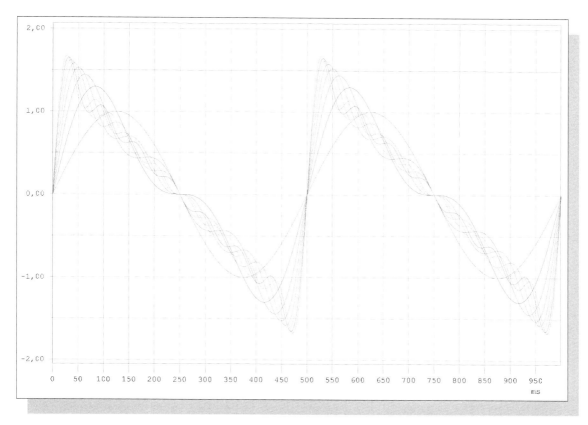

Illustration 26: ***FOURIER synthesis of the sawtooth oscillation***

It is worth looking very carefully at this picture. It shows all the cumulative curves beginning with a sinusoidal oscillation (N = 1) and ending with N = 8. Eight appropriate sinusoidal oscillations can "model" the sawtooth oscillation much more accurately than for example three (N = 3.) Please note - the deviation from the ideal sawtooth signal is apparently greatest where this oscillation changes most rapidly. First find the cumulative curve for N = 6

> *If it is known how a given system reacts to sinusoidal signals of different frequencies it is also clear how it reacts to all other signals because all other signals are made up of nothing but sinusoidal oscillations.*

Suddenly the entire field of communications engineering seems easier to understand because it is enough to to look more closely at the reaction of communications engineering processes and systems to sinusoidal signals of different frequencies.

It is therefore important for us to know everything about sinusoidal signals. As can be seen from Illustration 24 the value of the frequency f results from the angular velocity $\omega = \varphi / t$ of the rotating pointer. If the value of the full angle (equivalent to 360°) is given in rad, $\omega = 2\pi / T$ or $\omega = 2\pi f$ applies.

In total a sinusdoidal signal has three properties. The most important property is quite definitely the frequency. It determines acoustically the height of the tone.

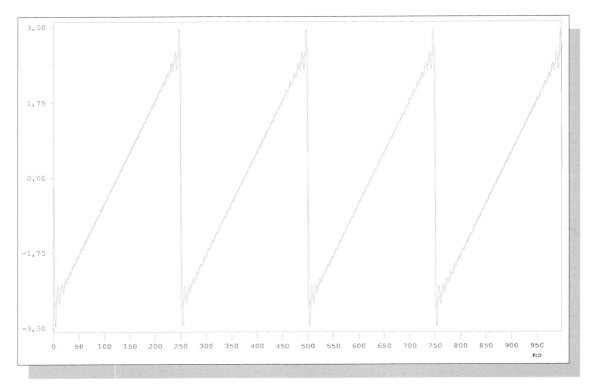

Illustration 27: **FOURIER synthesis: the more the better!**

Here the first N = 32 sinusoidal signals were added from which a sawtooth signal is composed. At the jump position of the "sawtooth" the deviation is greatest. The cumulative function can never change faster than the sinusoidal signal with the greatest frequency (it is practically visible as "ripple content"). As the "sawtooth" at the jump position can theoretically "change infinitely rapidly", the deviation can only have disappeared when the cumulative function also contains an "infinitely rapidly changing" sinusoidal signal (i.e. $f \to \infty$). As that doesn't exist, a perfect sawtooth signal cannot exist either. In nature every change takes time!

Terms such as "frequency range" or "frequency response" are well-known. Both concepts are only meaningful in the context of sinusoidal signals:

Frequency range: the frequency range which is audible for human beings lies in a range of roughly 30 to 20,000Hz (20 kHz). This means that our ear (in conjunction with the brain) only hears acoustic sinusoidal signals between 30 and 20,000Hz

Frequency response: if a frequency response for a bass loudspeaker is given as 20 to 2500 Hz this means that the loudspeaker can only transmit acoustic waves which contains sinusoidal waves between 20 and 2500 Hz.

Note: In contrast to the term *frequency range* the term *frequency response* is only used in connection with a system capable of oscillation.

The other two - also important properties - of a sinusoidal signal are:

- *amplitude* and
- *phase angle*

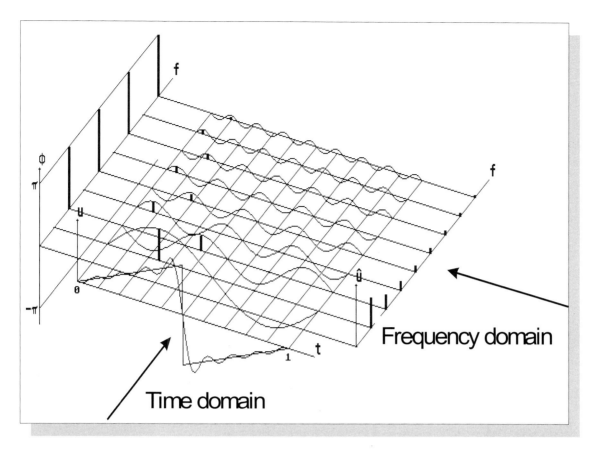

*Illustration 28:   **Picture-aided FOURIER transformation***

The illustration shows in a very graphic way for periodic signals (T = 1) how the path into the frequency range - the FOURIER transformation - arises. The time and frequency domain are two different perspectives of the signal. A "playing field" for the (essential) sinusoidal signals of which the periodic "sawtooth" signal presented here is composed serves as the pictorial "transformation" between the two areas. The time domain results from the addition of all the sine components (harmonics). The frequency domain contains the data of the sinusoidal signals (amplitude and phases) plotted via the frequency f. The frequency spectrum includes the amplitude spectrum (on the right) and the phase spectrum (on the left); both can be read directly on the "playing field". In addition the "cumulative curve" of the first eight sinusoidal signals presented here is also entered. As Illustration 26 and Illustration 27 show: the more sinusoidal signals contained in the spectrum are added, the smaller is the deviation between the cumulative curve and the "sawtooth".

The amplitude - the amount of the maximum value of a sinusoidal signal (is equivalent to the length of the pointer rotating in an anti-clockwise direction in Illustration 24) - is for example in acoustics a measure of volume, in (traditional) physics and engineering quite generally a measure of the average energy contained in the sinusoidal signal.

The phase angle φ of a sinusoidal signal is in the final analysis simply a measure of the displacement in time of a sinusoidal signal compared with another sinusoidal signal or a reference point of time (e.g. t = 0 s).

> As a reminder: The phase angle φ of the rotating pointer is not given in degrees but in "rad" (from radiant: arc of the unit circle (r = 1), which belongs to this angle).

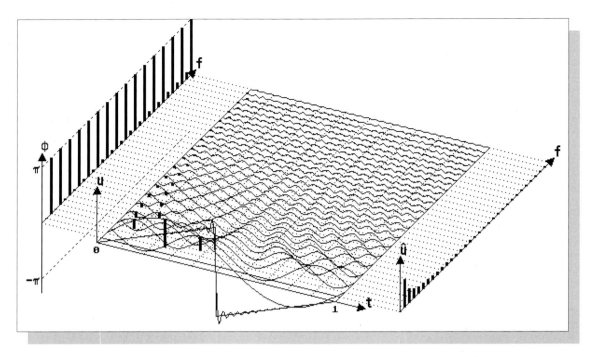

Illustration 29: **"Playing field" of the sawtooth signal with the first 32 harmonic**

The discrepancy between the sawtooth signal and sum curve is clearly smaller than in Illustration 28. See Illlustration 27.

- Circumference of the unit circle $= 2 * \pi * 1 = 2 * \pi$ rad
- 360 degrees are equivalent to $2 * \pi$ rad
- 180 degrees are equivalent to π rad
- 1 degree is equivalent to $\pi/180 = 0.01745$ rad
- x degrees are equivalent to $x = 0.01745$ rad
- for example, 57.3 degrees are equivalent to 1 rad

FOURIER -Transformation: from the time domain to the frequency domain and back

As a result of the FOURIER Principle all oscillations or signals are seen from two perspectives, i.e. :

the time domain and the

the *frequency domain*

In the *time domain* information is given on the values of a signal at any given time within a certain period of time (time progression of the values at any given moment).

In the *frequency domain* the signal is described by the sinusoidal signals of which it is composed.

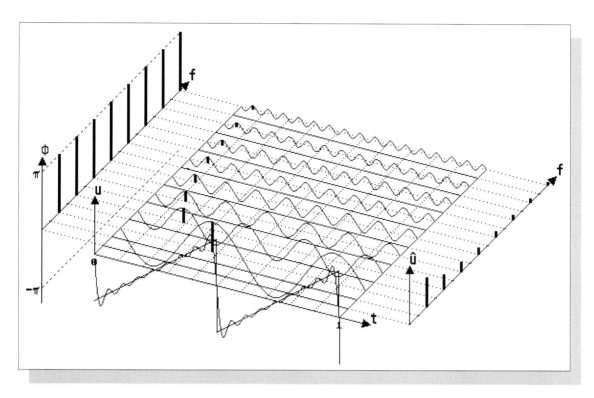

Illustration 30: **Doubling frequency**

Here the period length of the sawtooth signal is T = 0.5s (or for example 0.5 ms). The frequency of the sawtooth signal is accordingly 2 Hz (or 2 kHz). The distance between the lines in the amplitude and phase spectrum is 2 Hz (or 2 kHz). Note the changed phase spectrum. Although it is an oversimplification it is possible to say: our eyes see the signal in the time domain on the screen of the oscillograph but our ears are clearly on the side of the frequency domain.

As we shall see in the case of many practical problems it is sometimes more useful to consider signals sometimes in the time domain and sometimes in the frequency domain.

Both ways of presenting this are equally valid, i.e. they both contain all the information. However, the information from the time domain occurs in a transformed form in the frequency domain and it takes a certain amount of practice to recognise it.

Apart from the very complicated (analogous) "harmonics analysis" measurement technique there is now a calculating procedure (algorithm) to compute the frequency-based way of presentation - the spectrum - from the time domain of the signal and vice-versa. This method is called the FOURIER transformation. It is one of the most important signal processes in physics and technology.

> *FOURIER-Transformation (FT):*
>
> Method of calculating the (frequency) spectrum of the signal from the progression in time.
>
> *Inverse FOURIER Transformation (IFT)*
>
> Method of calculating the progression of a signal in time from the spectrum.

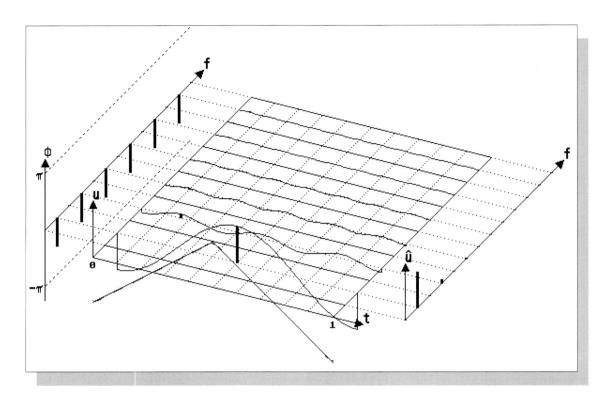

Illustration 31: **_Periodic triangle signal_**

The spectrum appears to consist essentially of one sinusoidal signal. This is not surprising in that the triangle signal is similar to the sinusoidal signal. The additional harmonics are responsible for subtle differences (see sum curve). For reasons of symmetry the even-numbered harmonics are completely absent.

The computer can work out the **FT** and the **IFT** for us. We are here only interested in the results presented graphically. In the interests of a clear illustration a presentation has been selected in which the time and frequency domain are presented together in a three-dimensional illustration.

The FOURIER Principle is particularly well illustrated in this form of representation because the essential sinusoidal oscillations which make up a signal are all distributed alongside each other. In this way the FT is practically described graphically. It can be clearly seen how one can change from the time domain to the spectrum and vice versa. This makes it very easy to extrapolate the essential transformation rules.

In addition to the sawtooth signals the cumulative curve of the first 8, 16 or 32 sinusoidal signals (harmonics) is included. There is a discrepancy between the ideal sawtooth and the cumulative curve of the first 8 or 32 harmonics, i.e. the spectrum does not show all the sinusoidal signals of which the (periodic) sawtooth signals consist.

As particularly Illustration 13 shows the following applies for all periodic signals:

> *All periodic oscillations/signals contain as sinusoidal components all the integer multiples of the base frequency as only these fit into the time frame of the period length T. In the case of periodic signals all the sinusoidal signals contained in them must be repeated after the period length T in the same manner!*

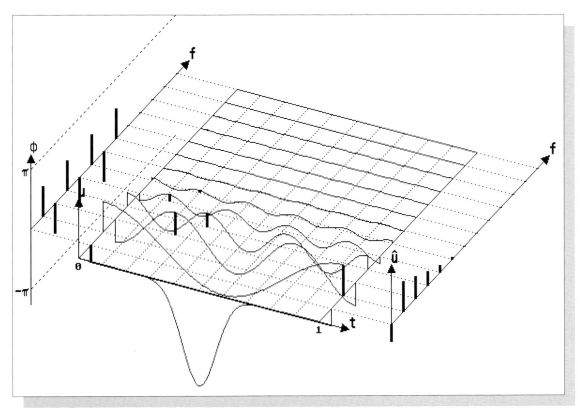

Illustration 32: ***Pulse form without rapid transitions***

Within the (periodic) sequence of GAUSSian pulses each pulse begins and ends gently. For this reason the spectrum cannot contain any high frequencies. This characteristic makes GAUSSian pulses so interesting for many modern applications. We will come across this pulse form frequently.

Example: a periodic sawtooth of 100 Hz only contains the sinusoidal components 100 Hz, 200 Hz, 300Hz etc.

The spectrum of periodic oscillations/signals accordingly always consists of lines at equal distances from each other.

> *Periodic signals have **line** spectra!*

The sawtooth and square wave signals contain steps in "an infinitely short space of time" from, for example 1 to -1 or from 0 to 1. In order to be able to model "infinitely rapid transitions" by means of sinusoidal signals, sinusoidal signals of infinitely high frequency would have to be present. Hence it follows:

> *Oscillations/signals with step function (transitions in an infinitely short period of time) contain (theoretically) sinusoidal signals of infinitely high frequency.*

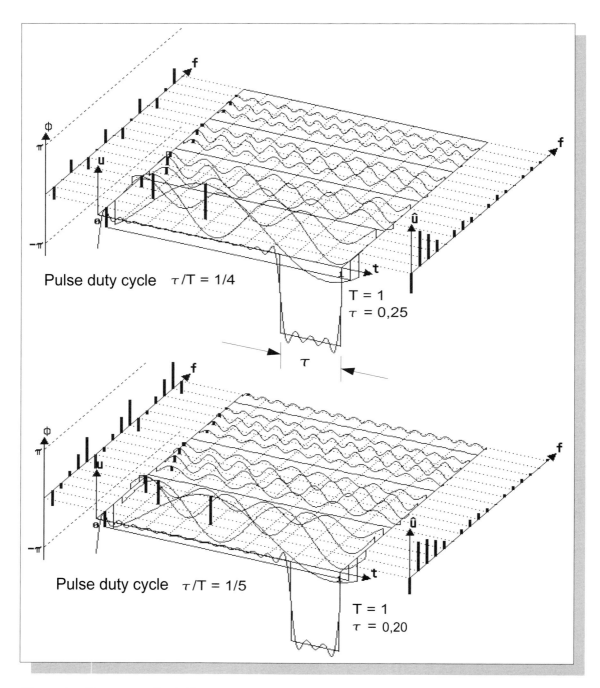

Illustration 33: ***Periodic square wave signals with different pulse duty factors***

This illustration shows how the information from the time domain is to be found in the frequency domain. The period length T is to be found in the distance 1/T of the lines of the frequency spectrum. As in this illustration T = 1s a line distance of 1Hz results. The pulse duration τ is 1/4 in the upper representation and in the lower 1/5 of the period length T. It is striking that every fourth harmonic above (4 Hz, 8Hz etc, and every fifth harmonic below (5 Hz, 10 Hz etc) has the value 0. The zero position is in each case at the point 1/τ. It is also possible to determine the period length T and the pulse duration τ in the frequence domain.

As from a physical point of view there are no sinusoidal signals "of a infinitely high frequency", in nature there cannot be signals with "infinitely rapid transitions".

> *In nature every change, including steps and transitions, needs time as signals/oscillations are limited as far as frequency is concerned.*

As Illustration 26 and Illustration 27 show, the difference between the ideal (periodic) sawtooth and the cumulative curve is greatest where the rapid transitions or steps are present.

> *The sinusoidal signals of high frequency contained in the spectrum serve as a rule to model rapid transitions.*

Thus, it also follows that

> *Signals which do not exhibit rapid transitions do not contain high frequencies either.*

Important periodic oscillations/signals

As a result of the FOURIER Principle it can be taken as a matter of course that the sinusoidal oscillation is the most important periodic "signal".

Triangle and sawtooth signals are two other important examples because they both change in time in a linear fashion. Such signals are used in measuring and control technology (for example, for the horizontal deflection of the electron beam in a picture tube).

They are easy to produce. For example, a capacitor switched into a constant current source is charged linearly.

Their spectra show interesting differences. In the first place the high frequency part of the spectrum of the triangle signal is much smaller, because - in contrast to the sawtooth signal - no rapid steps occur. While in the case of the (periodic) "sawtooth" all the even numbered harmonics are contained in the spectrum, the spectrum of the (periodic) "triangle" shows only odd-numbered harmonics (e.g. 100 Hz, 300 Hz, 500 Hz etc). In other words, the amplitudes of the even-numbered harmonics equal zero.

Why are the even-numbered harmonics not required here?

The answer lies in the greater symmetry of the triangle signal. At first, the sinusoidal signal looks very similar. This is why the spectrum only shows "small adjustments". As Illustration 31 shows, only sinusoidal signals can be used as components which exhibit this symmetry within the period length T and those are the odd-numbered harmonics.

Comparison of signals in the time and frequency domain

As a result of digital technology, but also determined by certain modulation processes, (periodic) square waves or rectangular pulses have a special importance. If they serve the purpose of synchronisation or the measurement of time they are aptly called clock signals. Typical digital signals are however not periodic. As they are carriers of (constantly changing) information they are not periodic or only "temporarily" so.

The so-called pulse duty factor, the quotient from the pulse duration τ and the period length T is decisive for the frequency spectrum of (periodic) rectangular pulses. In the case of the symmetrical rectangular signal $\tau/T = 1/2 = 0.5$. In this case there is symmetry as in the case of the (regular) triangle signal and its spectrum therefore contains only the odd-numbered harmonics. (see Illustration 34).

We can obtain a better understanding of these relationships by close examination of the time and frequency domains in the case of different pulse duty factors τ/T (see Illustration 33). In the case of the pulse duty factor 1/4 it is precisely the 4th, the 8th, the 12th harmonic etc which are missing, in the case of the pulse duty factor 1/5 the 5th, the 10th the 15th etc, in the case of the pulse duty factor 1/10 the 10th, 20th, 30th harmonic (see Illustration 35).

These "gaps" are termed "zero positions of the spectrum" because the amplitudes formally have the value of zero at these positions. Consequently, all the even-numbered harmonics are lacking in the case of the symmetrical rectangular signal with the pulse duty factor 1/2

It can now be seen that the core values of the time domain are "hidden" in the frequency domain:

> *The inverse ratio of the period length T is equivalent to the distance between the spectral lines in the spectrum. In this connection please again look carefully at Illustration 30. The frequency line distance ($f = 1/T$ equals the base frequency f_1 (1st harmonic).*

Example:

T = 20 ns results in a base frequency or a frequency line distance of 50 MHz.

> *The inverse ratio of the pulse duration τ is equivalent to the distance ΔF_o between the zero positions in the spectrum:*
>
> *Zero position distance $\Delta F_o = 1/\tau$*

This allows one to draw a conclusion about the fundamental and extremely important relationship between the time domain and the frequency domain.

Chapter 2 Signals in the time and frequency domain

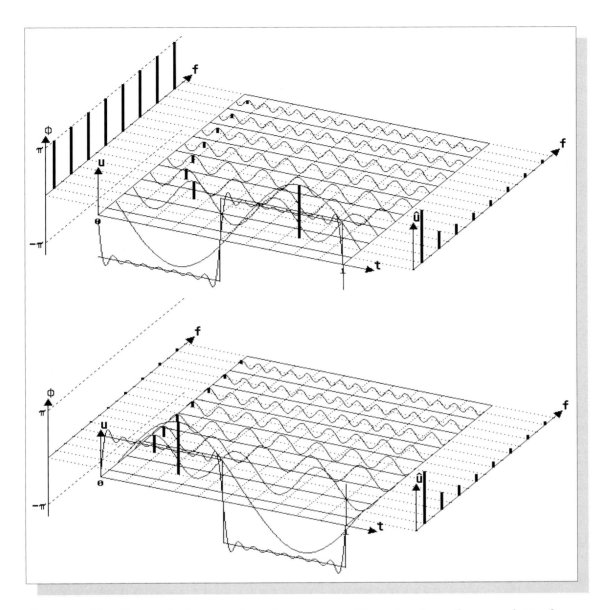

Illustration 34: **Symmetrical rectangular pulse sequence with varying time reference point t = 0 s**

In both representations it is the same signal. The lower one is staggered compared with the upper one by T/2. Both representations have a different time reference point t = 0 s. A time displacement of T/2 is exactly equivalent to a phase displacement of π. This explains the different phase spectra. On account of $\tau/T = 1/2$ all the even-numbered harmonics are lacking (i.e. the zero positions of the spectrum are 2 Hz, 4Hz etc).

> *All the **large** characteristic time values appear **small** in the frequency domain, all the **small** characteristic time values appear **large** in the frequency domain.*

Example: Compare period length T and pulse duration τ

The confusing phase spectrum

It is also possible to draw an important conclusion with regard to the phase spectrum. As Illustration 34 shows, the same signal can have different phase spectra. The phase spectrum depends on the time reference point t = 0.

By contrast, the *amplitude* spectrum is unaffected by time displacements.

For this reason the phase spectrum is more confusing and much less revealing than the amplitude spectrum. Hence in the following chapters usually only the amplitude spectrum will be demonstrated in the frequency domain.

Note:

- In spite of this, only the two spectral representations together provide all the information on the progression of a signal/oscillation in the time domain. The inverse FOURIER transformation IFT requires the amplitude and phase spectrum to calculate the course of the signal in the time domain.

- The property of our ear (a FOURIER analyzer!) which scarcely perceives changes in the phase spectrum of a signal is a particularly interesting phenomenon. Any important change in the amplitude spectrum is immediately noticed. In this connection you should carry out acoustic experiments with DASY*Lab*.

Interference: nothing to be seen although everything is there.

The (periodic) rectangular pulses in Illustration 33 have a constant (positive or negative) value during the pulse duration τ, but between pulses the value is zero. If we only considered these periods of time T - τ, we might easily think that "there cannot be anything there when the value is zero", i.e. no sinusoidal signals either.

This would be fundamentally erroneous and this can be demonstrated experimentally. In addition, the FOURIER Principle would be wrong (why?). One of the most important principles of oscillation and wave physics is involved here:

> *(Sinusoidal) oscillations and waves may extinguish or strengthen each other temporarily and locally (waves) by addition.*

In wave physics this principle is called *interference*. Its importance for oscillation physics and signal theory is too rarely pointed out.

Let us first off all look at Illustration 33 again. The cumulative curve of the first 16 harmonics has everywhere been - intentionally- included. We see that the sums of the first 16 harmonics between the pulses equal zero only in a very few places (zero crossings), otherwise they deviate a little from zero. Only the sum of an infinite number of harmonics can result in zero. On the sinusoidal "playing field" we see that all the sinusoidal signals of the spectrum remain unchanged during the entire period length T.

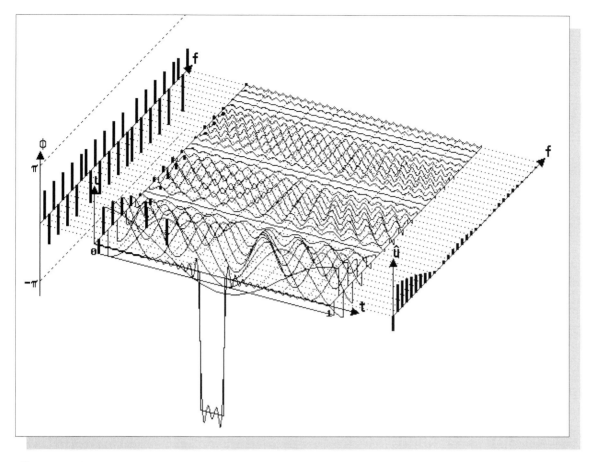

Illustration 35: ***An exact analysis of relationships.***

In this illustration the important relationships are to be summarised once again and additions made:

- *The pulse duty factor of the (periodic) rectangular pulse sequence is 1/10. The first zero position of the spectrum lies at the 10th harmonic. The first 10 harmonics lie at the position t = 0.5 s in phase so that in the centre all the "amplitudes" add up towards the bottom. At the first and every further zero position a phase step of π rad takes place. This can easily be recognised both in the phase spectrum itself and also on the "playing field". In the middle all the amplitudes overlay each other at the top and afterwards - from the 20th to the 30th harmonic towards the bottom again etc.*

- *The narrower the pulse becomes, the bigger the deviation between the sum of the first (here N = 32) harmonics and the rectangular pulse appears. The difference between the latter and the cumulative oscillation is biggest where the signal changes most rapidly, for example at or near the pulse flanks.*

- *Where the signal is momentarily equivalent to zero - to the right and left of a pulse - all the (infinite number of) sinusoidal signals add up to zero; they are present but are eliminated by interference. If one "filters" out the first N = 32 harmonics from all the others this results in the "round" cumulative oscillation as represented; it is no longer equivalent to zero to the right and left of the pulse. The ripple content of the cumulative oscillation is equal to the highest frequency contained.*

Even when the value of signals is equal to zero over a time domain Δt, they nevertheless contain sinusoidal oscillations during this time. Strictly speaking, "infinitely" high frequencies must also be contained because otherwise only "round" signal progressions would result. The "smoothing out effect" is the result of high and very high frequencies.

In Illustration 35 we see a value („offset") in the amplitude spectrum at the position f = 0. On the "the playing field" this value is entered as a constant function ("zero frequency"). If we were to remove this value -U – for instance by means of a capacitor – the previous zero field would no longer be zero but +U. Thus the following holds true:

> *If a signal contains a constant part during a period of time Δt the spectrum must theoretically contain "infinitely high" frequencies.*

In IIllustration 35 there is a (periodic) rectangular pulse with the pulse duty factor 1/10 in the time and frequency domain. The (first) zero position in the spectrum is therefore at the 10th harmonic.

The first zero position of the spectrum is displaced further and further to the right in Illustration 36 the smaller the pulse duty factor selected (e.g. 1/100). If the pulse duty factor approaches zero we have a (periodic) delta pulse sequence whereby the pulse duration approaches zero.

Opposites which have a great deal in common: sine and δ–pulse

Such needle pulses are called δ-pulses (delta-pulses) in the specialised theoretical literature. After the sinusoidal signal the δ-pulse is the most important form of oscillation or time function.

The following factors support this:

- In digital signal processing (DSP) number-strings are processed at regular time intervals (clock pulse frequency). These strings pictorially represent a sequence of pulses of a certain magnitude. Number 17 could for instance be equivalent to a needle pulse magnitude of 17. More details will be given later in the chapters on digital signal processing.

- Any signal can theoretically be conceived of as being composed of as a continuous sequence of δ–pulses of a certain magnitude following each other. See Illustration 37 in this connection.

- A sinusoidal signal in the time domain results in a "needle function" (δ–function) in the frequency domain (line spectrum). What is more - all periodic oscillations/signals result in line spectra that are equidistant (appearing at the same intervals) delta functions in the frequency domain.

- From a theoretical point of view, the δ-pulse is the ideal test signal for all systems. If a δ-pulse is connected to the input of a system, the system is tested at the same time with all frequencies and, in addition, with the same amplitude. See the following pages, especially Illustration 36.

- The (periodic) δ-pulse contains in the interval $\Delta f = 1/T$ all the (integer multiples) frequencies from zero to infinity always with the same amplitude.

Illustration 36: **Steps in the direction of a δ–pulse**

The pulse duty factor above is roughly 1/16 above and 1/32 below. Accordingly, the the first zero position above is at N = 16, and below at N = 32. The zero position "moves" towards the right with higher frequencies if the pulse becomes narrower. Below, the lines of the spectrum represented seem to have almost equally large amplitudes. In the case of a "needle" pulse or δ-pulse the width of the pulse tends towards zero, thus the (first) zero position of the spectrum tends toward infinity. Hence, the δ-pulse has an "infinitely wide frequency spectrum"; in addition, all the amplitudes are the same.

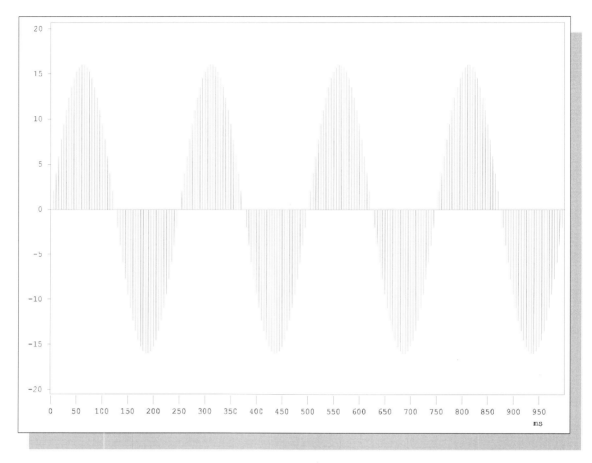

Illustration 37: ***Signal synthesis by means of δ-pulses***

Here a sine wave is "assembled" from δ-pulses of an appropriate magnitude following on each other. This is exactly equivalent to the procedure in "digital signal processing" (DSP). Their signals are equivalent to "strings of numbers" which, seen from a physical point of view, are equivalent to a rapid sequence of measurements of an analog signal; every number gives the "weighted" value of the δ-pulse at a given point of time t.

This strange relationship between sinusoidal and needle functions (uncertainty principle) will be looked at more closely and evaluated in the next chapter.

Note:

Certain mathematical subtleties result in the δ-pulse being theoretically given an amplitude tending to infinity. Physically this also makes a certain sense. An "infinitely short" needle pulse cannot have energy unless it were "infinitely high". This is also shown by the spectra of narrow periodic rectangular pulses and the spectra of δ-pulses. The amplitudes of individual sinusoidal signals are very small and hardly visible in the Illustrations, unless we increase the pulse amplitude (to extend beyond the screen of the PC).

For purposes of illustration we normally choose delta pulses of magnitude "1" in this book.

Illustration 38: **From the periodic signal with a line spectrum to the non-periodic signal with a continuous spectrum.**

On the left in the time domain you see sequences of periodic rectangular pulses from top to bottom. The pulse frequency is halved in each case but the pulse width remains constant. Accordingly the distance between the spectral lines becomes smaller and smaller ($T = 1/f$), but the position of the zero positions does not change as a result of the constant pulse duration.

Finally, in the lower sequence a one-time rectangular pulse is depicted. Theoretically it has the period length $T \to \infty$. The spectral lines lie "infinitely close" to each other, the spectrum is continuous and is drawn as a continuous function.

We have now gone over to the customary (two-dimensional) representation of the time and frequency domains. This results in a much more accurate picture in comparison to the "playground" for sinusoidal signals used up to now.

Non-periodic and one-off signals

In actual fact a periodic oscillation cannot be represented in the time domain on a screen. In order to be absolutely sure of its periodicity, its behaviour in the past, the present and the future would have to be observed. An (idealised) periodic signal repeated itself, repeats itself and will repeat itself in the same way. In the time domain only one or a few periods are shown on the screen.

It is quite a different matter in the frequency domain. If the spectrum consists of lines at regularly spaced intervals, this immediately signals a periodic oscillation. In order to underline this once again - there is at this moment only one (periodic) signal whose spectrum contains precisely one line - the sinusoidal signal.

We shall now look at the non-periodic signals which are more interesting from the communcations technology point of view. As a reminder: all information-bearing oscillations (signals) may have a greater information value the more uncertain their future course is (see Illustration 23).

In the case of periodic signals their future course is absolutely clear.

In order to understand the spectra of non-periodic signals we use a small mental subterfuge. Non-periodic means that the signal does not repeat itself "in the foreseeable future". In Illustration 36 we constantly increase the period length T of a rectangular pulse without changing its pulse duration until it finally tends "towards infinity". This boils down to the sensible idea of not attributing the period length $T \to \infty$ ("T tends towards infinity") to all non-periodic or one-off signals.

If however the period length becomes greater and greater the distance (f = 1/T between the lines in the spectrum gets smaller and smaller until they "fuse". The amplitudes ("end points of lines") no longer form a discrete sequence of lines at regular intervals but now form a continuous function (see Illustration 38).

> *Periodic oscillations/signals have a discrete line spectrum whereas **non-periodic** oscillations/signals have a **continuous** spectrum.*

A glance at the spectrum is enough to see what type of oscillation is present - periodic or non-periodic. As is so often the case the dividing line between periodic and non-periodic is not entirely unproblematical. It is occupied by an important class of signals which are termed *near-periodic*. These include language and music, for instance.

One-off signals are, as the word says, non-periodic. However, non-periodic signals which only change within the period of time under consideration, for instance a bang or a glottal stop, are also called non-periodic.

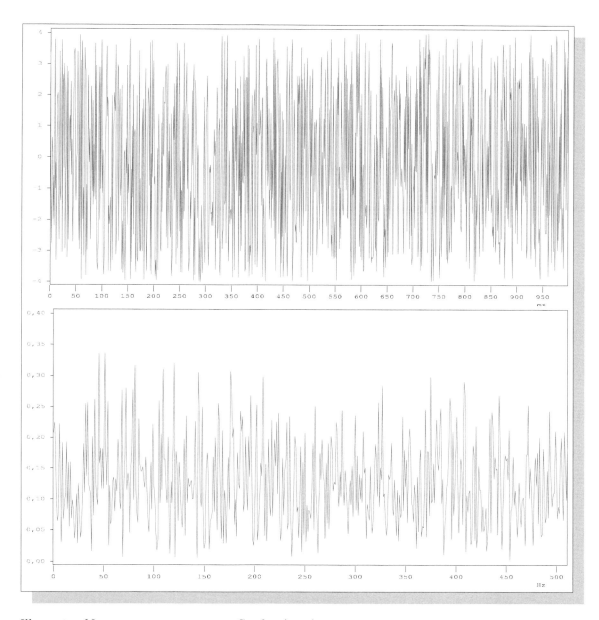

Illustration 39: ***Stochastic noise***

The upper picture shows stochastic noise in the time domain (for 1s) and below this the amplitude spectrum of the above noise. As the time domain develops randomly regularity of the frequency spectrum within the period of time under consideration is not to be expected (otherwise the signal would not be stochastic). In spite of many "irregular lines" it is not a typical line spectrum for otherwise the time domain would have to be periodic!

Pure randomness: stochastic noise

Noise is a typical and extremely important example of a non-periodic oscillation. It has a highly interesting cause, namely a rapid sequence of unpredictable individual events.

In the roar of a waterfall billions of droplets hit the surface of the water in a completely irregular sequence. Every droplet goes "ping" but the overall effect is one of noise. The applause of a huge audience may also sound like noise, unless they clap rhythmically to demand an encore (which simply represents a certain order, regularity or periodicity!)

Electric current in a solid state implies movement of electrons in the metallic crystal grid. The movement of an individual electron from an atom to the neighbouring atom takes place quite randomly.

Even though the movement of electrons mainly points in the direction of the physical current this process has a stochastic - purely random, unpredictable - component. It makes itself heard through noise. There is therefore no pure direct current DC; it is always accompanied by noise. Every electronic component produces noise, that is any resistance or wire. Noise increases with temperature.

Noise and information

Random noise means something like absolute chaos. It contains no "pre-arranged, meaningful pattern" - i.e. no information.

Stochastic noise has no "conserving tendency", i.e. nothing in a given time segment B reminds one of the previous time segment A. In the case of a signal, the next value is predictable at least with a certain degree of probability. If for example you think of a text like this, where the next letter will be an "e" with a certain degree of probability.

> *Stochastic noise is therefore not a "signal" because it contains no information bearing pattern - i.e. no information.*

Everything about stochastic noise within a given time segment is random and unpredictable, i.e. its development in time and its spectrum. Stochastic noise is the "most non-periodic" of all signals!

All signals are for the reasons described always (sometimes more or less or too much) accompanied by noise. But signals which are accompanied by a lot of noise differ from pure stochastic noise in that they display a certain conserving tendency. This is characterised by the pattern which contains the information.

> *Noise is the biggest enemy of communications technology because it literally "buries" the information of a signal.*

One of the most important problems of communications technology is therefore to free signals as far as possible from the accompanying noise or to protect or modulate and code the signals from the outset in such a way that the information can be retrieved without errors in spite of noise in the receiver.

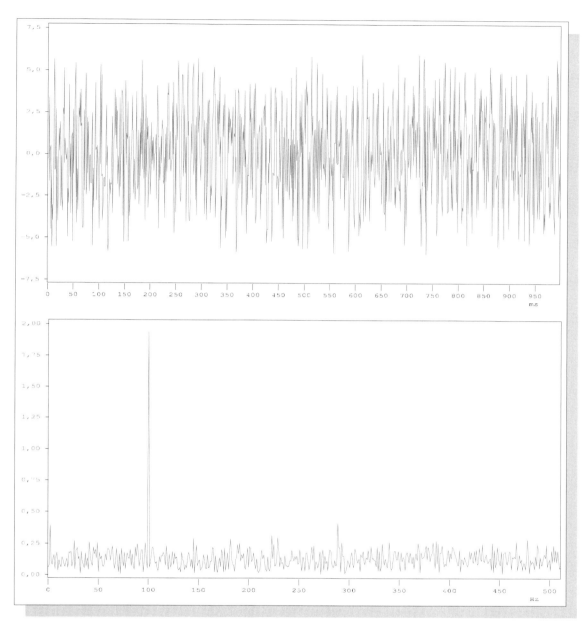

Illustration 40: ***Conserving tendency of a noisy signal***

Both illustrations - the time domain above, the amplitude spectrum below - describe a noisy signal, that is not pure stochastic noise, which displays a conserving tendency (influenced by the signal). This is shown by the amplitude spectrum below. A line protruding from the irregular continuous spectrum at 100 Hz can clearly be seen. The cause can only be a (periodic) sinusoidal signal of 100 Hz hidden in the noise. It forms the feature which conserves a tendency although it is only vaguely visible in the time domain. It could be "fished out" of the noise by means of a high-quality bandpass filter.

This is in fact the central theme of "information theory". As it presents itself as a theory formulated in purely mathematical terms, we shall not deal with it systematically in this book. On the other hand, *information* is the core term of information and communications technology. For this reason important findings of information theory turn up in many places in this book.

> *Signals are regularly non-periodic signals. The less their future development can be predicted, the greater their information value may be. Every signal has a "conserving tendency" which is determined by the information-bearing pattern. Stochastic noise is by contrast completely random, has no "conserving tendency" and is therefore not a signal in the true sense.*

We should, however, not completely denigrate stochastic noise. Since it has such extreme qualities, i.e. it embodies the purely random, it is highly interesting. As we shall see it has great importance as a test signal for (linear) systems.

Chapter 2   Signals in the time and frequency domain

Exercises for Chapter 2:

Exercise 1:

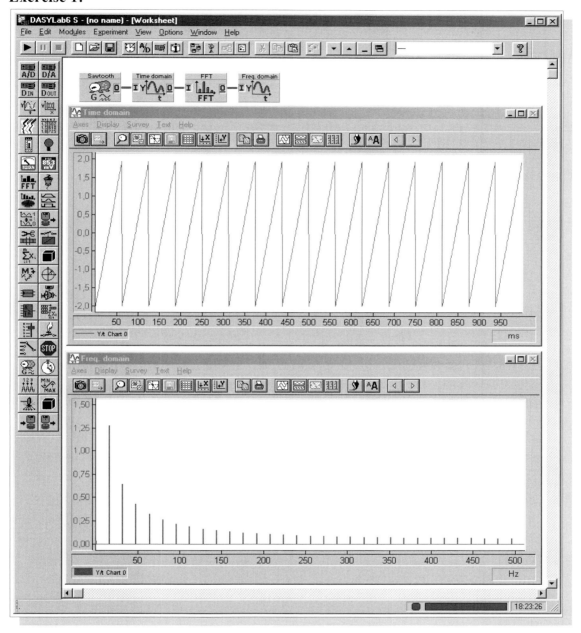

Illustration 41:   *Sawtooth in time and frequency domain*

Here you see the whole DASY*Lab* window displayed. By far the most important circuit for analysis and representation of signals in the time and frequency domain is to be found at the top of the picture.

(c)   Create this circuit and visualise - as above - a periodic sawtooth without a direct voltage offset in the time and frequency domain.

(d)   Measure the amplitude spectrum by means of the cursor. According to what simple rule do the amplitudes decrease?

(e) Measure the distance between the "lines" in the amplitude spectrum in the same way. In what way does this depend on the period length of the sawtooth?

(f) Expand the circuit as shown in Illustration 22 and display the amplitude spectra of different periodic signals one below the other on a "screen".

Exercise 2:

(a) Create a system using DASY*Lab* which produces the FOURIER synthesis of a sawtooth as in Illustration 25

(b) Create a system using DASY*Lab* which gives you the sum of the first n sinusoidal signals (n = 1,2,3,....9) as in Illustration 27

Exercise 3:

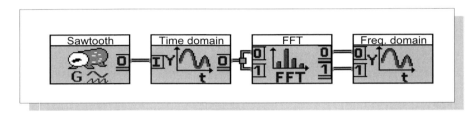

Illustration 42:   ***Block diagram: Amplitude and phase spectrum***

(a) Try to represent the amplitude spectrum and the phase spectrum of a sawtooth one directly beneath the other as in Exercise 1. Select amplitude spectrum on channel 0 in the menu of the module "frequency domain" and "phase spectrum" on channel 1. Select "standard setting" (sampling rate and block length = $1024 = 2^{10}$ in the A/D button of the upper control bar) and a low frequency (f = 1; 2; 4; 8 Hz. What do you discover if you choose a frequency whose value cannot be given as a power of two?

(b) Select the different phase modifications π (180°), $\pi/2$ (90°), $\pi/3$ (60°) and $\pi/4$ (45°) for the sawtooth in the menu of the generator module and observe the changes in the phase spectrum in each case.

(c) Do the phase spectra from Exercise 2 agree with the 3D representation in Illustration 28 ff.? Note deviations and try to find an explanation for the possible erroneous calculation of the phase spectrum.

(d) Experiment with various settings for the sample rate and block length (A/D button on the upper control bar, but select both values in the same size, e.g. 32, 256, 1024!)

Exercise 4:

Noise constitutes a pure stochastic signal and is therefore "totally non-periodic".

(a) Examine the amplitude and phase spectra of noise. Is the spectrum continuous? Do amplitude and phase spectra display stochastic behaviour?

(b) Examine the amplitude and phase spectrum of lowpass filtered noise (e.g. cutoff frequency 50 Hz, Butterworth filter 6th order). Do both exhibit stochastic behaviour? Is the filtered noise also "completely non-periodic"?

Exercise 5:

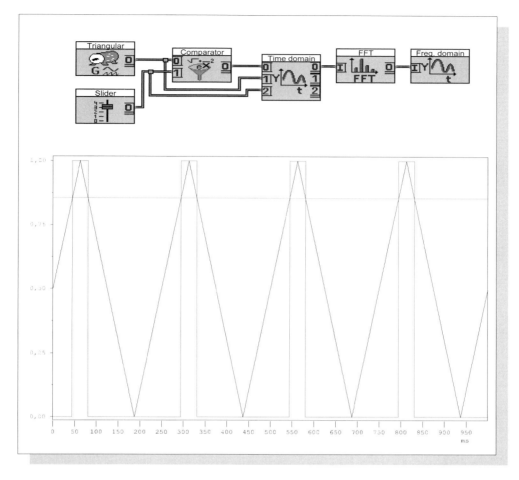

Illustration 43: ***Square wave generator with variable pulse duty factor***

(a) Design a square wave signal generator by means of which the pulse duty factor and the frequency of the periodic rectangular signal can be set as desired. If necessary use the enclosed illustration to help you.

(b) Interconnect (as above) your square wave signal generator with our standard circuit for the analysis and visualisation of signals in the time and frequency domain.

(c) Examine the amplitude spectrum by keeping the frequency of the square wave signal constant and making the pulse duration τ smaller and smaller. Observe particularly the development of the "zero positions" of the spectrum as shown in Illustration 33 ff.

(d) In the amplitude spectrum usually additional small peaks appear between the expected spectral lines. Experiment on ways of optically avoiding these, for instance by the selection of suitable scanning rates and block lengths (A/D setting in the upper control bar) and signal frequencies and pulse lengths. You will discover their cause in Chapter 10 (Digitalisation).

(e) Try to develop a circuit such as that used for the representation of signals in Illustration 38 - transition from a line spectrum to a continuous spectrum. Only the frequency, not the pulse length should be variable.

Exercise 6:

(a) How could one prove using DASY*Lab* that practically all frequencies - i.e. sinusoidal oscillations- are present in a noise signal. Try experimenting.

(b) How is is possible to ascertain whether a (periodic) signal is contained in an extremely noisy signal?

Chapter 3

The Uncertainty Principle

Musical notes have something to do with the simultaneous presentation of the time and frequency domains as they are to be found in the three-dimensional Illustration 28 ff (Chapter 2) of periodic signals. The height of the notes on the lines of the score gives the pitch of the tones; that is in the final analysis the frequency. The form of the notes gives their duration in time. Notes are written by composers as if the pitch and length could be determined quite independently of each other. Experienced composers have, however, long been aware of the fact that, for example, the low notes of an organ or a tuba must last a certain time in order to be felt to be sonorous. Sequences of such low notes can therefore only be played at reduced speed.

A strange relationship between frequency and time and its practical consequences.

It is one of the most important insights of oscillation, wave and modern quantum physics that certain quantities - such as frequency and time - cannot be measured independently of each other. Such quantities are termed complementary.

Illustration 44: **Simultaneous representation of the time and frequency domain in musical scores.**

Norbert Wiener, the world famous mathematician and founder of cybernetics, writes in his autobiography (Econ-Verlag - publishers): "Now let us look at what a musical score actually denotes. The vertical position of a note in the line system gives the pitch or frequency of a tone, while the horizontal position allocates the pitch to time. ..."Thus musical notation appears at first sight to be a system with which signals can be described in two independent ways, i.e according to frequency and duration". However, "things are not quite so straightforward. The number of oscillations per second which a note comprises is a piece of information which refers not only to the frequency but also to something which is distributed in time" ..."Beginning and ending a note involves a change in its frequency combination, which may be very small but which is very real. A note which lasts for only a limited period of time must be seen as a band of simple harmonic movements none of which can be regarded as the sole simple harmonic movement present. Precision in time implies a certain indefiniteness in pitch, just as precision of pitch involves vagueness in time".

Strangely, this aspect which is immensely important for signals is often disregarded. It is an absolute limit of nature which cannot be surmounted even with the most sophisticated technical equipment. Frequency and time cannot be measured accurately at the same time even with the most sophisticated methods.

The Uncertainty Principle **UP** follows from the FOURIER Principle **FP**. It represents the second column of our platform "Signals - Processing - Systems". Its characteristics can be described in words.

> *The more the duration in time Δt of a signal is restricted the wider its frequency band Δf automatically becomes. The more restricted the frequency band Δf of a signal (or a system) is, the greater the duration in time Δt of the signal must automatically be.*

Anyone who keeps this fact in mind will quickly understand many complex signal technology problems. We shall return to this constantly.

First, however, the **UP** is to be proved experimentally and assessed in its implications. This is carried out by means of the experiment documented in Illustration 45 and Illustration 46. First a (periodic) sine wave of, for example, 200 Hz is made audible via the sound card or amplifier and loudspeaker. As is to be expected there is only a single tone audible and the spectrum shows only a single line. But this is not ideal either and exhibits a slight spectral uncertainty in this case, for example, only 1 second was measured and not "infinitely long".

Now, step by step, we restrict the length of the "sinusoidal signal", which is actually no longer an ideal one.

The signals shown can be generated by means of the "Cut out" module and can be made audible via the sound card. The more the time section is reduced in size the more difficult it becomes to hear the original tone.

> Definition:
> An oscillation pulse consisting of a specific number of sine periods is called a *burst* signal. A burst is therefore a section from a (periodic) sinusoidal signal.

In the case of a longer burst signal many other tones can be heard alongside the "pure sinusoidal tone". The shorter the burst the more the tone becomes a crackle.

If the burst finally consists of very few (e.g. two) sine periods (Illustration 45, bottom) the original sinusoidal tone can hardly be heard for crackling.

The spectra on the right betray more specific details. The shorter the time duration Δt of the burst, the greater the bandwidth Δf of the spectrum. We must first however agree on what is meant by bandwidth.

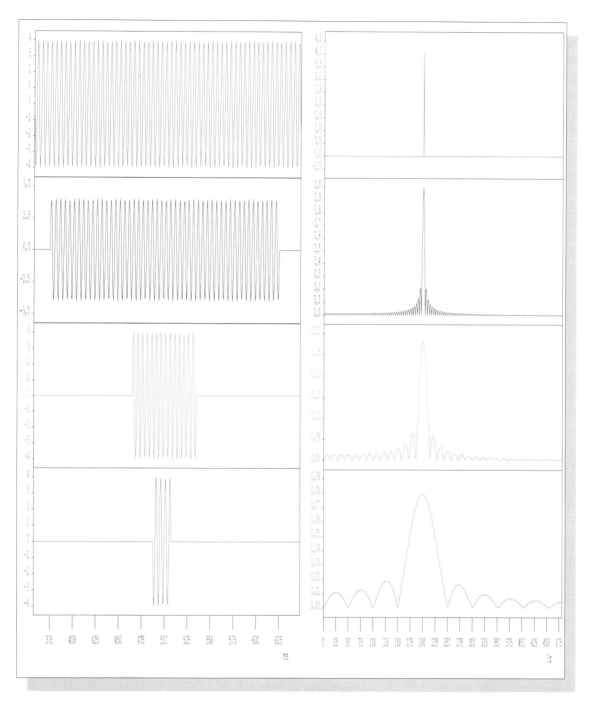

Illustration 45: **Restriction in time involves expansion of the frequency band**

As follows from the sequence of pictures from top to bottom, it is not possible to speak of a"time-limited sinusoidal signal" with one frequency. An oscillation pulse of this kind called a "burst" has a frequency band which gets wider and wider as the length of the burst gets shorter. The frequency of the sinusoidal signal in the upper picture is 40 Hz, the time segment of the sinusoidal signal in the upper sequence established by measuring was 1 s (only visible in extracts here). Fot this reason the spectrum in the upper sequence does not consist of a clear line.

Strangely, with increasing band width the spectrum apparently becomes more and more asymetrical (see bottom). In addition, the maximum point moves more and more to the left. We shall discuss the reasons later. Conclusion: there is every reason to speak of uncertainty.

In the present case the "total bandwidth" appears to tend towards infinity as - on closer inspection - the spectrum goes beyond the visible frequency domain. However, the amplitude tends very rapidly towards zero so that this part of the frequency band can be neglected. If we take "bandwidth" to mean the essential frequency, in the present case half the width of the average main maximum could be designated "bandwidth". It apparently follows that if time length Δt is halved, the bandwidth Δf is doubled. Δt and Δf are in an inversely proportional relationship to each other. Thus it follows:

$$\Delta t = K * 1/\Delta f \quad \text{or} \quad \Delta f * \Delta t = K$$

The constant K can be determined from the Illustrations, although the axes are not scaled. Assume that the pure sine wave has a frequency of 200 Hz. You can now produce the scaling if you remember that the period length is $T = 5$ ms where $f = 200$ Hz. N period lengths then represent the length of the burst $\Delta t = N*T$ etc. In this estimate roughly the value $K = 1$ results. Thus $\Delta f * \Delta t = 1$ follows. But as the bandwidth Δf is a question of definition (it does not usually agree entirely with ours) an inequation is formulated which permits an estimate. And that's really all we want to achieve.

> *Uncertainty Principle **UP** for time and frequency:*
>
> $\Delta f * \Delta t \geq 1$

An alert observer will have noticed that the maximum of the frequency spectrum moves more and more to the left - that is, towards the lower frequencies - the shorter the length of the burst. For this reason it would be a mis-interpretation to assume that the "correct frequency" of the burst was where the maximum is. This is ruled out by the **UP** and the spectrum shows that it is not possible to speak of one frequency in this case. It will be explained in Chapter 5 where this displacement or asymmetry comes from.

> Note: Do not attempt to outwit the **UP** by trying to interpret more than the **UP** permits. You can never give more precise information on the frequency than the **UP** $\Delta t * \Delta f \geq 1$ gives because it embodies an absolute natural boundary.

Illustration 46 shows how suitable it is to choose an inequation for the **UP**. A sine oscillation pulse is selected which begins and ends gently. Then the spectrum also begins and ends in the same way. How large is the time length Δt and how large is the bandwidth Δf of the spectrum? It would be possible to define uniformly the essential area of the time length Δt and the band width Δf to begin and to end where half of the maximum value is reached. In this case an evaluation results - which you should follow step by step - in the relationship $\Delta f * \Delta t = 1$

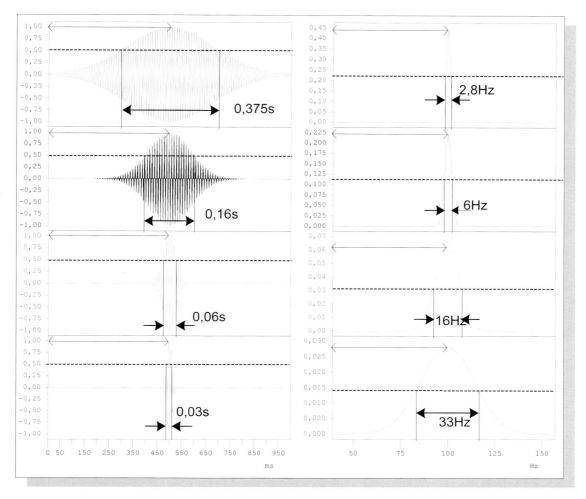

Illustration 46: **Bandwidth Δf, time length Δt and limiting case of UP**

Here a so-called GAUSSian oscillation pulse is more and more restricted in time. The GAUSSian function as revealing a "sinusoidal signal restricted in time" guarantees that the oscillation pulse begins gently and ends gently without any abrupt changes. As a result of this option selected the spectrum also develops according to a Gauss function; it also begins and ends gently.

The time duration Δt and the bandwidth Δf must now be defined as a Gauss pulse is also theoretically "infinitely long". If the time duration Δt and the bandwidth Δf relate to the two threshold values at which the maximum functional value (of the envelope) has dropped to 50%, the product of Δf * Δt is roughly 1, i.e. the physical limiting case Δf * Δt = 1.

Check this assertion using a ruler and calculation using the rule of three for the above four cases: e.g. 100 Hz on the frequency axis are x cm, the bandwidth Δf entered - marked by arrows - is y cm. Then the same measurement and calculation for the corresponding time duration Δt. The product Δf * Δt ought to be about 1 in all four cases.

Sinusoidal signal and δ-pulse as a limiting case of the Uncertainty Principle

In the "ideal" sinusoidal signal Δt –> ∞ (e.g. a billion) applies for the time duration. It follows that for the bandwidth Δf –> 0 (e.g. a billionth part) as the spectrum consists of a line or a thin stroke or a δ-function. In contrast, the δ-pulse has the time duration Δt –> 0. In contrast to the sine the bandwidth Δf –> ∞ (with a constant amplitude!) applies. Sine and the δ-function give the limiting values 0 and ∞ in the time and frequency domains, inverted in each case.

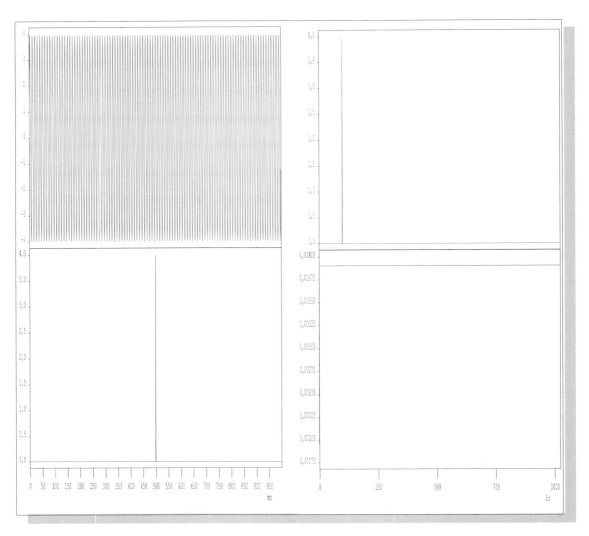

Illustration 47: ***δ-function in the time and frequency domain***

A δ-pulse in one of the two domains (Δt –> 0 and Δf –> 0) always implies an infinite extension in the complementary domain (Δf –> ∝ and Δt –> ∝).
*On closer examination it emerges that the spectral line of the sine (above right) is not a line in the true sense (Δf –> 0) but in a certain sense is blurred i.e. uncertain. The sine was also evaluated only within the segment illustrated from Δt = 1s. According to the Uncertainty Principle **UP** this results in Δf ≥ 1, i.e. a blurred stroke of at least 1 Hz bandwidth.*
A (one-off) δ-pulse produces an "infinite" bandwidth and Δf –> ∝ as a result of Δt –> 0. It contains all the frequencies with the same amplitude; see also Illustration 36. This makes the δ-pulse an ideal test signal from a theoretical point of view, because - see the FOURIER Principle - the circuit/ system is tested at the same time with all the frequencies (of the same amplitude).

Why ideal filters cannot exist

Filters are signal technology components which allow frequencies - i.e. certain sinusoidal signals within a frequency range to pass through (conducting state region) or block them (blocking state region). If only the low frequencies up to a certain limiting frequency are to be allowed to pass, this is called a lowpass filter. As we wish to demonstrate, the transition from a conducting state region to a blocking state region and vice versa must always take place with a certain uncertainty.

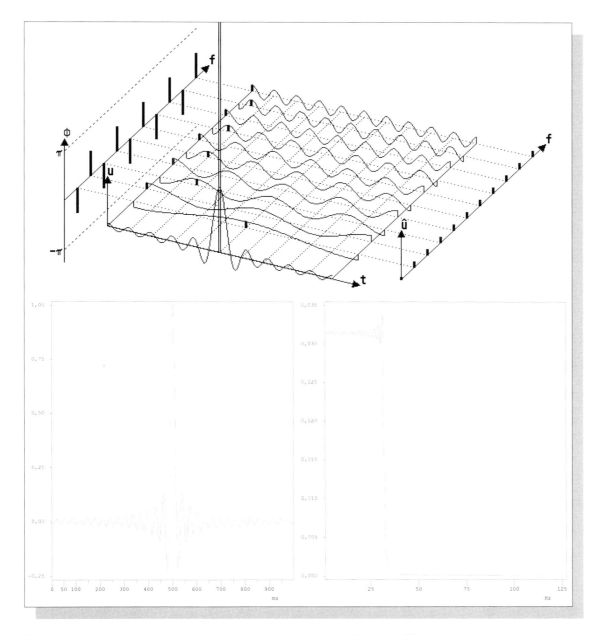

Illustration 48: **Pulse response of an ideal lowpass filter**

The upper FOURIER "playing field" shows a δ-pulse in the time and frequency domain. The sum of the first 10 sinusoidal signals is also entered in the time domain. If there were an ideal "rectangular" lowpass filter which let through (in this case) the first ten sinusoidal signals and then blocked all subsequent sinusoidal signals, precisely this cumulative curve would have to appear at the exit if a δ-pulse had been connected to the entrance.

In the middle illustration there is an indication that strictly speaking this cumulative curve reaches far into the "past" and the "future". This would in its turn mean that the initial signal would have to have begun before the arrival of the δ-pulse at the entrance to the filter. This contradicts the causality principle: first the cause then the effect. Such an ideal rectangular filter cannot therefore exist.

*If this δ-pulse response, which is termed a Si-function, is restricted to the segment of 1s represented here and a FFT is carried out, it results in rounded or rippled lowpass characteristics. All real pulse responses are limited in time; hence as a result of the **UP** there cannot be ideal filters with "rectangular" conducting state regions.*

Note:

Filters are also conceivable in the time domain. A "gate" such as that used in Illustration 45 to generate burst signals, could also be called a „time filter". Gates which filter out a certain signal field in the time domain are however generally called *windows*. An ideal lowpass filter with a limiting frequency of 1 kHz would allow all the frequencies from 0 to 1000 Hz to pass unattenuated but would block the frequency 1000.0013 completely (blocking state region). Such a filter does not exist. Why not? You can probably guess the answer. Because it contravenes the **UP**.

Please note Illustration 48 carefully in connection with the following explanation. Assuming we give a δ-pulse as a test signal to an ideal lowpass filter. What does the initial signal, the so-called pulse response (what is meant is the reaction of the lowpass to a δ-pulse) look like? It must look like the cumulative curve in Illustration 48, as this signal forms the sum of the first 10 harmonics, all other frequencies above the "limiting frequency" are disregarded as in the case of the lowpass filter.

This signal is reproduced on a completely different scale in Illustration 48 centre. This is the pulse response of an ideal lowpass to a one-off δ-pulse. At first its symmetry is clearly recognisable. It is very important that the pulse response of a lowpass of this kind is (theoretically) infinitely wide; it extends to the right and left from the illustrated segment. The pulse response would (theoretically) have had to begin in the past when the δ-pulse had not yet been given to the entrance. A filter of this kind is not causal ("first the cause, then the effect"), contradicts natural laws and can neither be imagined or produced.

If we limit this time response to the segment illustrated - this is done in Illustration 48 - and look which frequencies or which frequency spectrum it exhibits, this results in rounded, "wavy" lowpass characteristics and not in ideal, rectangular characteristics.

The **UP** can therefore be defined more precisely. As the example above shows, it is not just a question of time segments Δt and frequency bands Δf, but more precisely how rapidly the signal in the time segment Δt changes or how abruptly the frequency spectrum or the frequency response (e.g. of the lowpass) changes within the frequency band Δf.

The steeper the curve in the time segment Δt and within the frequency band Δf, the more extensive and marked the frequency spectrum Δf and the time duration Δt.

Time and frequency step transitions always produce extensive transients in the complementary frequency or time domain.

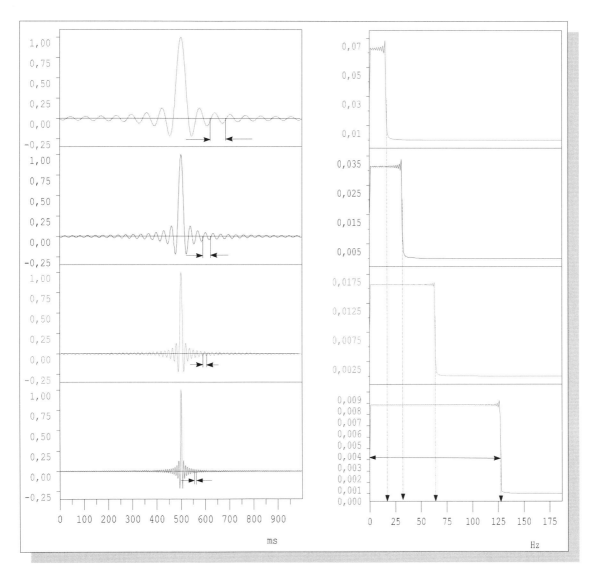

Illustration 49: **Pulse response (Si-function) with different lowpass bandwidths**

As already suggested, the lowpass filter has (at best) a rectangular progression. Up to now we have dealt mainly with rectangular progressions in the time domain. Now look closely at the Si-function in the time domain and compare it with the progression of the frequency spectrum of a rectangular pulse (see in this connection Illustration 48 bottom).

You will probably have noticed that with the Si-functions the time $T' = 1/\Delta f$ is entered which appears to describe something like the period length visually. But there cannot be a period length because the function is not repeated exactly after the time T'. However, each of the Si-functions represented have a different ripple content: it depends on the bandwidth Δf of the lowpass. This ripple content is equivalent to the ripple content of the highest frequency which passes through the lowpass. The pulse response can never change faster than the highest occurring frequency in the signal. The progression of the Si-function is determined precisely by this highest frequency.

The pulse response of an ideal "rectangular" lowpass filter (which - as already pointed out - is physically impossible) has a special importance and is called an Si-function. It is like a sine which is "compressed or bundled in time". For this reason it cannot consist of only one frequency because of **UP**.

The frequency of this invisible sine - if defined exactly, it is the ripple content of the Si-function - is exactly equivalent to the highest frequency occurring in the spectrum. This frequency which is the highest occurring in the spectrum determines how quickly the cumulative signal can change. See Illustration 49.

Frequency measurements in the case of non-periodic signals.

Up to now we have avoided non-periodic signals and the near- or quasi-periodic signals. The **UP** is, however, precisely the right tool to come to grips with these signals. So far we know:

> *Periodic signals have a line spectrum. The distance between the lines is always an integer multiple of the basic frequency $f = 1/T$.*
>
> *Non-periodic one-off signals have a continuous spectrum, that is for every frequency there are other frequencies in the tiniest most immediate neighbourhood.*
>
> *The question now remains: How can the frequencies contained in non-periodic signals, with their continous spectrum, be measured as accurately as possible?*

As a result of $\Delta t * \Delta f \geq 1$ the general answer is obvious: the longer we measure the more accurately can we determine the frequency.

What is the position with one-off i.e. non-periodic signals which only last for a short time? In this case the measuring time will be greater than the length of the signal simply to be able to better capture the whole process . What is decisive for the accuracy of measurement and the frequency resolution: the length of measurement or the length of the signal?

A corresponding experiment is documented in Illustration 50. If you interpret the scaled measurements for the time and frequency domain you should arrive at the following result:

> *If in the case of a one-off signal, the length of measurement is greater than the duration of the signal, the **duration of the signal** alone determines the frequency resolution.*

In the case of non-periodic signals which last a long time - such as language or music - it is only possible for technical or other reasons to analyse a time segment. It would not make sense to have the entire spectrum of a whole concert displayed. The spectral analyses must change as rapidly as the sounds as this is exactly what our ears do!

The only possibility is thus to analyse long-lasting non-periodic signals in segments. But how? Can we simply cut up the signal into several equal parts as if we were using a pair of scissors. Or are more intelligent methods necessary in this case to analyse segment by segment?

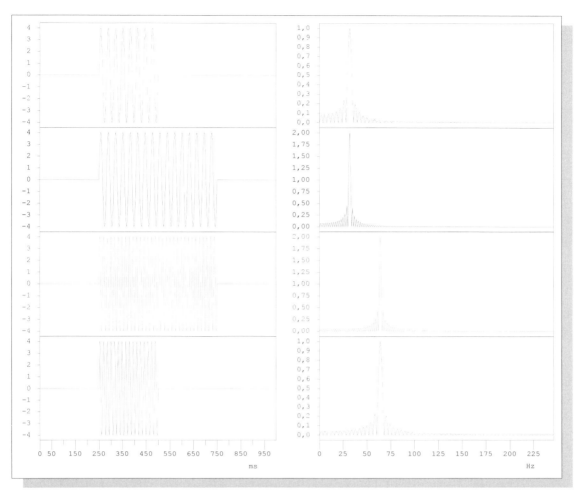

Illustration 50: **Does the frequency resolution depend on the measurement length or the signal length?**

Here are four different one-off burst signals. Two of the burst signals have the same length, two burst signals have the same medium frequency. The length of measurement and thus the length of analysis is in all four cases 1s. The result is quite clear. The shorter the length of the signal the more uncertain the medium frequency of the burst pulse! The uncertainty does not depend on the period of measurement, it depends purely on the length of the signal. This is to be expected as the whole of the information is only contained in the signal, and not in the length of measurement which can be varied at will.

Let us carry out an appropriate experiment. In Illustration 51 we use a lowpass filtered noisy signal as a test signal which in physical terms shows similarities to the production of speech in the throat cavity. The stream of air is equivalent to the noise, the throat cavity forms the resonator/filter. At all events it is non-periodic and lasts for an indefinite time. In this case a lowpass of superior quality (10th order) is selected which filters out practically all frequencies above 100 Hz.

The signal is first analysed as a whole (bottommost series). Above this four individual segments are analysed. The result is strange. The four segments contain higher frequencies than the lowpass filtered overall signal. The reason is easy to identify. The vertical segment has created steep transitions which have nothing to do with the original signal. However, according to the Uncertainty Principle steep transitions bring about a broad frequency band.

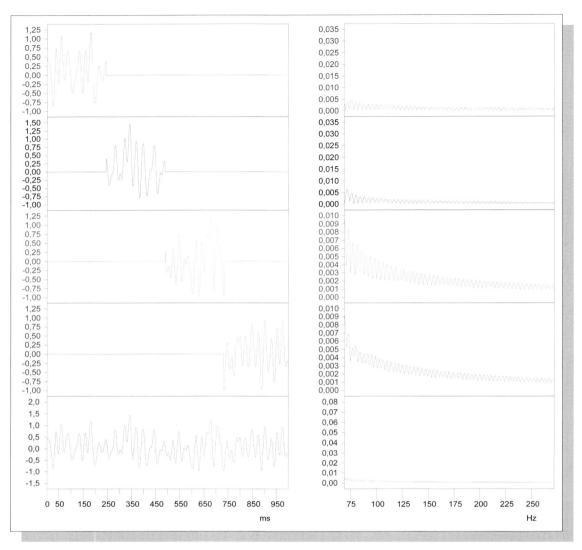

*Illustration 51:   **Analysis of a long-lasting, non-periodic signal.***

The non-periodicity is achieved here by using a noisy signal. This noisy signal is now filtered through a high quality lowpass (with steep edges) with the cut-off frequency 50 Hz. This does not mean that this filter does not let anything through above 50 Hz. These frequencies are damped to a greater or lesser degree depending on the filter quality.

Here we are looking at the "blocking state region" above 50 Hz, beginning at 70 Hz. The upper four signal segments contain considerably more or rather "stronger" frequency shares in this area than the overall signal (bottom). Thus, "cutting out" partial segments produces frequencies which were not contained in the original signal! And - the shorter the time segment the more uncertain the frequency band is. This can be clearly seen by comparing the spectra of the last but one signal segment - which lasts longer - with the four upper signal segments. Incidentally, here too the overall signal is analysed via the signal length (= length of measurement) 1s.

In addition the "link" between the individual segments of the signal which have been separated arbitrarily is lost. This may mean that information has been cut up. Information consists of certain "agreed" patterns - see Chapter 1 - and therefore lasts a certain time. In order to capture this information in its entirety the signal segments would as a precaution have to overlap.

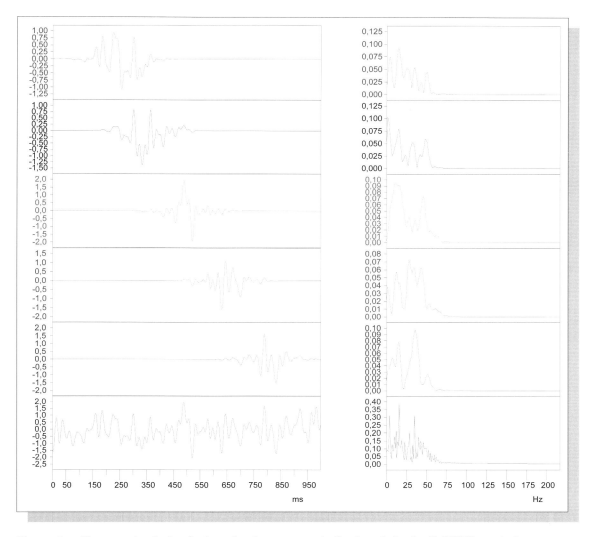

*Illustration 52: **Analysis of a long-lasting, non-periodic signal via the GAUSSian window.***

As in Illustration 51 the long, non-periodic signal here is also divided up into individual time segments. This so-called windowing is carried out by means of a time staggered GAUSSian window. The partial segments thus begin and end gently. In contrast to Illustration 51 the frequency domain of the time segments is not greater than the frequency domain of the signal as a whole.

This important signal technology process is called "windowing". This is intended to differentiate "cutting out" from "filtering" in the frequency domain.

From Illustration 46 you are already familiar with the trick of how to make a signal segment begin and end gently using the GAUSS-function. With this "time weighting" the central area of the signal segment is analysed precisely and the peripheral areas are analysed less precisely or not at all.

Illustration 52 shows what this solution which is in a relative sense the best looks like. The segments begin and end gently, thus avoiding steep transitions. Moreover, the segments overlap in the time domain. This lessens the danger of losing information. On the other hand, the signal is so "distorted" that only the central part is fully displayed and strongly weighted.

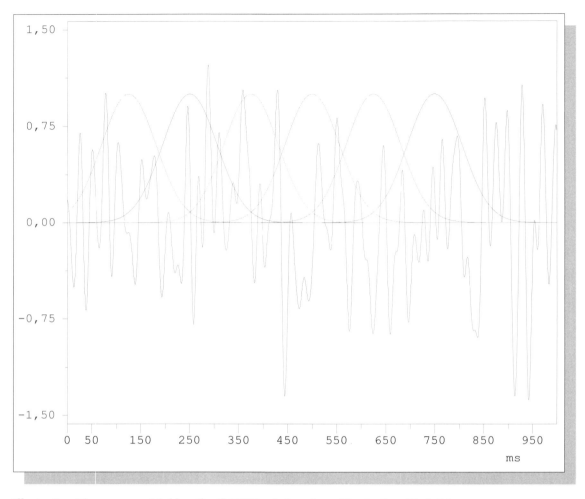

Illustration 53: ***Making the GAUSS-window from Illustration 37 visible***

This illustration precisely shows the 5 GAUSS windows which were used in Illustration 52 to divide the signal into meaningful segments. All the GAUSS-windows have the same form; the successive windows are displaced to the right by a constant value of 75ms. Cutting out the partial segment is mathematically equivalent to the multiplication of the signal with each window function. The overall signal illustrated here is not identical with Illustration 52.

As a result of the Uncertainty Principle there is no ideal solution, only a reasonable compromise. You should not assume that this is simply a technical problem. The same problems occur, of course, in the human production and perception of speech. We are simply used to dealing with them! After all, the Uncertainty Principle is a natural law!

Our ear and brain analyse in real time. Is a long-lasting signal - for instance, a piece of music - analysed constantly and simultaneously - by "windowing" in the time domain?

No, our ear is a FOURIER analyzer, functioning in the frequency domain. It works like a large number of very narrow bandfilters which lie alongside each other frequency-wise. As a result of the Uncertainty Principle **UP** the reaction time (build-up time) is greater the narrower the bandwidth of the filter. More details will be given in the next chapter.

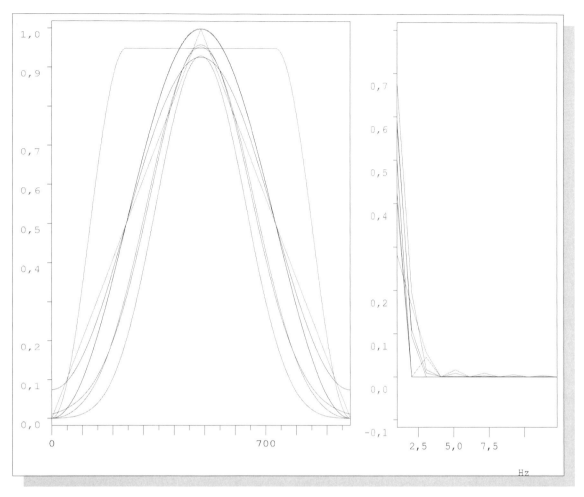

Illustration 54: ***Overview of the most important "window types"***

Here you see the most important examples of window types. With the exception of the triangular type and the "rounded rectangular window" they all look very similar and their spectrum hardly differs. With a duration of roughly 1 s they generate a frequency uncertainty of only roughly 1 Hz. The „rippled curves" in the frequency domain derive from the worst windows - triangle and "rounded rectangle".

"Windowing" always involves a compromise. As it is a very important process, a great deal of thought has been given to the ideal form of a time window. In principle they all do the same thing and with few exceptions resemble the GAUSS-function. They begin gently and end gently. The most important types of window are presented in Illustration 54 and their frequency-based effects are compared with each other. The worst is the triangular window, of course, as it displays fluctuations at the beginning, in the middle and at the end. The other windows are scarcely different from each other so that we shall continue to use the GAUSSian-window.

In the frequency-based analysis of long-lasting non-periodic signals - e.g. speech - these are divided up into several segments. The frequency-based analysis is then carried out for each individual segment.

> *These segments must begin gently and end gently and overlap in order to lose as little as possible of the information contained in the signal.*

> *The greater the time duration Δt of the time window selected the more precisely can the frequencies be established or the greater the frequency-based resolution.*

This process is called "windowing". This segment by segment dissection is equivalent from a mathematical point of view to the multiplication of the (long non-periodic) original signal with a window function (e.g GAUSS-function).

Ultimately, a long-lasting non-periodic signal is divided up into a multiplicity of individual events and analysed. The link between the individual events must not be lost. They should therefore overlap.

In the case of one-off, brief events which begin abruptly at zero and also end there (for instance, with a bang) a rectangular window should always be chosen, limiting the actual event in time. Thus the distortions are avoided which inevitably occur with all the "gentle" window types.

Near-periodic signals

Near-periodic signals form the ill-defined borderline area between periodic signals - which strictly speaking do not exist - and non-periodic signals.

> *Near-periodic signals are repeated over a given period of time in the same or a similar way.*

A sawtooth is selected as an example of a near-periodic signal which is repeated i the same way over variious different periods of time (Illustration 55). The effect is the same as in Illustration 50. In the case of the burst, the sine wave is repeated in the same way. In each case a comparison of the time and frequency domain with consideration of the Uncertainty Principle leads to the following results.

Near-periodic signals have more or less linear-like spectra (smudged or blurred lines) which include only the integer multiples of the basic frequency. The shorter the overall length the more blurred the lines. This is true of the line width:

$$\Delta f \geq 1/\Delta t \quad (\mathbf{UP})$$

Real, near-periodic signals or near-periodic phases of a signal are - as the following illustrations show - not always recognised as near-periodic in the time domain. This is successful at the first attempt in the frequency domain.

> *All signals which have "linear like" continuous spectra and in which these blurred lines can be interpreted as integer multiples of a basic frequency are defined here as **near-periodic**.*

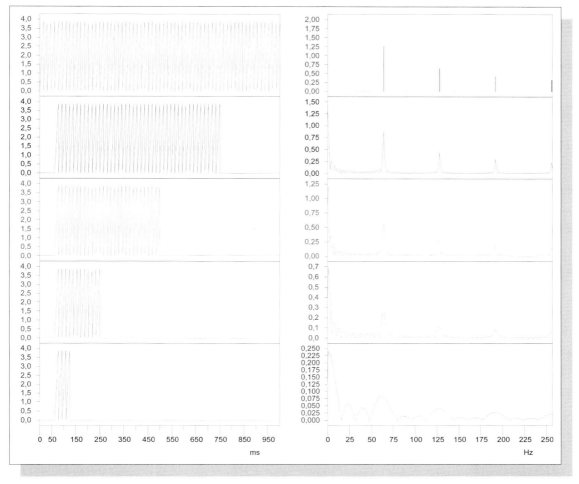

Illustration 55: ***On the spectrum of near-periodic sawtooth signals.***

This series of sawtooth signals illustrates very clearly how often signals should be repeated in order to be regarded as near-periodic. The upper series also contains a near-periodic signal, because this sawtooth was only recorded for 1s. Both the lower series include the transition to non-periodic signals.

In practice there are signals which have a linear-like spectrum whose "blurred" lines cannot in part be interpreted as integer multiples of a basic frequency. They are defined here as quasi-periodic. The reason why they exist will be described in the next section.

Tones, sounds and music

Up to now we have examined signals which were produced artificially such as rectangular, sawtooth and even noisy signals. We now come to the signals which are really meaningful; which are important in an existential sense because they impinge on our sense organs.

Strangely, in all theoretical books about Signals-Processes-Systems they are largely despised or ignored. They do not always fit into simple patterns, they are not just the one thing but also have characteristics of the other. We are talking here about tones, sounds and song, but above all about language.

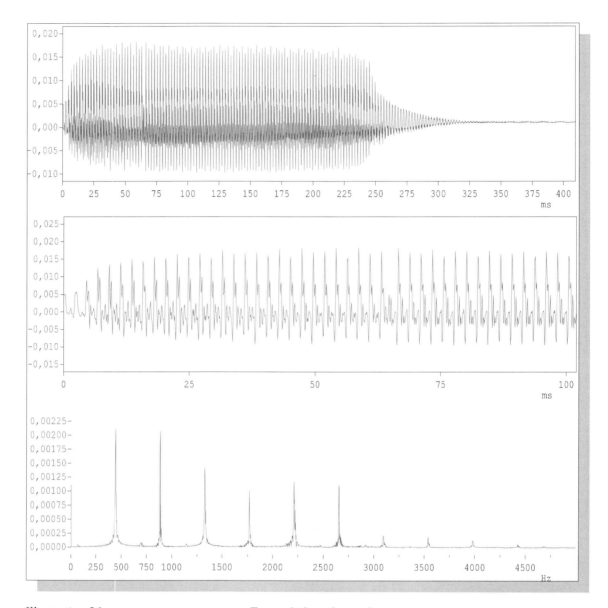

Illustration 56: **Tone, pitch and sound**

The near-periodicity of all tones is illustrated by means of a short clarinet tone (440 Hz = concert pitch "A"). In the time domain it is possible to perceive "similar events" at the same distance T from each other. Use a ruler to measure 10 T (why not simply T?), define T and calculate the inverse value $1/T = f_A$.

The result ought to be the basic frequency $f_A = 440$ Hz. As our ear is a FOURIER analyser - see Chapter 2 - we are able to recognise the (base) pitch. If you are not entirely unmusical you can also sing this tone after it has been played.

The "concert pitch" of a clarinet sounds different from that of a violin., i.e. every instrument has its own timbre. These two tones differ in the amplitude of the overtones and not in the basic pitch ($= f_A$). As a violin sounds "sharper" than a clarinet there are more overtones than in the spectrum of the clarinet.

A short tone/sound purposely wa chosen because it demonstrates a small "defect" in the near-periodic segment. The actual tone lasts roughly 250 ms and produces a near-periodic spectrum. One thus arrives at the following rule of thumb: every uniform tone/sound which lasts at least 1 s produces a practically periodic spectrum!

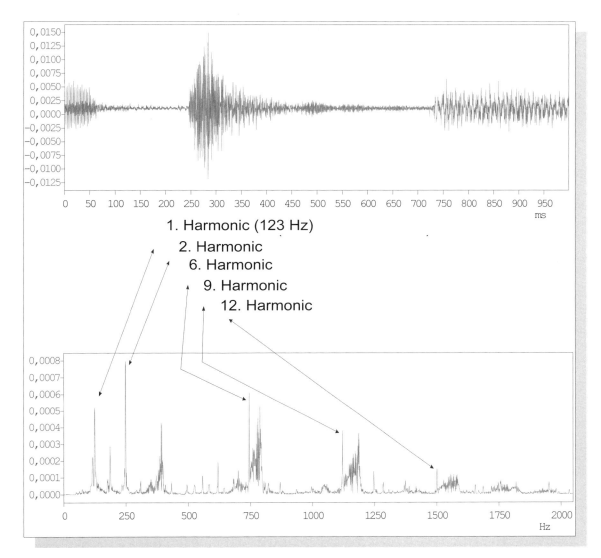

Illustration 57: **Sound as the superimposition of various different tones**

*Extract from a jazz recording (Rolf Ericson Quartet). At this moment trumpet and piano are playing. While the time domain betrays little of the near-periodic character of the music, the situation is quite different in the case of the frequency domain. The lines speak an unambiguous language. However, which lines belong together? In addition the spectrum does not contain any information as to when certain tones/sounds within the period of time under consideration were present. From the "width" of the lines, however, conclusions can be drawn as far as the length of these notes/sounds is concerned (**UP!**). In this connection note once more Illustration 45 and Illustration 55.*

We can now continue with simple experiments using our proven method. Thus, we now use a microphone as a "sensor" - as the source of the electrical signal.

The human ear perceives an acoustic signal as a tone or a sound if it succeeds in allocating it more or less clearly to a particular frequency. In addition, the signal is felt to be harmonic if all the frequencies are in a particular relationship to each other (they are equidistant from each other). As a result of the **UP**, this clear frequency allocation is as a only possible if the signal is repeated in a similar way over a longer period of time within the time segment observed.

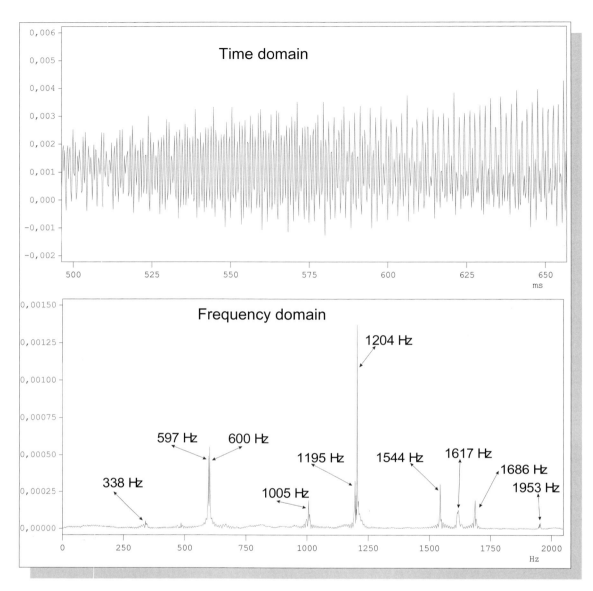

*Illustration 58: **Sound of a wine glass as a quasi-periodic signal***

In the time domain it is very difficult to detect periodicity. The signal appears to change constantly. Only the distance between the maximum values in the right-hand half appears to be practically constant.

On the other hand, the frequency domain displays clear lines. The frequencies were measured on the screen using the cursor. As you will easily establish, not all lines are integer multiples of a basic frequency. The signal is therefore not near-periodic. We term this type of signal quasi-periodic. The physical cause of quasi-periodic signals is the oscillations of a membrane for example. A wine glass is a kind of deformed membrane. Waves, so-called oscillation modes with certain wave-lengths or frequencies, are formed on the membrane depending on its size and shape. These frequencies then appear in the spectrum.

An analysis of this kind can be used, for example, in automation technology in the manufacture of glasses or roof tiles to find defective objects, for instance those with cracks. Their spectrum differs considerably from that of an intact glass or roof tile.

Tones or sounds must therefore last for a longer period of time in order to be recognised as such. For this reason tones/sounds are near-periodic or quasi-periodic!

The following rule of thumb results from the analysis presented in Illustration 55.

Every uniform note/sound which lasts at least 1s produces a practically periodic spectrum. Any practically periodic spectrum corresponds acoustically to a tone/sound which lasts at least 1s and which can be clearly identified as far as pitch is concerned.

Note: in ordinary language (and even among experts) the terms "tone" and "sound" are not clearly differentiated. People speak of the sound of a violin, but they also say the violin has a beautiful tone. We shall define and use the terms in the following way:

The pitch can be defined precisely in the case of a tone. It is thus a sinusoidal variation in pressure which is perceived by the ear.

- In the case of a tone the pitch can be defined precisely. A violin tone contains several audible frequencies; the lowest perceptible frequency is the basic tone and determines pitch. The others are called overtones and, in the case of near-periodic acoustic signals, are integer multiples of the basic frequency.

- A sound - for example a piano chord - generally consists of several tones. In this case it is not possible to identify a single pitch or even a clearly defined pitch.

- Every instrument and every speaker has a certain timbre. It is characterised by the overtones contained in the superimposed tones.

A clear differentiation of the terms "tone" and "sound" is hardly possible because they have been in use colloquially for considerably longer than the physical terms "tone" and "sound" in acoustics.

Tones, sounds and music are an unsurpassed mental stimulus for human beings. Only optical impressions can compete. In the evolution of human beings a certain sensibility for the superimposition of near-periodic signals - tones, sounds, music - seems to have prevailed.

Although the amount of information has to be limited as a result of its near-periodicity, it is music which we find particularly appealing. Speech also falls into this category. It has a great deal to do with tones and sounds. On the other hand it serves almost exclusively to convey information. The next chapter will therefore look at this complex in a case study.

Exercises on Chapter 3

Exercise 1

Design a circuit with which you can reproduce the experiments in Illustration 45. You will obtain the burst signals, using the "Cut out" module, by cutting out a periodic sinusoidal signal in the time domain using this module.

Exercise 2

In the "filter" module lowpass filters and highpass filters of different types and orders can be set.

(c) Direct a δ-pulse to a lowpass filter and examine the way in which the duration of the δ-pulse response h(t) depends on the bandwidth of the lowpass filter.

(d) Change the quality (steepness) of the lowpass filter (via menu) and examine its influence on the pulse response h(t).

(e) Put the δ-pulse and the pulse response on a screen and make sure that the pulse response can only begin after the δ-pulse has been directed to imput.

Exercise 3

The so-called Si-function is the pulse response of an ideal "rectangular" filter. It is also a practically ideal band-limited LF-signal which contains all the amplitudes up to cut-off frequency at (virtually) the same strength.

(a) Switch on DASY*Lab* and select the circuit in Illustration 48. A Si-function is produced and its spectrum displayed. Using the formula component change the form of the Si-function and the effect on the spectrum by experimenting.

(b) Make sure that the ripple content of the Si-function is identical with the highest frequency of the spectrum.

(c) You wish to measure the properties of a high-quality lowpass filter but you only have a normal oscilloscope to observe the Si-like pulse response. How can you deduce the features of the filter from this?

Exercise 4

Generate a speech-like signal for your experiments by lowpass filtering a noisy signal. Where is there a "noise generator" and a "lowpass filter" in the mouth and throat cavity?

Exercise 5

Why do near-periodic signals look "near-periodic" while quasi-periodic signals (see Illustration 58) do not look at all "near-periodic" although they have line spectra?

Exercise 6

(a) Develop a circuit with which you can represent the time window types of the "Data window" module graphically as in Illustration 54

(b) Compare the frequency curve of these various different time windows as in Illustration 54 on the right.

(c) Take a longish filtered noisy signal and try as in Illustration 52 to carry out the "windowing" by means of staggered overlapping GAUSSian windows.

(d) Present the spectrum of these signal segments in a time-frequency landscape.

Exercise 7

Examine the pulse response h(t) of various different lowpass filters in the time and frequency domain.

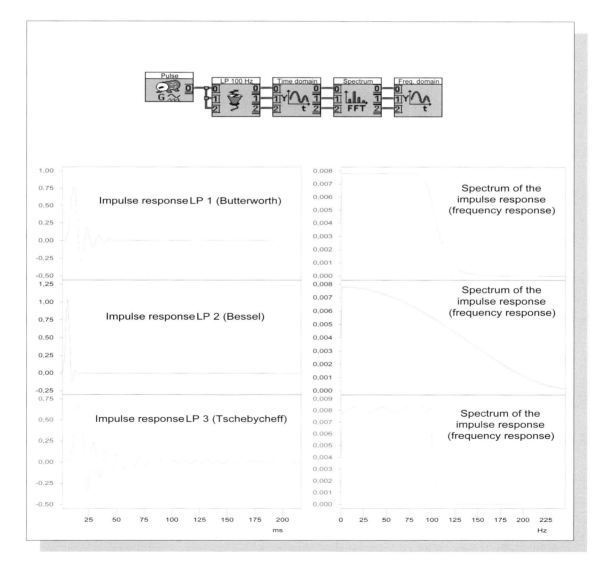

Illustration 59: ***Pulse response and transfer function***

A δ-pulse is directed at three different lowpass types with the same cut-off frequency (100 Hz) and of the 10th order. Remember that the one-off δ-pulse contains all the frequencies from 0 to ∞ with the same amplitude.

(a) What frequencies can the pulse responses only contain? What features of the filters can you deduce from the different pulse responses h(t)?

(b) The frequency domain of the pulse response apparently gives the "filter curve" and the frequency response of the filters. Why?

(c) Why is the length of the lower pulse response much greater than that of the other pulse responses? What does this mean from the point of view of the **UP**?

(d) Design this circuit using DASY*Lab* and carry out the experiments.

Chapter 4

Language as a carrier of information

It is always interesting to see what a short step it is from the fundamental principles of physics to their practical application. A knowledge of fundamental principles is absolutely essential. In this context, however, the word "theory" for many people has unpleasant associations, probably because it is mostly described by means of abstract mathematical models. This is not the case in this book.

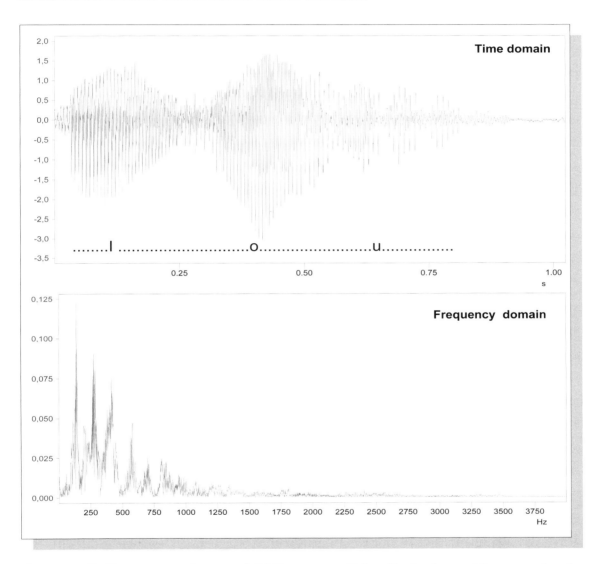

Illustration 60: ***The sequence „I owe you" (IOU) as a one-off signal in the time and frequency domain***

Even a short sequence such as "I owe you" has a very complicated structure in both the time and frequency domains. In the time domain 5 phases are recognizable. The initial "i", the transition phase from "i" to "o", the "o"-phase, the transition phase from "o" to "u", and the final, slightly longer "u"-phase. The initial transient phase and the final concluding phase could be added. The frequency spectrum shows us that it does not make sense to determine the frequencies over the entire length of the word. Many frequency patterns are interlocking. It is obviously more sensible - as shown in Illustrations51 -49 - to analyse individual sections of the word separately. Special techniques have been developed for this purpose ("Waterfall-analysis", see Illustration 64).

*Illustration 61: **"Waterfall"-representation of the sequence "I owe you"***

The entire sequence has been divided up into individual sections which have, as in Illustration 52, a gentle beginning and ending (weighting with a GAUSSian window). The word begins in the uppermost line and ends at the bottom. The "i"-phases, the "o"-phase and the transition phases between them can be clearly discerned. The frequency spectrum of each section is calculated (Illustration 63).

Before we have a closer look at how language is actually generated and above all how it is perceived we should analyse it with regard to its time and frequency domain.

Two words suffice as an example to be able to draw general conclusions on the possibilities and the speed with which information is transmitted by means of language.

The first sequence, "I owe you" (IOU), is known in everyday-life. It consists mainly of 3 vowels. If this sequence is pronounced slowly, five different phases can be discerned: the vowel "i" - the transition from "i" to "o" - the vowel "o" - the transition from "o" to "u" - the vowel "u".

Illustration 60 presents the overall event as a one-off signal in the time and frequency domain. If one looks at the event through a magnifying class, as in Illustration 61, near-periodic segments of the signal can be detected. These are clearly the vowels (Illustration 62). The spectrum in Illustration 60 also shows a number of "lines" which hint at near-periodic elements. As I have already pointed out, it is, however, not possible to detect the sequence of these near-periodic sectors in the spectrum/frequency domain. It is thus an obvious possibility to analyse the signal segment by segment by means of windowing and to examine its frequencies.

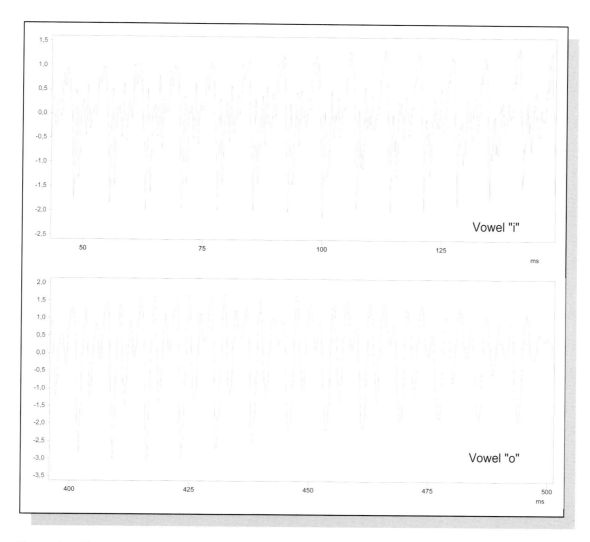

Illustration 62: ***Near-periodic nature of vowels***

These are segments of the time domain of the vowels "i" and "o" from Illustration 60 which explain the expression "near-periodic" better than words could. The segments clearly show in what shape signals can repeat themselves in a similar way.

This is carried out in Illustration 63 where a particularly impressive technique - the "waterfall analysis" - is employed which presents a landscape-like, three-dimensional picture of the frequency-time landscape of this word. If windows overlap - which, as I have demonstrated in Illustrations 51 and 52, is the case here - the 3D-landscape must contain all the information of the overall signal "I owe you" because the connection between the individual segments has not been disrupted. Hence, the first important results are:

> *A vowel can be characterised as a **near-periodic segment** of a "word signal". In other words, it also has a "line-like" spectrum. The longer it takes to pronounce, the more clearly it can be identified. The shorter the period of pronunciation, the less clearly it can be identified (**UP!**). An extremely short vowel would just be like a crack.*

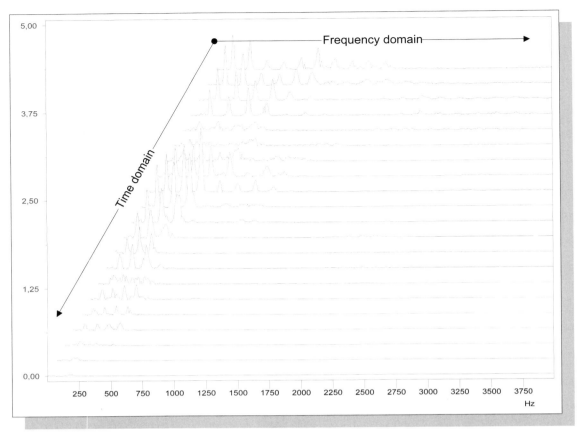

Illustration 63: ***Waterfall analysis: frequency-time landscape of the sequence „I owe you"***

This is an Illustration which corresponds to Illustration 61 in the frequency domain. At the top there are the spectra of the beginning of the word, at the bottom those of the end of the word. The "i" spectra at the top and the „u" spectra at the bottom are clearly identifiable. In the middle there are the "o" spectra with a higher frequency than the "u" spectra and the blurred transition spectra.

*An ideal frequency-time-landscape would differ slightly from the one presented here. It would be composed of several partly parallel disc-shaped pinnacles stretching from the foreground into the background with deep ravines between the peaks. This perfect frequency-time-landscape would present in an ideal way the connection between length of time and bandwidth and thus the **UP** for this event. See also Illustration 66.*

> The way experienced lecturers and speakers stress a particular word contributes to its intelligibility. They stress the vowels.
>
> Those who know how to stress words well ought in theory to be good singers as well because a stre-e-e-ssed word is basically a sung word with a clear pitch.

As you will know, language consists of a sequence of vowels and consonants. Vowels are near-periodic. But this is not true of consonants. They are like noises, some of which use the noise of the air stream (try and pronounce a very long "s") some are like small explosions, such as "b", "p", "k", "d", "t". These consonants have a continuous spectrum without near-periodic elements.

The second example is the English word "history" because it consists clearer of an alternating sequence of vowels and consonants.

This sequence limits the speed of human speech and thus the speed of information. Language is a rapid succession of vowels and consonants. The shorter the vowels the less comprehensible they (and speech generally speaking) become.

There are people who - in order to demonstrate how fast they can think or how intelligent they are - talk so fast that it is extremely difficult to follow them. It is not, however, difficult because they think so fast but because our brain has to work very hard to assign the linguistic elements to individual vowels. Often the whole context is needed if one wants to get the message. And that is a very stressful business.

Let me now point out the most important results so far: owing to the fact that the human ear can only hear sinusoidal signals - I will come back to this point later - the following statement can be made:

> *Every tone, sound and vowel have some characteristic frequencies which - just like a fingerprint - are virtually unmistakable.*

The acoustic recognition of patterns by means of our ears takes place in the frequency domain because glasses, coins and other solids produce a certain sound when they are clinked, i.e. they emit - after a transient phase - near- or quasi-periodic signals. The same is true of vowels. The spectra of these sounds are almost exclusively characterised by only a few specific frequencies - lines - and thus produce very simple patterns which serve as an "identity tag".

> *The acoustic recognition of patterns in nature and in technology takes place mainly in the frequency domain.*
>
> *Frequency patterns of near-periodic and quasi-periodic signals - e.g. vowels - are particularly simple because they merely consist of several blurred peaks of different height.*
>
> *Frequency-time landscapes of sounds and vowels resemble **beds of nails** with nails of different height.*

The human acoustic system - ears and brain - functions in very much the same way as presented and described here in the Illustrations. The human brain does not wait for a piece of music to end before it begins its frequency analysis but analyses the music on a continuous basis. Otherwise we would not be able to hear sounds continuously. As I have already shown in the last chapter, this real-time operation is not carried out through many successive time windows (windowing) but basically in the ear through many parallel frequency windows (filters, bandpass filters). The human ear is a FOURIER analyser, i.e. a system, which is organized on a frequency basis. Illustration 72 describes the structure of this chain of filters and its location.

Besides, the human brain must have something like a library or database where the numerous "beds of nails" are stored as references. How else could we recognize a particular sound or piece of music?

The acoustic processes in the human ear are much more complex than described here. So far, it is largely unknown how signals are processed in the brain. It is well-known which areas of the human brain are responsible for specific functions, but a precise model of these functions has not been elaborated yet.

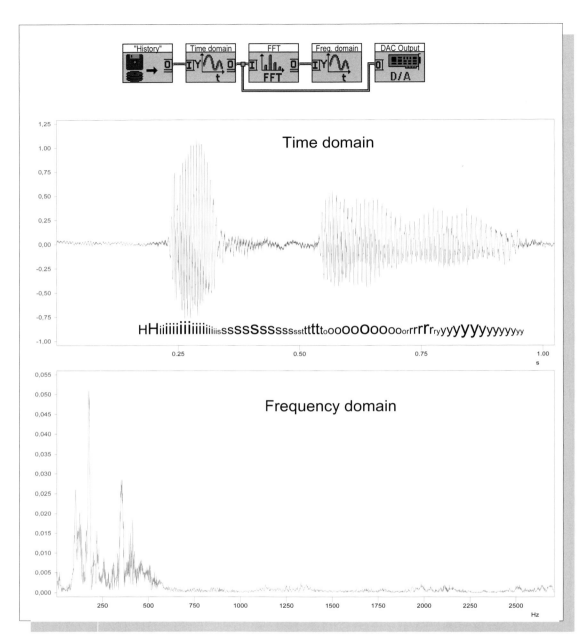

Illustration 64: *The word "history" as a sequence of vowels and consonants*

In the time domain vowels and consonants can be distinguished quite easily. The shape of the plosives (consonants) are clearly recognisable.

The frequency domain of the complete word "history", however, does not provide any relevant information. Again, a waterfall representation or frequency-time landscape as shown in Illustrations 65 and 66 is needed.

But no matter what the details behind these processes are, one thing is certain, they are biophysical processes, or to be more precise, phenomena from the field of the physics of waves and oscillations. As nature is the great preceptor of natural sciences such as communications engineering I should like to discuss briefly the principles of language generation and perception.

Chapter 4   Language as a carrier of information   Page 95

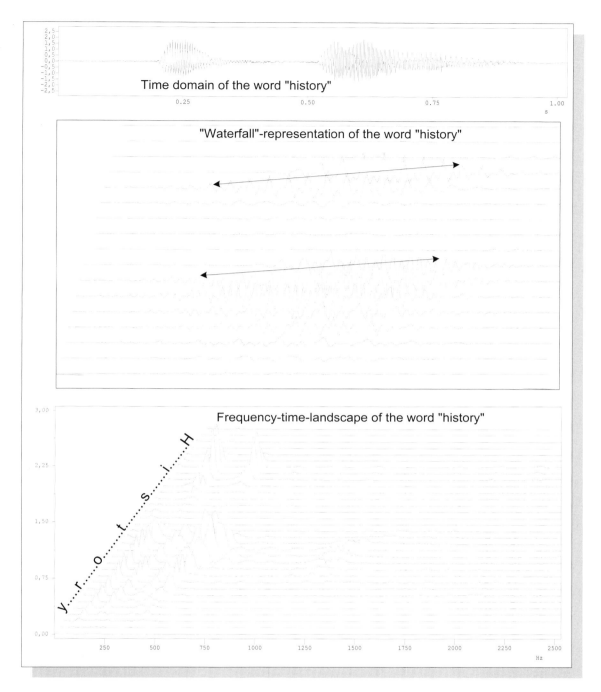

Illustration 65:   ***The word "history" as a waterfall representation and frequency-time landscape***

At the top you can see the complete sequence of the word in the time domain, at the bottom the waterfall representation (time domain). Please note the overlapping sections of signals in the waterfall representation. They are marked by the two arrows. The sequence starts with the plosive "h", followed by the vowel "i". The vowels "i"and "o" are clearly distinguishable since they both have a distinctive near-periodic structure. The sibilant "s" can also be clearly recognized in time domain, but hardly in frequency domain..

The waterfall representation is used for a frequency analysis in order to resolve the non-periodic and near-periodic characteristics more effectively. These are clearly recognizable in the frequency-time landscape.

This type of signal analysis should be used in connection with acoustic pattern and voice recognition in order to be able to identify a specific word by its characteristic sequence of consonants and vowels.

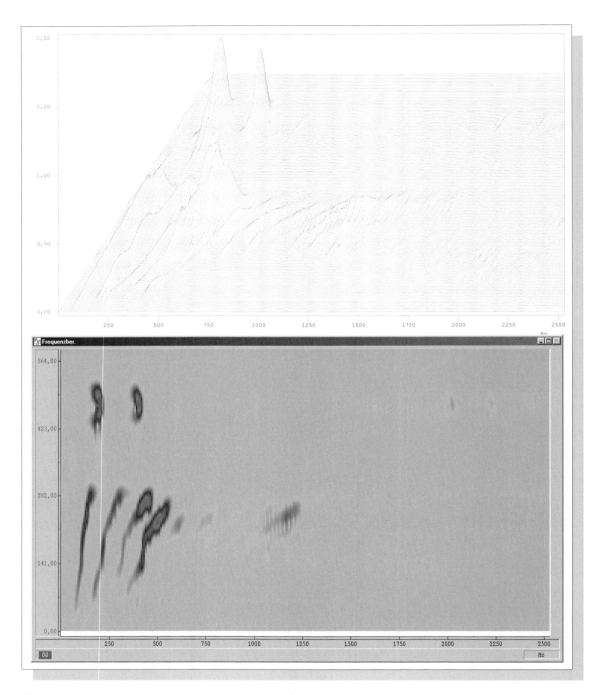

Illustration 66: **The sonogram as a specific form of the frequency-time landscape**

Above you see a variant of the perfected time-frequency landscape of the word "history". In contrast to Illustration 65 the FFT window was only displaced minimally compared with the previous window.

A total of 100 spectra were recorded successively. Now the time-frequency landscape looks like a proper landscape. The vowels are represented by "discs" which are the same distance from each other.

The sonogram at the bottom represents the vertical pattern of the graph at the top. The amplitude level is divided into a few vertical layers. Each layer has a different colour, which of course you cannot see on the black and white print (the electronic document shows the colours on the screen).

Sonograms are used in voice and language research to recognize acoustic characteristics of voices by means of graphic patterns.

How speech, tones and sounds are generated and perceived

There are two reasons why I would like to make a short excursion into human anatomy and, in the same context, into the physics of waves and oscillations:

- Nature is the greatest preceptor of natural sciences such as communications engineering. In the course of evolution nature has developed mechanisms of perception which have enabled the various different species to survive. Human sense organs are highly sensitive and are precise sensors which far surpass technological imitations. A good example is the human ear: even the weakest acoustic signals which are just slightly above the noise level caused by the bombardment of the eardrum with air molecules are perceived.

- These physiological correlations are the basis of acoustics and of any form of spoken communication.

Engineers for some reason often shy away from physics and above all from biophysics although engineering is nothing more than the sensible and responsible application of the laws of natural sciences. The theorists in natural sciences, however, have so far failed to produce a mathematical model of sensory perception such as hearing, and anything which cannot be defined mathematically is widely ignored in textbooks on the theory of communications engineering.

My objective is to demonstrate how sounds and language are generated in the speech organs and how the acoustic perception in the human ear works. As a highlight we are going to develop a case study of a simple computer-aided voice recognition system.

We will have to content ourselves with a very simple model of the human ear because the acoustic perception and above all the processing of signals by the human brain are so complex - as I have already pointed out - that so far relatively little is known about these processes.

In any case, it has all got to do with the physics of oscillation and waves of which a little knowledge is necessary to be able to understand these processes. Please remember: there are no processes in nature or in engineering which would contradict the laws of natural science - e.g. the laws of physics. So let us begin with these laws.

If we want to do research in the field of the generation of tones, sounds and speech we will have to ask ourselves what is required to generate them. What we basically need are oscillators and cavity resonators as well as energy to trigger the process.

> Definition: a mechanical oscillator is a vibrating system which - once it has been given a mechanical pulse - oscillates in its characteristic natural frequency. The oscillation can be maintained if the relevant energy is supplied. The oscillation can be particularly effectively maintained if the spectrum of the energy supplied contains in its signal the natural frequency of the oscillator. An example of a mechanical oscillator is a tuning fork, another example is the reed in the mouthpiece of a clarinet which is kept vibrating by means of an airstream.

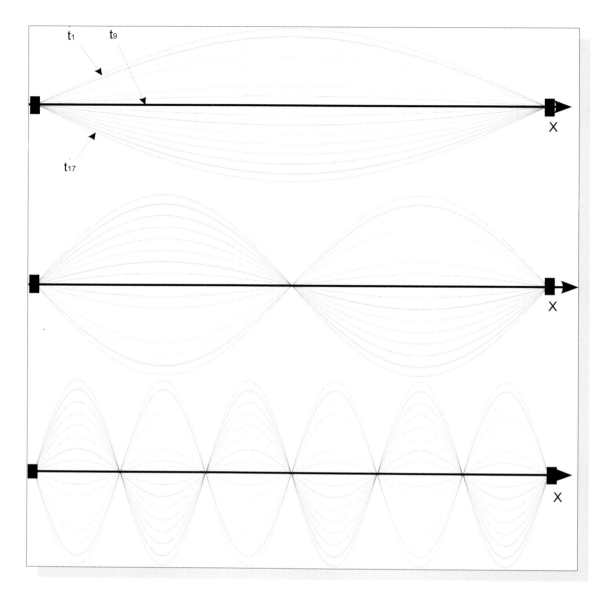

Illustration 67: ***Standing sinusoidal wave on the string of an instrument***

The sinusoidal deflection of the string is presented in a highly exaggerated way. In the above depiction the length of the string is λ/2. The top run of the curve is valid for the moment t_1, the run below that for the moment t_2 etc. Moment t_9 shows that the string is temporarily immobile. With moment t_{17} the bottom run is reached. After that the whole process reverses. At t_{25} the string would be immobile again and at t_{33} the same situation as at t_1 would be reached. This means that the standing wave is a dynamic, i.e. time-variable, condition.

Those areas of the string which are permanently immobile are called nodes, the areas of greatest deflection are called antinodes.

*In the middle row the frequency is twice as high but the wavelength half as long as at the top. As you can see only integer multiples of the base frequency (top) can form a standing wave on a string. That is why the string of a guitar produces a (near) periodic sound with a characteristic timbre. The sinusoidal basic "conditions" depicted above overlap each other (as you have already seen in connection with the FOURIER Principle **FP**) when the string is plucked to produce a sawtooth or triangular deflection.*

A (mechanical) cavity resonator is, for example, any type of flute, organ pipe or the resonant cavity of a guitar.

Depending on the shape and size of the air volume, particular frequencies or frequency ranges are amplified in a resonator, others are attenuated. The frequency ranges are amplified for which standing waves can be produced in the cavity.

Transitions are fluid because a flute can be defined as an oscillator or a resonator. In either case it is a vibrating system.

Let us first concentrate on a one-dimensional vibrating system, a string of a guitar. At either end it is firmly fixed and can thus not be deflected there. When the string is plucked it always produces the same tone or at least the same pitch. Why is that?

A fixed vibrating string is a one-dimensional oscillator/resonator. When it is plucked - which is the supply of energy - it produces free-running oscillations in its characteristic frequencies (natural frequencies; free-running oscillations).

If the string were stimulated in a periodic sinusoidal way with varying frequencies it would oscillate in a controlled way. In that case the frequencies leading to an extreme deflection - in the form of standing sinusoidal waves - are called resonance frequencies. The natural frequencies are identical with the resonance frequencies but are theoretically slightly lower because the attenuation leads to a delay in the oscillation process.

Plucking the string means stimulating it in all frequencies because a one-off short pulse contains basically all the frequencies (see Illustration 47). But all the sinusoidal waves running up and down the string - reflected at the ends of the string - delete each other, except those which are amplified by interference and which produce standing waves with nodes and antinodes. The nodes are those points of the string which are permanently immobile, i.e. which are not deflected. These natural frequencies are always integral multiples of a base frequency, i.e. a string oscillates in a near-periodic way and thus produces a harmonic tone. The wavelength can be calculated quite easily by taking the length of the string L.

Definition:

The wavelength λ of a sinusoidal wave is the distance covered by the wave in the period length T. For the velocity of the wave c the following is true:

$$c = \text{distance} / \text{time} = \lambda / T \text{ and as } f = 1 / T \text{ it follows}$$

$$\boxed{c = \lambda * f}$$

Example:

The wavelength λ of a soundwave (c = 336 m/s) of 440Hz is 0.74 m.

For the standing wave on the string of an instrument only sinusoidal waves for which the length of the string is an integer multiple of $\lambda/2$ are possible. This is to be seen clearly in Illustration 67. All the other sinusoidal waves either peter out alongside the string or eliminate each other.

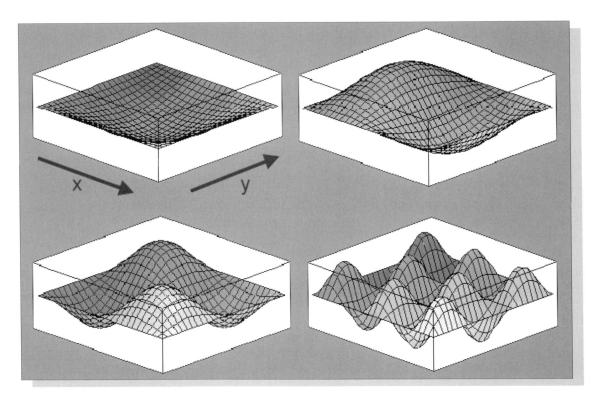

Illustration 68: ***Sinusoidal oscillation modes of a rectangular membrane***

Above left a snapshot of a standing (basic) wave with λ/2 in x and y direction. Note that in the subsequent moments the amplitude decreases, the membrane is briefly flat and is then deflected downwards (in a sinusoidal shape).
Above right: x =λ and y = λ/2; bottom left: x =λ and y =λ ; bottom right: x = 2λ and y = 3/2λ.. If x and y differ both directions also have different wavelengths.

The propagation speed of the wave of the basic tone can thus be easily established experimentally. In the case of the basic tone the length of the string is λ/2 (see Illustration 67 above). Now the frequency f of the basic tone is established by using a microphone and the propagation speed c is calculated by the above formula.

Let us now turn to the two-dimensional oscillator/resonator. A firmly anchored rectangular membrane is selected as an example. If it is struck certain modes of oscillation form on it. Oscillation modes arise in their turn by the formation of standing waves, in this case however in a two-dimensional form. The basic forms of these modes of oscillation are sinusoidal. Now carefully examine Illustration 68 and you will understand which oscillation modes can arise and the way in which they depend on the length and width of the rectangular membrane. Above left you see a case in which the membrane oscillates in the direction of x and y with λ/2.

A membrane - and this is what oscillating coins and glasses are - can be seen as a "two-dimensional" string. If for example the length of a rectangular membrane is not in an integer relationship to the width, natural oscillations or frequencies are possible which are not in an integer relationship to the basic oscillation but which together form a (somewhat blurred) line spectrum. These are quasi-periodic signals. They often do not sound "harmonic". The sound of a drum for example has a noise-like percussion sound which is typical of this.

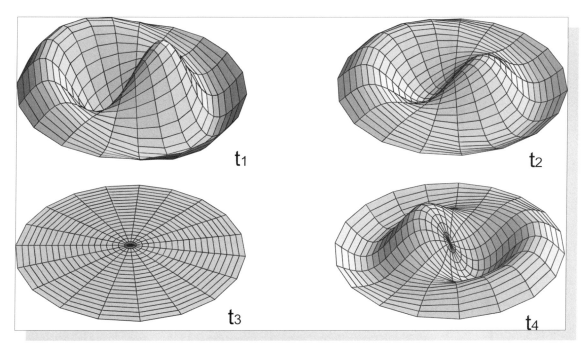

Illustration 69: ***Membrane oscillation of a drum***

Using a program ("Mathematica" from Wolfram Research) the deformation of a drum membrane after being struck by a drumstick was calculated for several consecutive moments on the basis of the laws of wave physics. Each illustration represents the superimposition of possible oscillation modes (natural frequencies). The oscillation modes of a circular fixed taut membrane are far more complex and difficult to understand than those of a rectangular membrane.

Membranes typically produce quasi-periodic signals. An example of this is a drum. For this reason a drum beat sounds quasi-periodic and not "harmonic" (near-periodic), although the spectrum consists of nothing but "lines". In this connection see the reverberating wine glass in Illustration 58.

Finally, we come to the three-dimensional oscillator/cavity resonator. An acoustic guitar has a resonance chamber or cavity resonator. This is what gives a guitar its characteristic timbre. Depending on the shape and size of this resonance chamber three-dimensional standing acoustic waves may be formed as also, for example, in any organ pipe or wooden wind instrument. Three-dimensional standing waves cannot be satisfactorily represented in graphic form (a two-dimensional standing wave is also drawn three-dimensionally). For this reason no illustration is given here.

Every string instrument with a resonance chamber works in the following way: the string is plucked and produces a sound with a near- periodic spectrum. The resonance chamber/ cavity resonator amplifies the frequencies where standing waves can be formed inside the cavity and weakens those for which this is not the case. From an information technology point of view the resonance chamber is thus equivalent to a weighting filter.

Let us now try to apply these insights to the human vocal organs (see Illustration 70). The source of energy is the air stream from the lungs. If pressure builds up at the vocal chords - they are something between a string and a membrane and in this case represent the oscillator - they open, the pressure is suddenly released, then they close again etc etc. In the case of vowels this is near-periodic, in the case of consonants non-periodic.

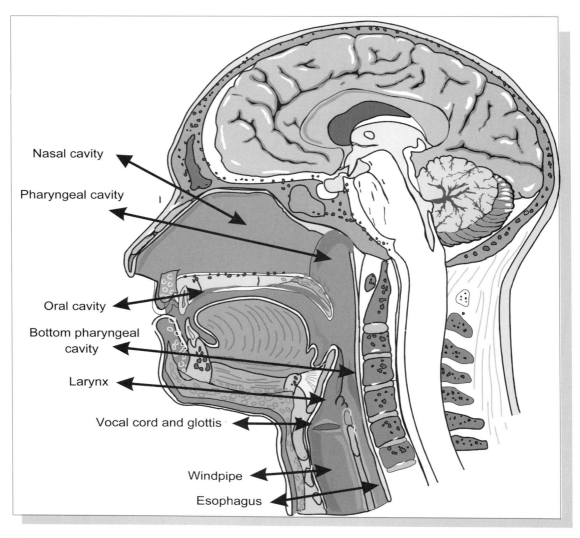

Illustration 70: **The human speech-tract as a deformable cavity resonator**

The human vocal organs consists of an energy store - the lungs - , an oscillator - the vocal chords - , and an adjustable cavity resonator - the mouth and throat cavity. The lungs mainly serve to generate an air stream with sufficient pressure. The air flows through the glottis, the space between the two vocal chords at the lower end of the larynx. If pressure builds up the vocal chords are pressed apart in a fraction of a second. As a result the pressure is suddenly released, the vocal chords close again, and in the case of a vowel or in singing this is near-periodic. If the vocal chords vibrate near-periodically the vocal chord spectrum is then practically a line spectrum.

Illustration 70 explains the role of the cavity resonator mouth and throat. The vocal chord spectrum - as described for the guitar - is influenced as far as frequency is concerned by the cavity resonator. The resonance chamber acts like a "filter bank" with several parallel bandpass filters on the near periodic signal produced by the vocal chords. In this way all the frequencies which lie near formant frequencies are ultimately emphasised. To put it another way - a frequency evaluation/weighting of the vocal chord signal takes place. A vowel belongs to each combination of the four to five formants. Vowels must therefore be harmonic, i.e. near-periodic because the resonance chamber carries out a controlled near-periodic signal, in contrast to the free oscillation (natural frequencies) of a drum membrane. The frequencies of the vocal chord spectrum which lie in these resonance areas are clearly "emphasised" or amplified. All the others are subdued.

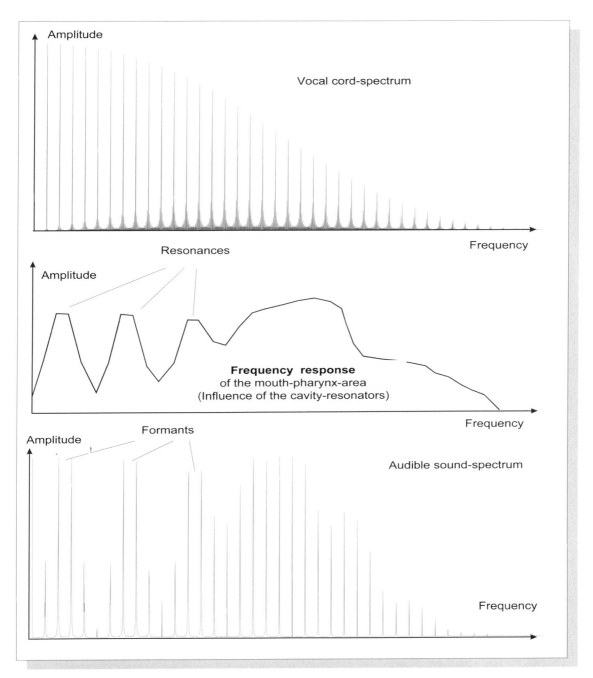

Illustration 71: ***Resonances and formants***

The "emphasis" or the resonance-based amplification of various frequency areas produces formants. Their combination corresponds to the different vowels. The resonance chamber can as directed by the brain be reduced or increased in size in many different ways. If this happens at one place only the formants change or shift in different ways. There are three visible possibilities of changing the cavity resonator - with the jaws, the back of the tongue and the tip of the tongue.

The true miracle is the systematic forming of the mouth and throat cavity as a cavity resonator and the control of the air stream by the individual in such a way that acoustic signals are produced which make possible communication between human beings. The human child needs many years to learn to produce sequences of sounds and to attribute specific meanings to them.

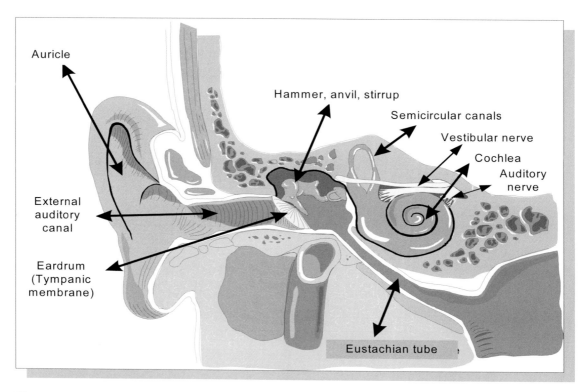

Illustration 72: **Structure of the hearing organs**

Acoustic fluctuations in pressure are transported to the eardrum. This is effected by the mechanics of the ossicles between the cochlea and the eardrum. The ossicles produce an ingenious travelling wave in the cochlea which is filled with liquid.

This wave should be seen in principle as a sweep signal - see Chapter 5 "System Analysis". The whole spectrum of the frequencies present in the fluctuation of pressure is spread over the length of the cochlea canal, i.e. certain areas of sensory hair cells are responsible for specific frequencies.

What is the role of hearing in this process? It can be shown that the actual pattern recognition is carried out here or at least is prepared here. Illustration 72 shows the structure of the hearing organs.

The acoustic fluctuations in pressure of the sound wave reach the eardrum from the auricle (funnel effect) or from the outer auditory canal. These fluctuations in pressure are conveyed by a mechanical system of ossicles - hammer and stirrup - to the most important "sub-system", the cochlea.

This is intended as a simplified description of these processes. The cochlea has the overall shape of a rolled-up tapering funnel and is filled with liquid. Along this funnel "sensory hairs" are distributed which are connected to nerve cells - neurones. These are the actual signal sensors.

The fluctuations in pressure create a travelling wave which subsides towards the end of the cochlea. This takes place in a particular way which is called frequency dispersion. High-frequency components in the pressure fluctuation align themselves at a different part of the wave than low-frequency elements. Thus well-defined parts of the cochlea with their sensory hairs are responsible for low, medium and high frequencies.

According to Helmholtz this phenomenal system functions in the final analysis like a reed frequency gauge. This consists of a chain of small tuning forks. Each tuning fork can only vibrate with a single frequency, i.e. each tuning fork oscillates sinusoidally! The natural frequency of these tuning forks increases continuously from one side to the other. Hence the following is true:

> *Our ear can only perceive sinusoidal signals, i.e. our ear is a FOURIER analyser.*
>
> *Our hearing organs transform all acoustic signals into the frequency domain.*

As can be seen from Illustrations 56 and 58, the spectral pattern of tones, sounds and vowels consists of a few lines in contrast to the progression in time. This contains an extremely effective simplification of patterns or, in modern terms, an efficient compression of data. Our brain needs "merely" to attribute a relatively simple line pattern to a certain concept.

> The transformation of acoustic signals into the frequency domain by means of our hearing organs at the same time implies a very effective simplification of patterns or data compression.

Case study: a simple system for voice recognition

The basic physiological principles described above are to be imitated technologically in a small project. We shall use DASY*Lab* for this. We would be overstretching our resources if we selected too many words.

For this reason, the exercise is as follows: In a warehouse with tall shelving the fork-lift trucks are to be controlled via a microphone. This requires the words "up", "down", "left", "right" and "stop". Thus, we need a microphone connected with a PC soundcard.

Alternatively, a professional multi-function board with a microphone attached could be used. In order to create the link with DASY*Lab* driver software is required. For this you would need the industrial version of DASY*Lab*.

In the first instance the language samples should be recorded and analysed accurately using an experimental circuit in the time and frequency domain which is as simple as possible. As described above the recognition or comparison of patterns is much easier in the frequency domain than in the time domain. The important thing is to establish the features of each of these five frequency patterns.

Planning and initial experimental phase:

- The microphone should respond only above a certain sound level, otherwise background noise would be directly recorded as soon as the circuit is activated and before the words were spoken. A trigger and relay module (Illustration 73) are required for this. This trigger level depends on the microphone and must be carefully tried tested.

- None of these words lasts longer than a second. In contrast to the flow of language we do not initially plan a "waterfall" analysis or an analysis of time-frequency landscapes.

- The settings in the module "FFT" allow the settings power spectrum, power density spectrum, FOURIER analysis and the logarithmic representation of the spectrum in db (decibels) apart from the amplitude spectrum which (with the phase spectrum) we have so far used exclusively.

- Note: The applications "complex FFT of a real or complex function" will be dealt with in Chapter 5.

- Experiment several times with each word. Note how small divergencies in emphasis make themselves felt. In order to be quite sure it is essential to establish several "frequency patterns" for each word to avoid regarding a random component as typical. A reference spectrum with which the spectrum of each spoken word should be compared, should then be selected from these patterns.

The object of these experiments is the *establishing of reference spectra.* The spectra of the words spoken into the microphone will later be compared with these reference spectra or the similarity of the measured and reference spectra will be established. But how is similarity to be measured?

There is also a suitable module for this purpose: *Correlation*. It determines the mutual relationship (correlation) or similarity between - in this case - two spectra by means of the correlation factor. The figure of 0.74 means practically a similarity of 74%, whatever this may refer to.

We use this module - in order not to interrupt the case study unnecessarily - in a simplified way initially without analysing more closely the way it functions. This will be done in the following section "Pattern recognition".

If we succeed in achieving a more or less distinct identification of the word spoken with subsequent digital display of the correlation factor using the module *correlation* our case study will be practically completed.

Since, however, every spectrum of a word spoken has some similarity with every reference spectrum it is important to examine what process gives the clearest safety margin between the word spoken and identified and the other words.

The error level or certainty will probably become more critical if the person who speaks the words is different from the originator of the reference spectra. Do processes exist whereby the system can be adapted to different speakers?

There is thus considerable scope to experiment. The findings ought to show that we are dealing here with one of the most complex systems far removed from run of the mill school science. Not for nothing are outstanding groups of researchers working worldwide on a reliable solution of this commercially viable killer application. What is more: our own acoustic system seems unbeatable.

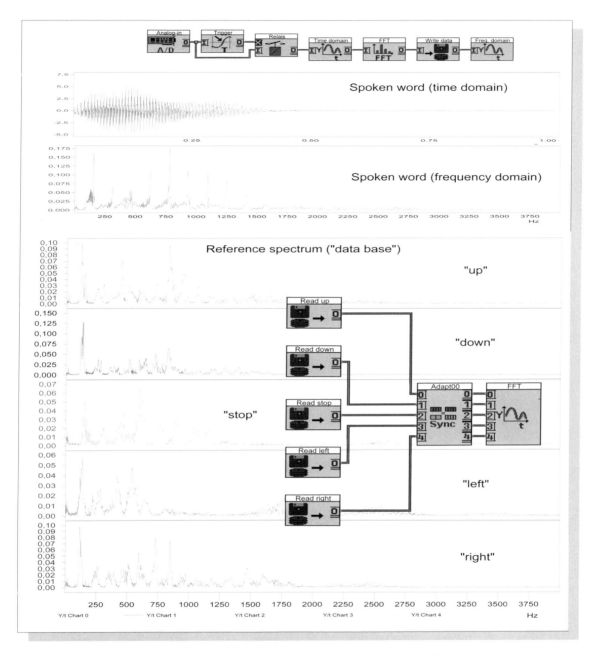

Illustration 73: ***Technical measurement of the reference spectra (amplitude spectra)***

At the top you see a suitable circuit for presenting and saving the reference spectra. Here the word "up" is displayed in the time and frequency domain. Make sure before storing a reference spectrum you designate the file correctly and set the appropriate path in the menu of the module "Reading Data". It is essential to proceed carefully.

In order to be able to compare the spectra qualitatively and quantitatively you should include all five reference spectra in one representation. For this purpose design the lower circuit. The module "Adaptation" serves to synchronise the data and to check whether all the signals were recorded with the same block length and sampling rate. In this case, a block length and sample rate of 4096 was set in the menu bar (A/D). You will see that surprisingly very little happens above 1000 Hz. Select the lens function to examine this area of the spectrum more closely.

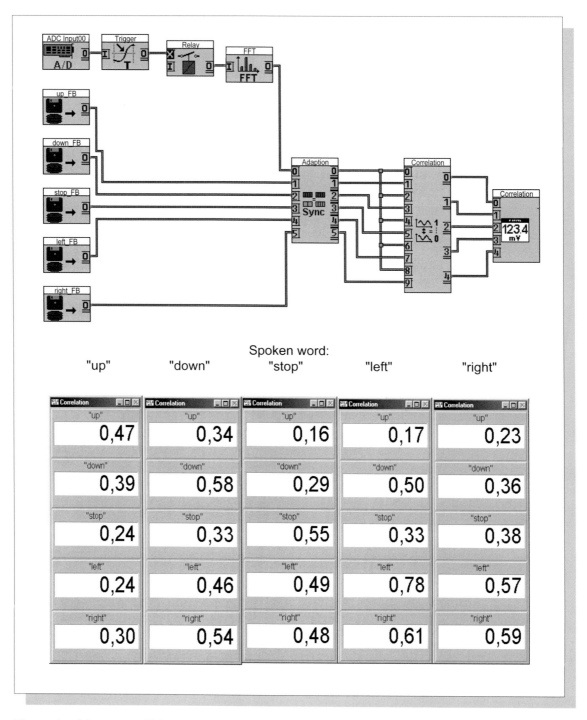

Illustration 74: ***Voice recognition by measurement of the correlation factor***

The words were spoken one after the other from top to bottom and the correlation factors for all the reference spectra were measured. The word "right" was recognised least reliably. The safety margin is here very small. Diagonally you see the correlation factors of each spoken word. The system is already functioning reasonably.

In the experiment you will see how important the module "adaptation" is. The reference spectra are entered immediately and the modules signal EOF (end of file). The spectrum of the word spoken takes longer to appear because it has first to be calculated. Only then are the five correlation factors calculated.

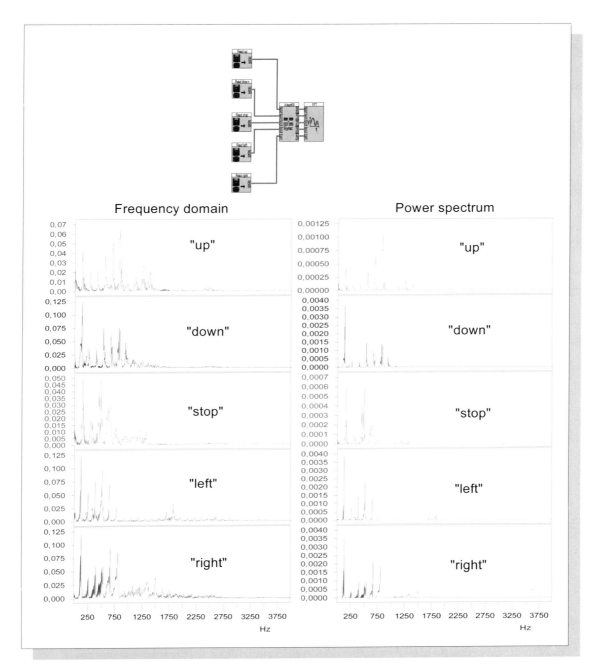

Illustration 75: **Amplitude spectrum and power spectrum**

In the left hand column you see the amplitude spectra and in the right hand column the power spectra for the same signal. You obtain a power spectrum by calculating the square of the amplitudes. What is the sense in this?

In the amplitude spectrum you recognise the "lines" of the characteristic frequencies which come from the vowels and which rise above the other less typical frequencies. By squaring the amplitudes of this spectrum the characteristic frequencies are given even more weight, the less important ones are so downgraded as to be negligible (see right hand side).

You should now examine whether it would not be better for this reason to carry out voice recognition by means of the power spectra. The power spectrum is of great theoretical importance (Wiener-Khintchine-theorem) in the field of pattern recognition.

Refinement and optimisation phase

The system shown in Illustration 74 already works in principle even though it is not perfect. If you experiment with it you will see how easily it can be "deceived". What might be the causes of this and what alternative possibilities are there?

- A frequency range up to 2000 Hz results according to our measurements from a sample rate and block length of 4096. The frequency resolution is roughly 1 Hz as you can easily check using the cursor (the smallest step $\Delta f = 1$ Hz; see Sampling Principle in Chapter 9. Any fluctuation in the voice in the pronunciation of the five words involves a shift in the characteristic frequencies in relation to the reference spectrum. In the case of this high frequency resolution the correlation factor fluctuates because the characteristic frequencies are no longer identical.

- One should therefore consider how for example the areas of characteristic frequencies could be perceived in a more "uncertain" way.

- In Illustration 75 the difference between the amplitude spectrum and the power spectrum can be seen clearly. By squaring the amplitudes in the power spectrum the characteristic frequencies with their high amplitudes - deriving from the vowels - are disproportionately amplified whereas the less relevant spectral elements of the consonants are disproportionately suppressed. A (slight) improvement in voice recognition by the correlation of power spectra should therefore be examined.

- Instead of the module "correlation" steep edge filters could be used which filter out the expected characteristic frequencies. A "filter comb" with the module "cut" could check for each of the five words whether the characteristic frequencies with the corresponding amplitude are present (Illustration 76). However, this procedure seems rather homespun compared with the correlation because it involves a lot of trouble setting the filter.

- Illustration 77 shows a further alternative. This involves the pre-processing of a signal. The signal is first limited in frequency by means of a high pass filter (e.g. $f_{HP} = 50Hz$) and lowpass filter (e.g. $f_{LP} = 1$ kHz). A waterfall representation is prepared by means of the module "data window". However, by means of the following module "separation" a time segment of part of a vowel is selected. This time segment could be so short that the new block length for example is only 128 (128 of 4096 corresponds to roughly ($\Delta t = 1/32$ s). On account of **UP** this implies a frequeny resolution of roughly ($\Delta f = 32$ Hz. As a result the frequency pattern would be much less clearly defined, i.e. *much more tolerant towards fluctuations in the voice*. Of course, all the reference spectra would have to be obtained according to the same separation method.

- Now, you might object, one cannot just take the vowels into account as the "i" and the "o" both occur in two words so that they could not be distinguished. However, as a precise analysis shows, the same vowels may differ considerably in their spectrum depending on the word. The "o" in "stop" is therefore not necessarily identical with the "o" in "down". This system is not happy with fluctuations over time. The vowels might no longer lie in the separated block.

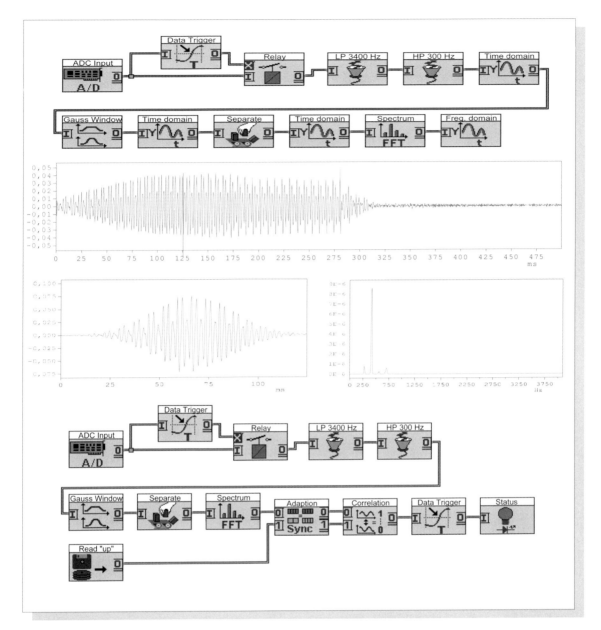

Illustration 76: **_Refined method of voice recognition_**

The method shown in Illustration 73 has among other things the disadvantage of analysing frequencies so accurately that the smallest vocal changes in pitch which appear as small changes in frequency may influence the correlation factor considerably.

Thus we ought to use the **UP** in order to be able to correlate more "tolerantly". To this end a waterfall representation is selected with only one block separated containing the segment of a vowel - here "u" in "up". The time segment is also reduced, for example to 1/32 thus increasing the frequency uncertainty by the factor 32 according to the **UP**. It is as simple as this when one is familiar with the basic principles of signal processing. Or would you have thought of it?

At the top you see the basic test circuit, also intended for the recording of the reference spectra. In the middle you see how easy it is possible to "hit" a vowel if the way of speaking is kept constant. It is then fascinating how simple the frequency pattern looks. Finally, at the bottom you see the completed circuit for the measurement of the correlation factor. Recognition is controlled via a trigger level which is to be defined. Tip: carry out the programming and test the system.

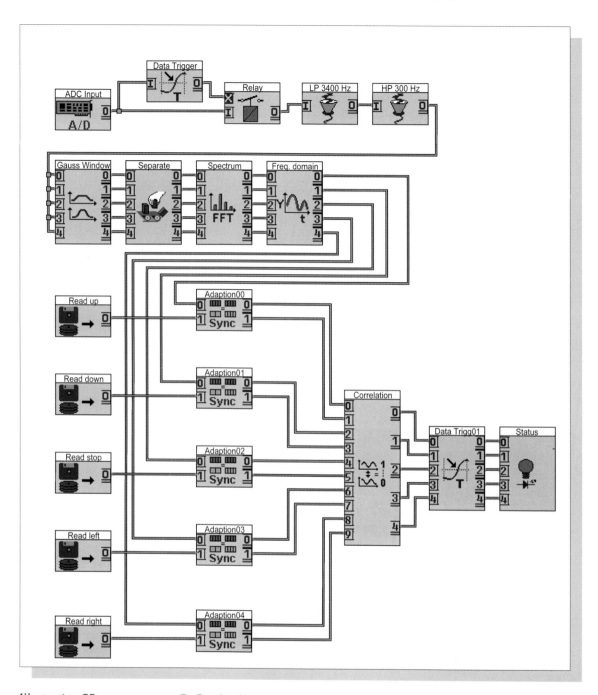

Illustration 77: **_Refined voice recognition for the stacker truck_**

Illustration 74 was completed to form a complete system which - given a certain "speech store" - can recognise five words. However, no parameters are defined. Try, therefore, to program and set the system represented here. To design the connections so that they are clear you should switch off the autorouter function under the menu point "options" and design the connection setup yourself. You obtain the reference spectra by means of the circuit in Illustration 73 (top). If you succeed in creating a demonstrable system with an acceptable level of accuracy, it is evidence that you have an advanced grasp of the subject "Signals - Processing - Systems"!

Thus, there are scarcely any limits to your creativity and inspiration and DASY*Lab* makes it possible to check almost any idea with a few mouse-clicks. A successful solution, however, requires you to proceed systematically. Experimenting at random will not produce results.

It is important to proceed according to scientific method. First, a number of possibilities based on an idea which can be explained in physical terms must be considered. They must be carefully tried out experimentally and reported.

A perfect solution is not possible on this basis, only the best in a relative sense. Our brain uses additional extremely effective methods, for instance, and recognises the word uttered from the context. This is where even DASY*Lab* has its limitations.

Pattern recognition

Basically, in introducing correlation we have taken up the fundamental phenomenon of communication: pattern recognition. A transmitter cannot communicate with a receiver unless a store of patterns conveying meaning exists or has been agreed. It doesn't matter whether it is a technical modulation process or a holiday abroad with its attendant language problems.

So that you do not have to use the module "correlation" in an uninformed way we will demonstrate here how straightforward pattern recognition can be (though it is not always so straightforward!). How is the correlation factor - that is the "similarity of two signals given as a percentage" - calculated by the computer?

Illustration 78 provides the basis of the explanation. The top half is intended to remind you that the computer in reality processes strings of figures arithmetically and does not represent continuous functions. The strings of figures may be represented graphically as a sequence of measurement data of a certain level.

The correlation factor for the two lower signals is now to be determined. The lower signal is to be the reference signal. We are not interested in whether it is the frequency or time domain. For the sake of simplicity we limit the number of measurement data to 16 and quantize the signal, by allowing only 9 different integer values from 0 to 8.

Now we shall multiply the measurements one below the other by each other. All the products are then totalled. This results in

$$2*6 + 2*7 + 2*8 + 3*8 + 3*8 + 4*8 + \ldots + 7*1 + 7*1 + 0*7 + 0*8 = 273$$

This figure already says something about the similarity; the greater it is the more agreement there should be.

But how do we "scale" this value so that it lies between 0 and 1? As the lower signal was taken as the reference signal the „similarity between the reference signal and itself" is determined in the same way. This gives

$$2*2 + 2*2 + 2*2 + 3*3 + 3*3 + 4*4 + \ldots + 7*7 + 7*7 + 7*7 + 8*8 = 431$$

At 431 the agreement would be 100% or 1.0. By dividing the upper by the lower figure (calculation using rule of three) we obtain $271/431 = 0.63$, i.e. a similarity of 63% between the two signals or signal segments.

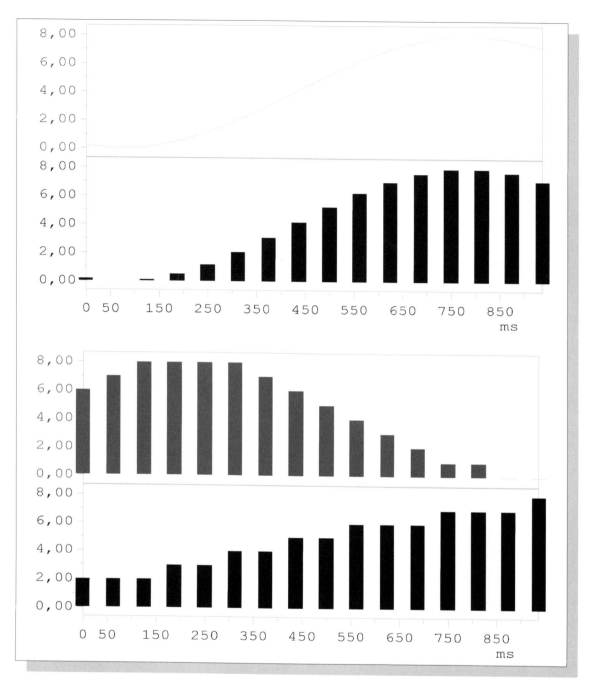

Illustration 78: ***Calculation of the correlation or correlation factor***

DASYLab represents signals as continuous functions by linking the measurement points with each other. If you look more closely at the top you will see a straight line between two measurement points. Although this enhances clarity it suggests deceptively that we are in an analog world.

What does similarity consist in here? On the one hand (here) all the measurements are positive. In addition they are of similar dimensions. Within the two small segments the signals develop in roughly the same way.

Exercises on Chapter 4

In order to be able to tackle the following exercises you require a multi-function board or a sound card with the appropriate DASY*Lab* driver. The DASY*Lab* S version (educational version), where the acoustic signals can be directly entered via the sound card is the most favourable. You could have access where necessary to saved speech files (*.ddf).

Exercise 1

(e) Try to recognise on the screen the segments of vowels and consonants as in Illustration 64 via the *.ddf file of a spoken word.

(f) Develop a circuit by means of which you can analyse these various segments more or less separately (tip: Illustration 76).

(g) Record a certain word several times at longer intervals and compare the spectra.

(h) Record a certain word spoken by various people and compare the spectra.

(i) Under what aspects are the reference spectra to be selected?

(j) What might be the advantages of selecting the setting *power* spectrum rather than amplitude spectra? Which frequencies are emphasised and which suppressed?

Exercise 2

(a) Design a system for the creation of frequency-time landscapes of speech (see Illustration 65.

(b) Divide the signal up into increasingly small, overlapping windowed blocks and observe the changes in the frequency-time landscape.

(c) Represent the frequency-time landscape as a bed of nails pattern by selecting the bar representation of the signal (see Illustration 78).

(d) Represent the frequency-time landscape as a sonogram (see Illustration 66). Set the colour scale in such a way that the areas of different amplitudes are represented in an optimal way.

(e) Do some research (e.g. on the Internet) to find out what sonograms are used for in science and technology.

(f) How can vowels and consonants be distinguished in sonograms?

Exercise 3

(a) Try to design a system for the acoustic recognition of people ("door openers" in the field of security).

(b) What technical possibilities can you think of to outwit a system of this kind?

(c) What measures could you take to prevent this?

Exercise 4

Carry out the experiment relating to Illustration 76 and 77 and try to develop a refined voice recognition system for five words. Think creatively and try to find novel solution where possible.

Chapter 5

The Symmetry Principle

Symmetry is one of the most important structural elements in nature. Space is symmetrical - i.e. no direction is given preference in a physical sense - and we demand and find for practically any elemental particle (e.g. the electron with its negative elemental charge) a "mirror-image" object (e.g. the positron with its positive elemental charge). For reasons of symmetry there is matter and anti-matter.

For reasons of symmetry: negative frequencies

A periodic signal does not begin at $t = 0$ s. It has a past and a future and both are practically symmetrical to the present.

However, the spectrum (up to now) always begins at $f = 0$ Hz. The equivalent to the past would be "negative frequencies" in the frequency domain. Negative frequencies seems not to make any sense because it does not appear possible to interpret them physically. But we should take nature's symmetry principle seriously enough to look for effects where negative frequencies might provide the key to understanding.

Illustration 79 describes an effect of this kind. To arouse your curiosity look carefully at Illustration 45 once again. The progression of the spectrum became more and more asymmetrical as the signal duration Δt was increasingly restricted. This requires an explanation as there is no equivalent to this in the time domain.

We intend to prove experimentally that asymmetry can be explained and offset, as it were, if negative frequencies and negative amplitudes are allowed. Physically negative and positive frequencies would have to work jointly, each representing half the energy.

> Note:
> Information on the direction in which measurements were taken (e.g. how high after low (downward gradient) or low after high (upward gradient) is of significance (physically). For this reason it seems sensible to accept the following rule: we obtain the frequency of a (periodic) sinusoidal signal by means of the inverse value of the period length $f = 1/T$. If the period length is measured from left to right i.e. in the direction of the positive time axis, a positive value results. If it is measured from right to left - in the direction of the past - a negative value results. Both measurements must be of equal value. Why should the positive direction be preferred? Since the positive and negative frequency both describe the same sinusoidal signal up to now we have limited ourselves to the positive frequency absolute value. However, in this way we have contravened the symmetry principle. And we shall see that we will get on better with these "two" frequencies.

Proof of the physical existence of negative frequencies

Illustration 79 provides the first indication of the physical existence of negative frequencies. By means of a trick - "convolution" of an Si-function (see Chapter 10) - the two frequency bands of a given signal are displaced more and more to the left in the direction of $f = 0$ Hz. What will happen when $f = 0$ Hz is finally exceeded?

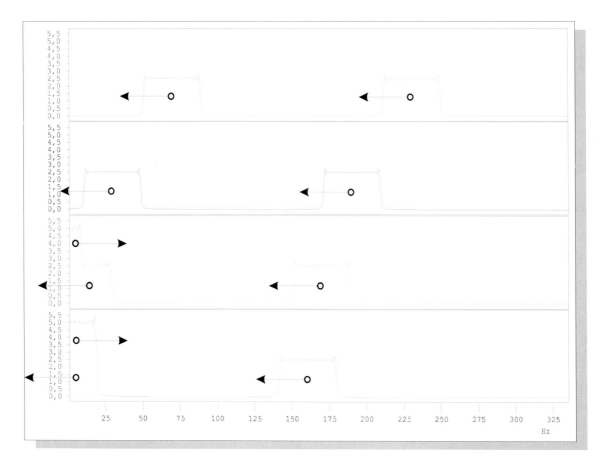

Illustration 79: ***Frequency reflection?***

An extremely complex signal - the curve of which in the time domain does not interest us here - produces a spectrum which consists of two separate, equally steep-edged and broad frequency bands. By skilful manipulation of the signal the two frequency bands are displaced step by step to the left in the direction f = 0 Hz - without changing their width or the distance between them. What happens when the lower frequency band rises above f = 0 Hz? In the three lower illustrations it can be seen that the "negative spectrum" appears in the positive area as though reflected and "overlaps" i.e. is added to the positive area. When the frequency is further displaced the "positive frequency band" "moves" further to the left, the "negative frequency band" on the other hand moves further to the right. The formerly lower frequencies are now the higher ones and vice versa. The most interesting illustration is without doubt the one at the bottom. In this case the positive and negative areas overlap in such a way that on the left lowpass characteristics are in evidence. Thus a lowpass is to a certain extent a bandpass filter with the medium frequency f = 0 Hz, the virtual - and as we shall show - the physical bandwidth is double the visible bandwidth here.

If there were no negative frequencies the frequency band would be gradually more and more cut off and finally disappear!

This is, however, certainly not the case. The part of the frequency band reaching into the negative area above f = 0 appears on the vertical axis as though it were again "reflected" in the positive area. Has the direction of movement of the frequency band turned round and are all the previously low frequencies now high and vice versa? Or is the absolute value of the negative frequency still taken into account? Or does a frequency band which is symmetrical to the positive domain like a mirror image move from the negative to the positive domain and vice versa?

The bottommost picture in 79 is of course the most interesting one. In the case of the left-hand frequency band it is a lowpass characteristic the bandwidth of which is exactly half the original bandwidth, that is, it corresponds to the bandwidth of the right-hand band. On the other hand the amplitude curve is twice as high. A lowpass filter thus seems to have a "virtual amplitude curve" which is twice as large as the visible bandwidth in the spectrum with positive frequencies. It is possible to show that this "virtual" bandwidth is the actual physical bandwidth. This follows from the **UP**. Hence, if the filter range began at f = 0 Hz the lowpass filter would at that point have an infinitely great edge steepness. But precisely this is ruled out by the **UP**. This is also revealed by the time domain (see Chapter 6 under the heading "Transients").

The ingenious signal from Illustration 79 was generated in the time domain by means of the Si-function. This is obvious if you look once again at Illustrations 48 and 49. This form of signal seems more and more important and the question arises whether the Si-function also exists in the negative frequency range. The answer is "yes" if we admit negative frequencies and negative amplitudes. Thus, the Symmetry Principle between the time and frequency domain is clearly demonstrated.

In Illustration 80 at the top we see once again the 3D-spectrum of a narrow (periodic) rectangular pulse. Look carefully at the "playing field" of the sinusoidal signals, particularly where the rectangular pulse is symmetrical to t = 0.5 s. As the pulse duty factor τ/T is roughly 1/10 the first 0 lies at the 10th harmonic. The amplitudes of the first 10 harmonics at t = 0.5 s on the "playing field" point upwards, those from 11 to 19 point downwards, then upwards again etc. It would be better to enter the amplitudes of the amplitude spectrum in the second (fourth etc.) sector (11 to 19) pointing downwards instead of upwards.

In Illustration 80 (centre) you see the continuous amplitude spectrum of a one-off rectangular pulse. If the curve is drawn downwards in the 2nd, 4th, 6th etc. sector it ought to begin to dawn on you. What you see is the right-hand symmetrical half of the Si-function. Last but not least if we describe the Si-function symmetrically towards the left in the negative frequency range we obtain the total Si-function (to be exact, however, only half as high because half the energy is allocated to the negative frequencies).

> *Background knowledge*:
>
> What we have discovered here by means of "experiments" - the simulations carried out here with a virtual system would in the real world lead to exactly the same results using suitable measuring instruments - is provided automatically by mathematics if calculations are carried out in the suitable scale (GAUSSian plane i.e. complex numbers).
>
> *Why do the mathematics (of the FOURIER transformation) provide this result?*
>
> If the calculation is based from the outset on a correct physical premise with real marginal conditions all further mathematical calculations produce correct results because the mathematical operations used are essentially free of contradictions. However, not every mathematical operation can necessarily be interpreted in a physical sense.

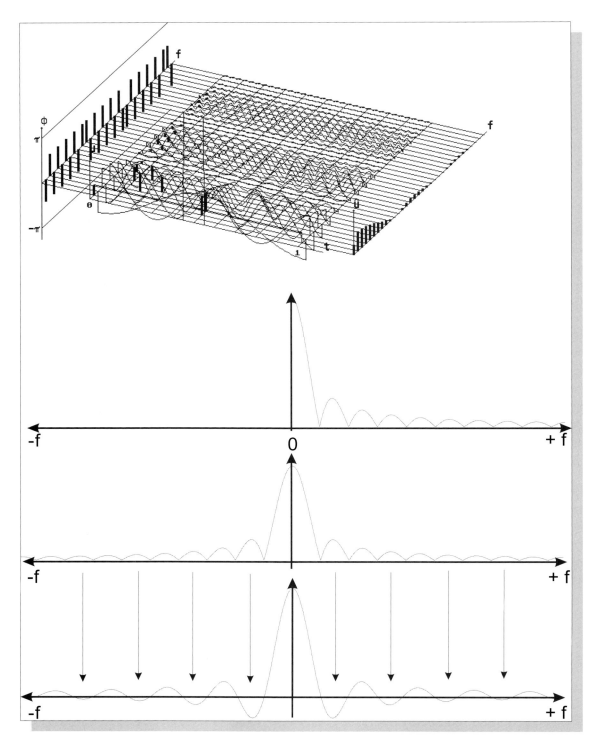

Illustration 80: **Symmetrical spectrum**

Top: representation of a (periodic) rectangular pulse with the pulse duty factor t/T = 1/10 in the time and frequency domain. The amplitudes of the first 9 harmonics point upwards on the "playing field" of sinusoidal signals at the centre of the pulse (t = 0.5 s) in the direction of the pulse, the next 9 harmonics by contrast point downwards etc. Bottom: representation of the amplitude spectrum (of a one-off, i.e. non-periodic rectangular pulse) with negative frequencies and amplitudes. A "symmetrical FOURIER transformation" from the time to the frequency domain and vice versa now results. Note that now the phase information is also contained in the amplitude spectrum.

A rectangular function in the time domain/frequency domain thus produces an Si-function in the frequency domain/time domain if for reasons of symmetry negative frequencies and negative amplitudes are allowed and given equal weight alongside the positive values.

All the frequencies and frequency bands occurring in the purely positive frequency domain appear symmetrically in mirror image form in the negative frequency domain. For reasons relating to energy the curve is only "half as high".

> *Symmetry Principle **SP**:*
>
> *The results of the FOURIER transformation from the time to the frequency domain and from the frequency domain to the time domain are largely identical when negative frequencies and amplitudes are admitted. As a result of this method of representation the representation of the signal in the frequency domain corresponds largely to that of the time domain.*

According to this only one single "signal" has a single spectral line in the symmetrical frequency spectrum: DC voltage (offset) where $f = 0$ Hz applies so that $+f$ and $-f$ coincide. On the other hand any sine has the two frequencies $+f$ and $-f$ which are symmetrical to $f = 0$ Hz. In the case of the sine a negative amplitude necessarily belongs to one of the two frequencies for reasons of symmetry. In the case of the cosine - i.e. a sine with a phase displaced by $\pi/2$ rad or displaced $T/4$ in time - the symmetry is more perfect: both lines point in one direction.

> *The frequency band half of a lowpass filter lying in the negative frequency range is termed "lower sideband". It is a "mirror image" of the "upper sideband" in the positive range*

It is now possible to explain how the "symmetry distortions" in Illustration 45 (bottom) arose. The negative frequency range protrudes into the positive frequency range and is added to the positive frequencies. This is most noticeable in the vicinity of the zero axis.

Conversely, the positive frequency range also protrudes into the negative one and overlaps with it resulting in two completely mirror image symmetrical spectral fields at the zero axis.

By means of the Symmetry Principle **SP** we have arrived at a simplified and unified description of the time and frequency domains. In addition this method of representation also contains more information. Thus "negative amplitudes" point to the development of phases i.e. to the relative positions of the sinusoidal signals.

To tell the truth: it cannot always be assumed that there is perfect symmetry with regard to negative amplitudes. With the FOURIER transformation it is restricted to signals which develop - like sine functions and rectangular functions - symmetrically into the "past" and "future" (see Illustration 84).

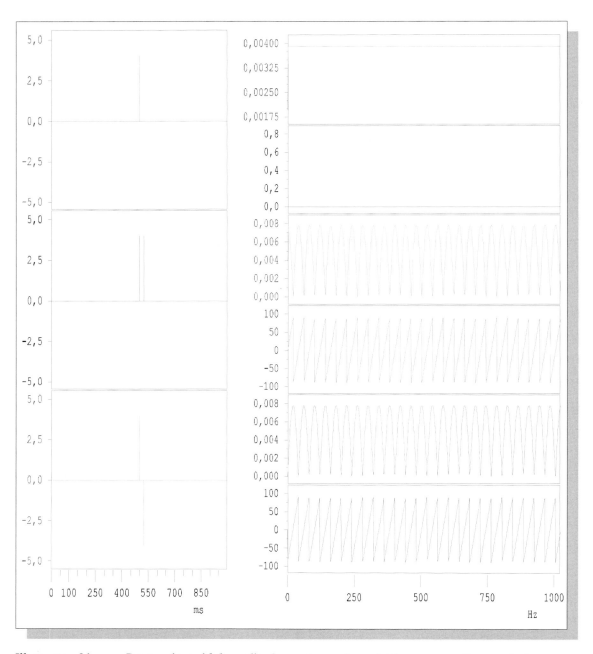

Illustration 81:   **Does a sinusoidal amplitude spectrum also exist for reasons of symmetry?**

An interesting question which may be regarded as a test case for the symmetry principle.

Top series: One line in the time domain (e.g. at t = 0) results in a constant progression of the amplitude spectrum just as a single line in the amplitude spectrum at f = 0 results in a "DC voltage" in the time domain.

Centre and bottom series: two lines (e.g. at t = -20 ms and t = +20 ms) produce a cosine shaped or sinusoidal progression of the amplitude spectrum (assuming that negative amplitudes are admitted!), just as two lines in the amplitude spectrum (e.g. at f = -50 Hz and f = +50 Hz) guarantee a cosine shaped or sinusoidal progression in the time domain. The phase step from π, or more precisely from $+\pi/2$ to $-\pi/2$, to the zero positions of the (sinusoidal) amplitude spectrum is proof that every second "half wave" ought to lie in the negative range.

With regard to the FOURIER transformation the time and frequency domains are largely symmetrical if, as already mentioned, negative frequencies and amplitudes for the spectral area are admitted.

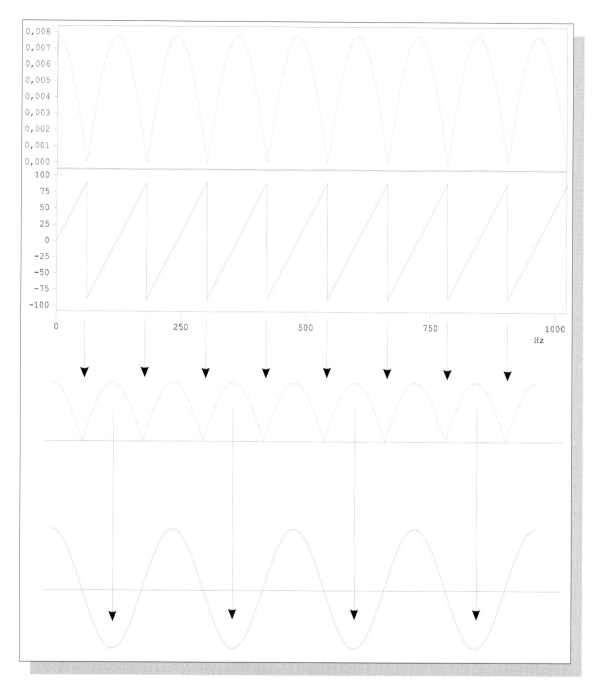

Illustration 82: **The phase spectrum provides the explanation for the sinusoidal curve**

You may have had the following idea: if one δ–pulse results in a constant spectral curve (since all frequencies have the same amplitude) two δ-pulses should theoretically result in a constant spectral curve which is twice as high.

This idea is not entirely wrong but it applies only to those points where the sinusoidal pulses with the same frequencies of both δ-pulses are in phase. This only applies for the zero points of the phase spectrum. At the shift points of the phase spectrum the sinusoidal pulses with the same frequencies of both δ–pulses are dephased exactly by π and thus eliminate each other (interference). Between these two extreme points they amplify or decrement each other depending on their relative positions in the phase.

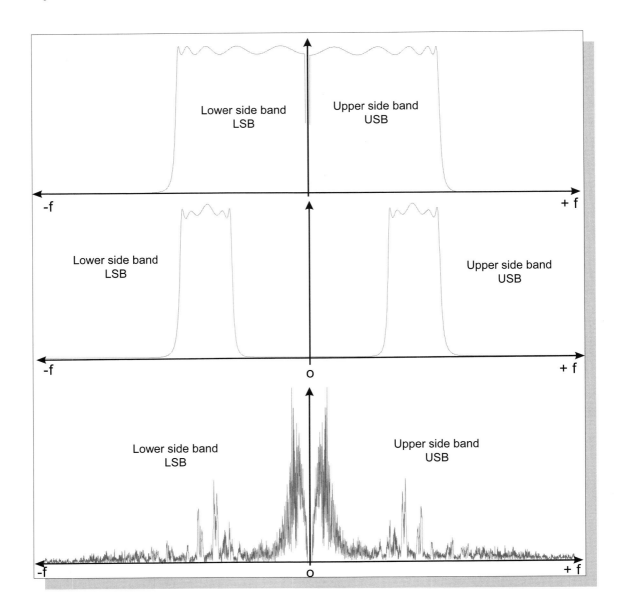

Illustration 83: **Upper and lower sidebands as a characteristic of symmetrical spectra**

*Top series: symmetrical spectrum of a lowpass. Every lowpass is, as it were, a bandpass with the medium frequency f = 0 Hz. Both halves of the spectrum are symmetrically identical. They contain above all the same information. However, the lowpass cannot start at f = 0 Hz because it would then have an infinitely steep-sloped filter edge. This would not be in accordance with the **UP**.*

Centre series: a band pass also has a symmetrical counterpart. Please note that the expressions "upper and lower cutoff frequency" ought to be used with care in connection with the upper and lower sideband.

Bottom series: symmetrical spectrum (with regular and inverse position) of a detail of an audio signal ("speech"). Some computer programs always present the symmetrical spectrum in the frequency range. The symmetrical spectrum is also automatically provided by mathematics (FFT). In that case, however, half the screen would be used up for redundant information.

Unlike (periodic) rectangles, there is no symmetry in the direction of past and future in connection with "asymmetrical" audio signals. Therefore there is no such simple symmetry in the frequency domain with positive and negative amplitudes etc.

Symmetrical spectra do not lead to misinterpretations of spectral curves as in Illustration 45. In addition, non-linear signal processes (Chapter 7) such as multiplication, sampling or convolution which are of great importance for digital signal processing (DSP) can be more easily understood.

Periodic spectra

Remember, periodic signals have line spectra. The lines are equidistant and are integer multiples of a basic frequency.

As a result of the symmetry principle the following should hold:

> *Equidistant lines in the time domain ought also to produce periodic spectra in the frequency domain.*

This is investigated experimentally in Illustration 85. In the first series you see the line spectrum of a periodic sawtooth function.

In the second and third series you see periodic δ-pulse sequences of different frequencies - both in the time and frequency domains! This is the special case in which both occur simultaneously in the time and frequency domains: periodicity and lines.

The fourth series shows a one-off, continuous function - part of an Si-function which has a relatively narrow continuous spectrum. If this signal is digitalized - i.e. presented as a chain of numbers - from a mathematical point of view this is equivalent to the multiplication of this continuous function by a periodic δ-pulse sequence (in this case by the δ-pulse sequence of the third series). This is shown in the bottom series.

Every digitalized signal consists as a result of a periodic but "weighted" δ-pulse sequence. Every digitalized signal also consists therefore of (equidistant) lines in the time domain and must therefore have a periodic spectrum.

> *The essential difference between continuous-time analog signals and time discrete digitalized signals is to be found in the frequency domain: digitalized signals always have **periodic** spectra!*

Periodic spectra are therefore by no means a theoretical curiosity, rather they represent the normal case because digital signal processing DSP has long since gained the upper hand in communications technology and signal processing. As already described in Chapter 1, analog technology is being increasingly restricted to places where it is physically necessary - the source or drain of a communications system (e.g. microphone - speaker in (digital) radio transmission) and the actual transmission path.

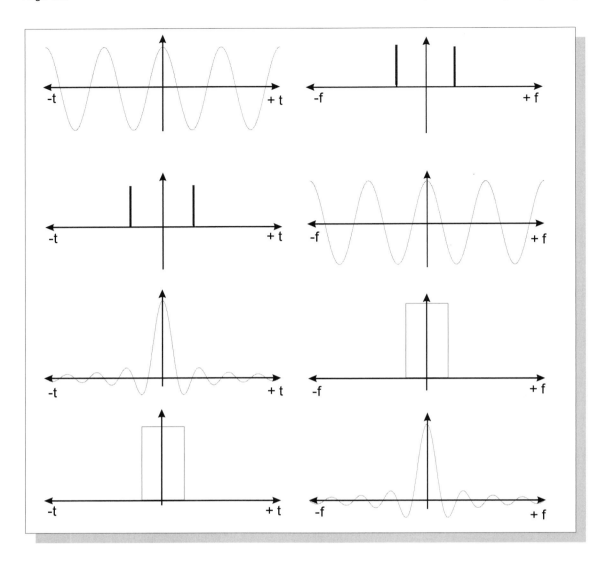

Illustration 84: **Symmetry balance**

The diagrams once again summarize the symmetry features for two complementary signals - a sine and δ–pulse and an Si-function and rectangular function. The important insight is: signals in the time domain may occur in the same form in the frequency domain (and vice versa) if negative frequencies and also negatives amplitudes are admitted. The time and frequency domain represent two "worlds" in which similar figures projected into the "other world" produce identical copies.

To be precise, this statement is only completely true for signals which - as shown here - have a mirror-image curve in the direction of the "past" and the "future". These signals consist of sinusoidal oscillations whose phases are not modified at all or are modified by π; or in other words which have a positive or negative amplitude.

*You should also note that idealized signals have been chosen as examples here. From a physical point of view rectangular functions do not exist either in the time or frequency domain. Because the Si-function is the result of a FOURIER transformation of the rectangular function, it cannot exist in nature either. Like the sine above, it also extends an infinite distance to the left and to the right, that is in the time domain an infinite distance into the past and into the future. Finally, lines cannot exist either because the sinusoidal signal would have to last for an infinitely long time as a result of **UP**.*

In nature every change requires time and everything has a beginning and an end. Idealized functions (processes) are examined in order to discover what real and near- ideal solutions look like.

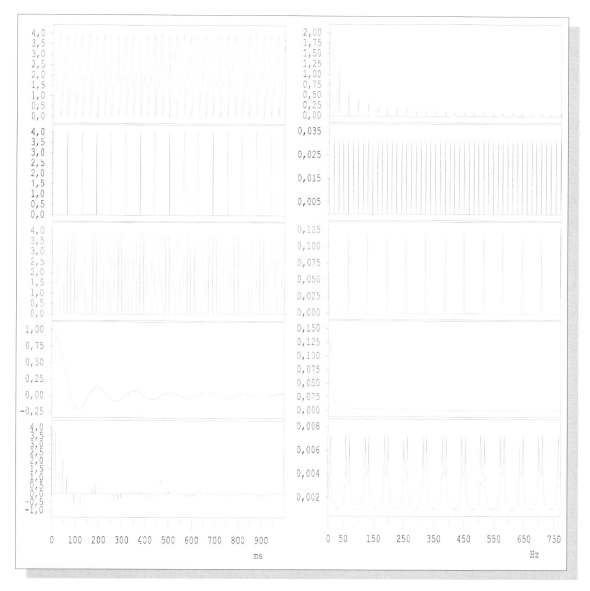

Illustration 85: **_Periodic spectra_**

This Illustration explains the most important conclusion from the Symmetry Principle **SP** for modern digital signal processing DSP.

The top series shows once again that periodic functions in the time domain - here a sawtooth signal - have a line spectrum in the frequency domain. This always consists of equidistant lines.

The Symmetry Principle implies the reverse proposition: Equidistant lines in the time domain should produce periodic spectra.

The second and third series present a special case in which you were already familiar with this phenomenon, which you, however, disregarded: the periodic δ–pulse sequence! Lines and periodicity are present to the same extent both in the time and frequency domains. As a consequence of the equidistant lines every digitalized signal must be regarded as a "periodic but weighted δ–pulse sequence". It arises by multiplication of the analog signal by a periodic δ–pulse sequence. In the spectrum both characteristics - that of the analog signal and that of the periodic δ–pulse sequence - must be present. The periodic result is to be seen bottom right.

Please note that only the positive frequencies are represented here. The spectrum of the analog signal looks therefore exactly like every (periodic) part of the periodic spectrum.

Inverse FOURIER transformation and GAUSSian plane

The moment a technical signal problem is seen from both perspectives of the time and frequency domains, not only the FOURIER Principle **FP** and the Uncertainty Principle **UP** but also the Symmetry Principle **SP** are involved.

Whereas up to now in this chapter we have mainly dealt with the phenomenon of "symmetry", we shall now deal with the application and its visualization by means of measuring techniques. As the highlight of the symmetry principle it ought to be possible to get from the frequency domain into the time domain by means of the same operation i.e. FOURIER transformation.

This can be deduced directly from Illustration 84 as the time and frequency domains are interchangeable here under the conditions given. In addition, the possibility of the FOURIER transformation in the other direction from the frequency domain to the time domain - the inverse FOURIER transformation IFT - was indicated in Chapter 2.

In chapter 2 under the heading "The confusing phase spectrum" it was pointed out that only the amplitude and phase spectrum together provide complete information on the signal in the frequency domain.

The perspectives which this moving backwards and forwards between the time and frequency domain offers are fascinating. They are exploited more and more in modern digital signal processing DSP. An example of this are high-quality filters which provide virtually rectangular filter functions except for the unavoidable limits imposed by the **UP**.

The signal to be filtered is first transformed in the frequency domain. There the unwanted frequency range is cut out, i.e. the values in this range are simply set at 0. Then by means of the IFT we go back to the time domain and the filtered signal is complete (see Illustration 88, bottom). The obstacle here is, however, "real time processing". The computational process described must be carried out so quickly that even in the case of a long-lasting signal no unwanted loss of information occurs.

The possibilities of this moving backwards and forwards are now to be explored and explained experimentally by means of DASY*Lab*.

Let us first look at the possibilities provided by module "FFT" to get into the frequency domain. Up to now we have only used the function group "real FFT of a real signal" and the amplitude and/or the phase spectrum in it.

In the previous Chapter we added the power spectrum. The FOURIER analysis, however, comes first in the menu. We now select this. Now construct the circuit according to Illustration 86 and set the values as they are given in the caption text.

Surprisingly, we see the Symmetry Principle **SP** realized to some extent as there is a mirror-image symmetry in relation to the vertical central line. In the new version of DASY*Lab* and in the educational version symmetry can be selected by clicking on to the menu point "symmetrical axis".

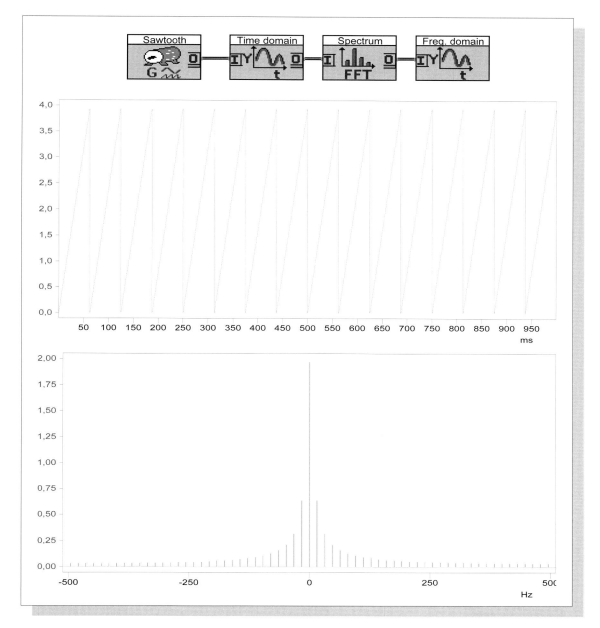

Illustration 86: ***FOURIER analysis and symmetrical spectrum***

In the FFT module the "Real FFT of a real signal" was selected as before and then the setting "FOURIER-Analysis". Our standard setting block length = sampling rate = 1024 is selected. The sawtooth frequency is 16 Hz.

If the frequency values are not a power of two - for example 21 Hz - , additional spectral lines result which will be dealt with in Chapter 9.

The result is an symmetrical spectrum measured by DASYLab. However, not with positive and negative frequencies at the top as previously shown, but at all events symmetrical. In the new 6.0 version and in the educational version there is now the possibility of selecting the lower representation with a positive and negative frequency range by means of the setting (bottom right) "symmetrical axis".

However, the negative amplitudes are lacking and there is no information on the phase spectrum. The FOURIER-analysis does not therefore provide all the information that we would need in order to obtain a clear representation of the signal in the frequency domain. For this reason we shall now examine what other variants of the FFT module there are (Illustration 87). There are two forms of the "complex FFT" which have two outputs. We shall now use these.

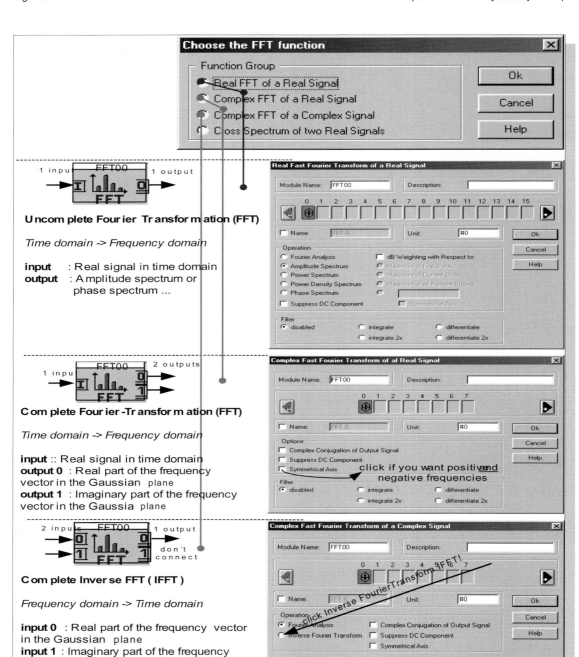

Illustration 87: **Selection of FFT functions**

The FFT module allows various different variants. This is most strikingly apparent in the number of inputs and outputs of this module (see top left). Up to now we have only made use of the "real FFT...". In this case, too, there are various different alternatives (see centre).

As we shall see, the Symmetry Principle **SP** is exploited in the two forms of the "complex FFT..." in order to achieve the "back and forwards" between the time and frequency domains (see bottom). For the path from the time to the frequency domain (FT) we require the module with one input and two outputs and for the reverse path (IFT) the module with two inputs and outputs. It is very important to select the setting "FOURIER synthesis" because you want to put together the time signal from the sinusoidal oscillations of the spectra.

Before we examine this type of FOURIER transformation more closely by means of experimentation, it should be shown to work.

In Illustration 88 you see an appropriately block diagram. If you set the parameters according to the representation and caption text of Illustration 87 the left-hand input signal does in fact re-appear at the upper output of the (inverse) FFT module. Clearly, all the information from the time domain was transferred to the frequency domain so that the inverse FOURIER transformation in the reverse direction back into the time domain via the complete set of all the necessary information was also successful.

Now it will be shown how easily it can be manipulated in the frequency domain. By adding the "cut out" module we have the possibility of cutting out any frequency range. We see this step has been successful in the lower half: a virtually ideal lowpass filter with the borderline frequency 32 Hz which it was not possible to realize up to now. This circuit will prove to be one of the most important and sophisticated in many practical applications which we shall be dealing with.

The next step is to establish experimentally how the whole thing works and what it has to do with the **SP**. First we represent three simple sinusoidal signals with 0, 30 and 230 degrees and 0, π/6 and 4 rad phase displacement by means of the "complex FFT" of a real signal. The result in Illustration 89 are two different - symmetrical - line spectra for each of the three cases. However, it does not seem to be a question of absolute value and phase, as only positive values are possible in the case of the absolute value. The lower spectrum in each case cannot be a phase spectrum as the phase displacement of the sinusoidal signal is not identical with these values.

We continue to explore and add an x-y-module (Illustration 90). Now you see a number of "frequency vectors" on the plane. Each of these frequency vectors has a mirror-image symmetrical "twin" in relation to the horizontal axis. In the case of the sinusoidal signal with 30 degrees and π/6 phase displacement the two frequency vectors which each have a 30 degrees and π/6 rad phase displacement *in relation to the vertical line* running through the central point (0;0) fall into this category. The phase displacement in relation to the line running horizontally through the point (0;0) is accordingly 60 degrees and/or π/3 rad.

For the time being we will call the vertical *sine axis* because both frequency lines with a phase displacement of 0 degrees or 0 rad lie on this. We will call the horizontal the *cosine axis* because the frequency lines with a phase displacement of the sine of 90 degrees or π/2 - this corresponds to the cosine - lie on this.

On the other hand, a sine displaced by π/6 rad is nothing other than a cosine displaced by -π/3. If you now compare the axis sections with the values of the line spectra the values of the upper spectrum belong to the cosine axis and the values of the lower spectrum to the sine axis.

The two frequency lines apparently possess the characteristics of vectors which in addition to their absolute value also have a certain direction. We shall see that the length of both frequency vectors reflects the amplitude of the sine and the angle of the "frequency vectors" in relation to the vertical and horizontal line reflects the phase displacement of the sine and cosine at the point of time t = 0 s.

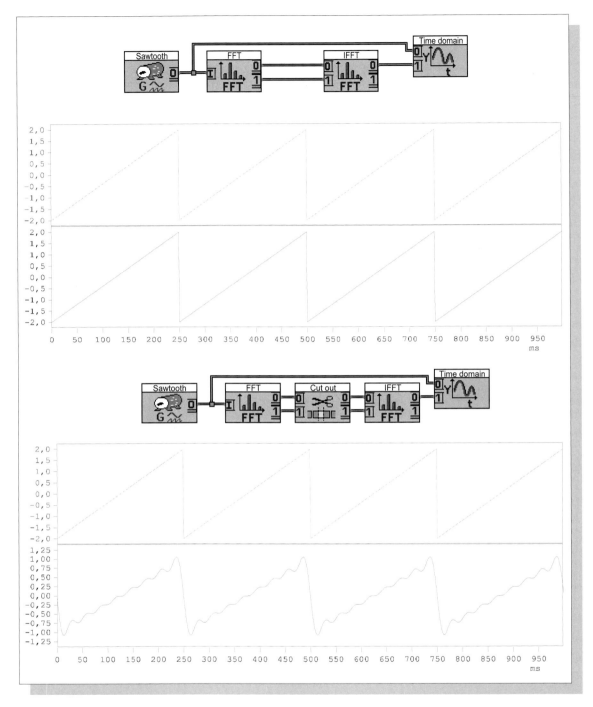

Illustration 88: **FT and IFT: symmetrical FOURIER transformation**

A periodic sawtooth signal with four Hz was selected as a test signal. You could, however, just as well take any other signal, for example noise. Set as usual the sampling frequency and the block length at 1024 using the menu option A/D.

In the module "cut out" the frequencies 0 - 32 Hz were allowed to pass (in the case of the settings selected the sample value is practically equivalent to the frequency). You see the highest frequency of 32 Hz as the ripple content of the sawtooth: this "ripple" sine covers 8 periods with any sawtooth; in the case of 4 Hz sawtooth frequency the highest frequency allowed to pass) has the value 4 * 8 = 32 Hz.

*It is very important in this connection to set exactly the same "cut out" on **both** channels (frequency domain).*

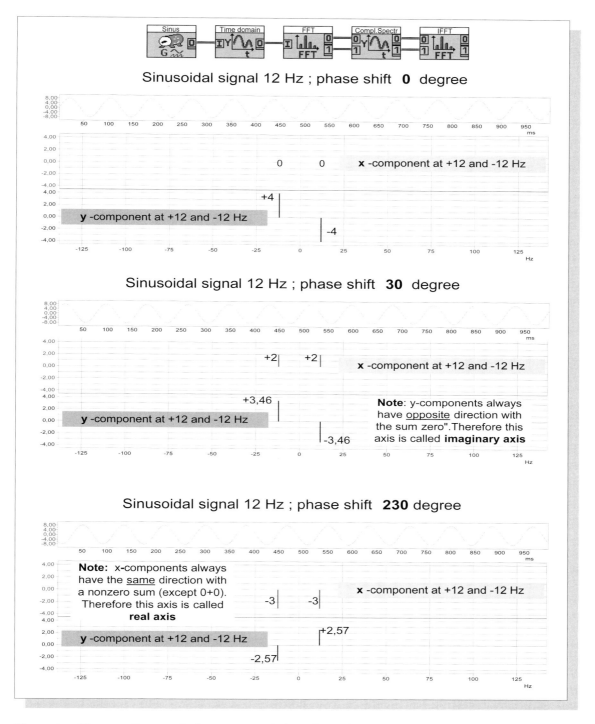

Illustration 89:   **Symmetrical spectra consisting of x and y components (x-y-representation)**

We must now clarify what information appears at the two outputs of this FFT module. Giving the amplitude and phase, apart from frequency, is part of every sinusoidal signal. One might therefore suspect that the amplitude and phase of the positive and negative frequency + and - 12 Hz will appear at the two outputs. The diagrams give a different result.

A reminder: Illustration 24 links the sinusoidal signal with a rotating pointer. If we pursue the idea further the rotating pointer can be represented like a vector by x and y components which may change over time. Following this idea in Illustration 90 an x-y-module is selected in order to visualize the two channels. Result: two pointers rotating in the opposite direction!

As Illustration 90 shows the direction of rotation of the two "frequency vectors" is opposite if the phase shift of the sine increases or decreases. The frequency on the left in Illustration 89 which we represented as a negative frequency in Illustration 86 (bottom) rotates in anti-clockwise direction, the positive frequency in a clockwise direction whereby the positive phase shift increases.

How do the instantaneous values of the three sinusoidal signals conceal themselves at the point of time t = 0 s in the plane? Compare carefully the symmetrical spectra of Illustration 89 with the plane of the x-y-module in Illustration 90. You should take into account that the frequency lines are vectors for which quite specific rules apply. Vectors - e.g. forces - can be divided up into parts by projection on to the horizontal and vertical axes which are entered here as markings.

For the sinusoidal signal with the phase shift of 30 degrees or π/6 rad we obtain the value 2 by projection on to the cosine axis. The sum is 4 (instantaneous value at the point of time t = 0s). The projection on to the sine axis gives the value 3.46 or -3.46, that is the sum adds up to 0.

For this reason the resulting vectors of all (symmetrical) "frequency vector" pairs always lie on the cosine axis and represent the real instantaneous values at the point of time t = 0, which can be measured. For this reason the so-called *real part* is represented on the cosine axis.

By contrast, the projections of the frequency vector pairs on the sine axis always lie opposite to each other. Their sum is therefore always 0 independent of the phase position. The projection on to the sine axis has therefore no counterpart which can be measured. Following mathematics of complex calculations in the so-called GAUSSian plane the projection on to the sine axis is referred to as the *imaginary part*. Both projections have an important physical sense. This is explained by Illustration 91. The projection reveals that every phase-displaced sinusoidal signal can always consist of a sine and a cosine oscillation of the same frequency. Important consequences result from this:

All signals can be represented in the frequency domain in three ways:

- as an *amplitude and phase* spectrum
- as the spectrum of the *frequency vectors in the GAUSSian plane*
- as a spectrum of *sine and cosine signals*

The symmetrical spectra from Illustration 89 (bottom) are revealed to be the last type of representation of a spectrum. This is proved by Illustration 91.

The Illustrations that follow deal with the spectra of periodic and non-periodic signals in the representation as a symmetrical "frequency vector pair" in the GAUSSian plane of complex numbers. You will find additional information in the caption text.

Complex numbers refers to numbers in mathematics which contain a real and an imaginary part. It would be quite tempting to demonstrate that calculating with complex numbers is far from "complex", and on the contrary is much easier than calculating with real numbers. But the original approach is to do without mathematics.

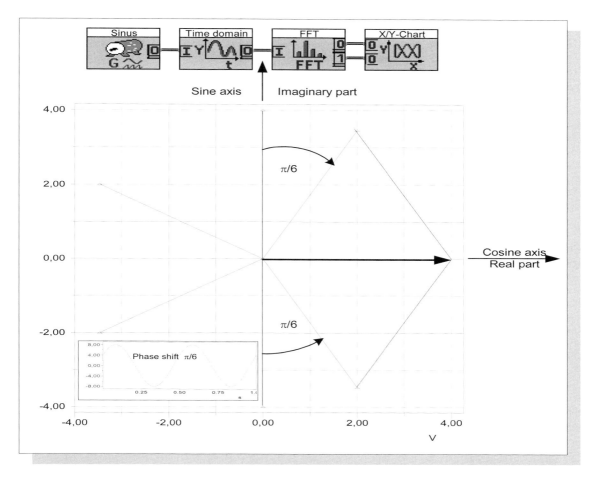

*Illustration 90: **Representation of the "frequency vectors" in the complex GAUSSian plane***

By means of the x-y-module all the information on the two spectra from Illustration 89 can be brought together on one level. Each of the three sinusoidal signals is here in the form of a "frequency vector pair" which is always symmetrical in relation to the horizontal axis. Instead of the usual vector arrowheads we use a small triangular form. The length of all the "frequency vectors" is in this case 4 V, that is, half the amplitude of the sinusoidal signal is allotted to each of the two frequency vectors.

*It is most difficult to recognize the sinusoidal signal **without** phase shift: this pair of frequency vectors lie on the vertical axis which passes through the point (0;0) and which for this reason we call the "**sine axis**". In the case of a phase displacement of 90 degrees or $\pi/2$ rad - this corresponds to a cosine - both frequency vectors lie above one another on the horizontal axis. We therefore call this the "**cosine axis**". In a phase displacement of 30 degrees or $\pi/6$ rad we obtain the two frequency vectors of which the angle in relation to the sine axis is entered. As you can now see a sine with a phase displacement of 30 degrees or $\pi/6$ rad is simply a cosine of -60 degrees or $-\pi/3$ rad. A phase displaced sine thus has a sine and cosine part!*

*Careful! The two equally large cosine parts of a frequency vector pair add up, as you ought to check in Illustration 91, to a quantity which is equal to the instantaneous value of this sinusoidal signal at the point of time t = 0s. By contrast, the sine-parts always add up to 0 because they lie **opposite** to each other. Because there are quantities along the cosine axis which are measurable in a real sense, we call this the **real part**. Because on the sine axis everything cancels each other out and nothing remains, we choose the expression **imaginary part** following the mathematics of complex calculations.*

We shall show in the next Illustration that the sinusoidal signal which belongs to the "frequency vector pair" can be produced from the addition of the sinusoidal signals which belong to the real part and the imaginary part.

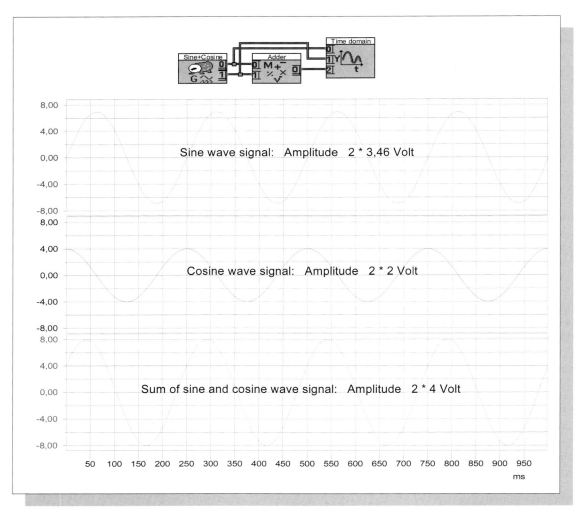

Illustration 91: **_Spectral analysis of sine and cosine components_**

This Illustration proves that the three different kinds of spectral representations or representations of the frequency areas are consistent in themselves. The relevant sine and cosine components result from the "frequency vectors" in the GAUSSian plane of complex numbers as projections on to the sine and cosine axes (imaginary and real components).

Let us first take a look at the sinusoidal oscillation (top) with the amplitude $2 * 3.46 = 6.92$ V. In the numerical plane of complex numbers it results in a "frequency vector pair" which is located on the sine axis. A vector with a length of 3.46 V points in the positive direction, its "twin" in the negative direction. Their vector sum equals 0.

Let us now take a look at the cosine oscillation with the amplitude 4 V. The relevant "frequency vector pair" can be located on the cosine axis pointing in the positive direction. Each of these two vectors has a length of 2. Thus their sum is 4.

Thus, everything is in accordance with Illustration 74. Please note that the sinusoidal oscillation the phase of which is displaced by 30 degrees or $\pi/6$ rad also has an amplitude of 8 V. This is also the result of appropriate calculations using the right-angled triangle: $3.46^2 + 2^2 = 4^2$ (Pythagorean Theorem).

The representation in the so-called GAUSSian plane is of great importance because in principle it combines all three ways of spectral representation: amplitude and phase correspond to the length and angle of the "frequency vector". The sine and cosine components correspond to the breaking down of a phase-shifted sine into **pure** sine and cosine forms.

So far, we can only see one disadvantage: we cannot read the frequency. The position of the vector is independent of its frequency.

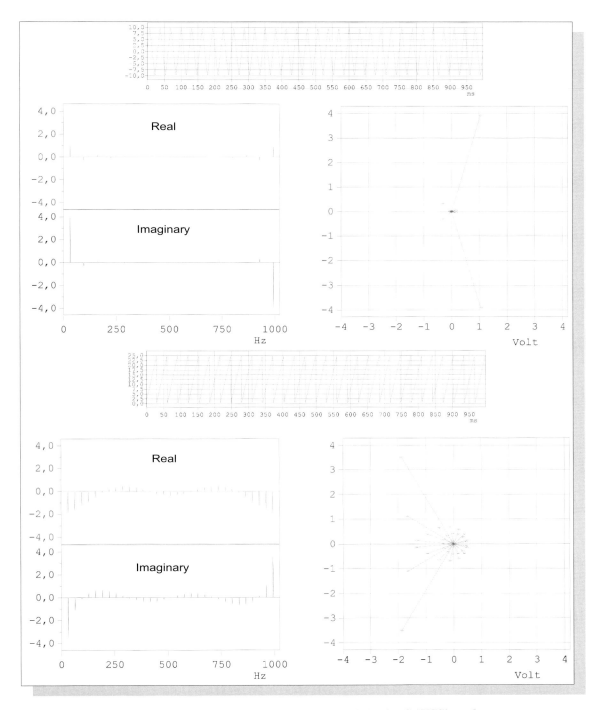

Illustration 92: **Spectral representation of periodic signals in the GAUSSian plane**

Periodic signals contain the common multiples of their basic frequency only in their spectra. In contrast to previous cases we now expect many "frequency vector pairs".

At the top you can see a periodic rectangular oscillation with a phase shift of 30 degrees or $\pi/6$ rad. In 31 you can see how rapidly the amplitudes decline with increasing frequency. In this case the same is true: the lower the amplitude the higher the frequency. A classification with regard to frequency is thus already possible provided we know the basic frequency.

The same is true of sawtooth oscillations with a phase shift of 15 degrees or $\pi/12$ rad. In this case the amplitudes change (see Illustrations 25 - 30) in accordance with a very simple law: $\hat{U}_n = \hat{U}_1/n$. The second frequency has therefore only half the amplitude of the first frequency etc.

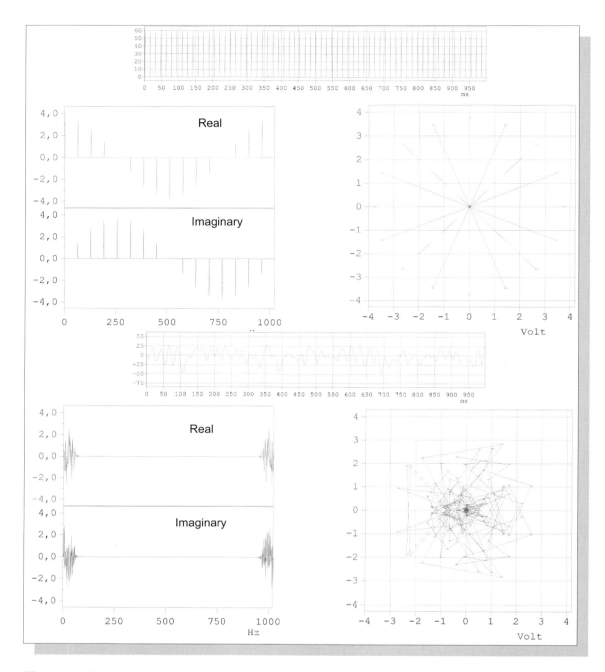

Illustration 93: ***Periodic and non-periodic spectra in the GAUSSian plane***

At the top you see - slightly spoiled by the grid - a periodic sequence of δ–pulses. Please note the sine and cosine-shaped form of the line spectrum of the real and imaginary components. If you transfer these sine and cosine components to the GAUSSian plane you will find the first "frequency vector pair" on the horizontal cosine axis in the positive direction, the second pair with twice the frequency at a π/8-angle to the cosine axis, the next pair at double the angle etc. The amplitudes of all frequencies are identical with a δ–pulse; the result is therefore a star-shaped symmetry.

Below that you can see a lowpass-filtered noise (cutoff frequency 50Hz), i.e. a non-periodic signal. This type of signal does not produce any law with regard to amplitude and phase as it is of a purely random - stochastic - nature. You can clearly see the symmetry of the "frequency vector pairs". The one higher and one lower frequencies are directly connected to each other, i.e. one line leads to the lower frequency, the other one to the higher frequency. This muddle makes it very difficult to find the beginning and the end of the general line. So how do we find out the frequency value of each pair?

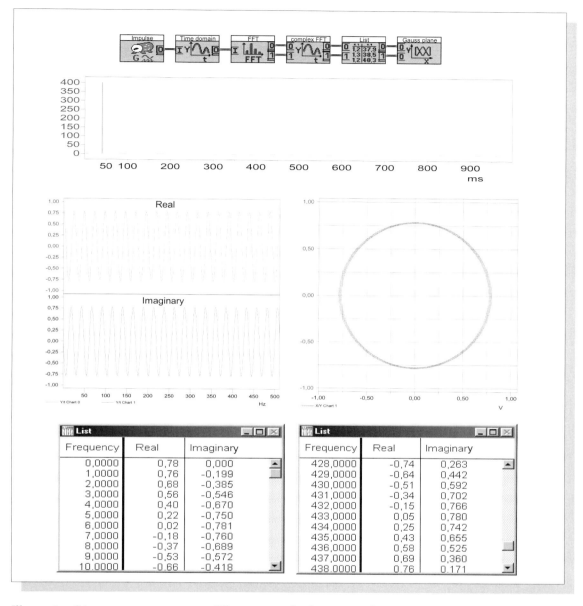

Illustration 94: *"Locus curve" of a one-off δ-pulse*

The frequencies which a one-off signal contains closely follow each other. The "frequency vector pairs" in this Illustration are all on one circle because all the frequencies of a δ–pulse have the same amplitude. However, as the cosine and sine-shaped curve of the real and the imaginary component shows, the phase varies considerably from frequency to frequency so that the neighbouring "frequency vector pairs" are arranged in a star-shaped symmetry as in Illustration 93.

The measured signal has a duration of a total of 1 s at a scanning rate of 1024 and a block length of 1024. As a result the frequency uncertainty is roughly 1 Hz (**UP**). The complex FOURIER transformation provides a spectrum of 0 to 1023, i.e. 1024 "frequencies". That means 512 "frequency vector pairs" all on this circle. The result of the number of the "periods" of the sine and cosine-shaped spectrum (roughly 42) is that the connecting chain of all frequencies orbits the circle roughly 42 times. The angle difference between neighbouring frequencies is thus just under (42*360)/1024 = 15 degrees or π/24 rad. A straight line is drawn between two neighbouring points. As these 1024 lines overlap the circle line in this Illustration appears thicker.

Using the cursor it is easy to indicate the relevant real and imaginary components. The relevant frequency can be determined using the chart module to a maximum accuracy of 1 Hz.

Exercises on Chapter 5

Exercise 1

(d) How can the sequence of spectra in Illustration 79 be explained by means of the Symmetry Principle **SP**?

(e) Draw a symmetrical spectrum for the two bottom spectra.

Exercise 2

(a) Try to make out the relevant block diagram for Illustration 81. You can produce the two δ-pulses using the module "cut out" from a periodic sequence of δ-pulses. For the bottom illustration it is possible to start by producing two delayed δ-pulses, then to invert (*(-1) one of them and finally to add them up. Another possibility is to use the "formula interpreter".

(b) Examine the effect which the time interval of the two δ-pulses has on the sinusoidal spectrum. Before you do that consider what the effect according to the Symmetry Principle ought to be.

(c) Do the δ-pulses (of +4V and -4V) illustrated at the bottom produce a sinusoidal or a cosinusoidal spectrum, i.e. a spectrum displaced by $\pi/2$?

(d) Cut out three or more δ-pulses which are very close to each other and observe the curve of the (periodic!) spectrum. What function determines the curve?

Exercise 3

Why is it impossible for an audio signal to have a perfectly symmetrical spectrum with negative amplitudes?

Exercise 4

(a) Summarize the importance of periodic spectra for digital signal processing DSP.

(b) Find an explanation for the fact that periodic spectra always consist of mirror-image components.

Exercise 5

(a) What would be the symmetrical counterpart of *near-periodic* signals. How could near-periodic spectra be created?

(b) Are there any *quasi-periodic* spectra?

Exercise 6

Show the different signals in the frequency domain in the following variants:

(a) Amplitude and phase spectra

(b) Real and imaginary components

(c) "Frequency vector pairs" in the GAUSSian plane.

Exercise 7

How can the frequency in the GAUSSian plane be measured?

Chapter 6

System analysis

Gradually we are able to reap the fruits of our basic principles (FOURIER, the Uncertainty and Symmetry Principles).

An important practical problem is measuring the properties of a circuit, component or system from outside. You will be familiar with test reports which, for instance, compare the features of different amplifiers. At issue is always the technical behaviour of transmission ("frequency response", "distortion factor" etc). Let us first look at the frequency-dependent behaviour of a system to be tested.

This is easy as long as we do not forget the **UP**: any frequency-dependent behaviour necessarily triggers a certain time-dependent reaction. The FOURIER Principle says even more precisely that the time-dependent reaction can be completely deduced from the frequency-dependent behaviour and vice-versa.

The technical signal test of a circuit, a component or system is generally carried out by comparing the output signal u_{out} with the input signal u_{in}. In the first instance it is immaterial (see above) whether the comparison of the signals is carried out in the time or frequency domain.

> Note:
> It is however, pointless to look at the signal from your TV aerial (roof) on the screen of a (rapid) oscilloscope. All you can see is a chaotic picture. All the radio and television transmitters are broadcast in staggered frequencies. For this reason they can only be represented separately on the screen of a suitable spectrum analyzer (See Chapter 8: Traditional modulation procedures).

The standard procedure is based on the direct implementation of the FOURIER Principle:

> *If it is known how a given (linear) system reacts to sinusoidal signals of different frequencies it is also clear how it reacts to all other signals ...because all other signals are composed of nothing but sinusoidal signals.*

This procedure is widely practised in school laboratories. The necessary equipment is:

- Sine wave generator with adjustable frequency or sweep mode.

- 2-channel oscilloscope

The properties in the time and frequency domain are to be established by comparing u_{out} and u_{in}. Then both signals should be represented simultaneously on the screen. For this reason u_{in} is connected not only to the input of the circuit but also to channel A of the oscilloscope. The output signal reaches the screen via channel B.

By means of a function generator and oscilloscope it is possible - using time-consuming measurement procedures, recording the measurement values, and calculation by means of a pocket calculator - to determine the representation of the frequency response according to amplitude ($\hat{U}_{out}/\hat{U}_{in}$) and phase ($\Delta\varphi$) between u_{out} and u_{in}.

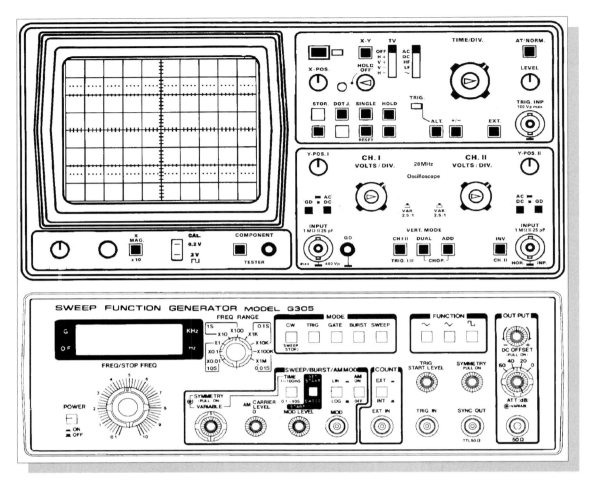

Illustration 95: **Function generator and oscilloscope**

These two instruments are never lacking in a traditional laboratory. The function generator produces the (periodic) test signal u_{in}. The following are available as standard signals: sine, triangle and rectangle. In addition, more sophisticated instruments make it possible to generate "sweep" signals, burst signals and "one-shot" signals (caused by a trigger process only a period of the signal set is produced). The "analog" oscilloscope can only make periodic "standing" signals visible on the screen. Digital storage oscilloscopes are required for one-off signals. (see above).

The days of these two traditional analog measuring instruments are numbered. Computers with an appropriate periphery (PC multi-function cards for the input and output of analog and digital signals) can be used for a specific application by means of graphic interfaces for any tasks in measuring and automatic control engineering or any form of signal processing (e.g FT)

The determination of frequency-selective properties - frequency response and transfer function - of a (linear) system is limited in principle to two different questions or measurements.

- To what extent are sinusoidal signals of different frequencies allowed to pass? In this context the amplitudes of u_{out} and u_{in} within the frequency range in question are compared with each other ($\hat{U}_{out}/\hat{U}_{in}$).

- How great is the time delay between u_{out} and u_{in}? This is determined via the phase difference $\Delta\varphi$ between u_{out} and u_{in}.

Note:
In the case of all *non*-sinusoidal test signals and modern analysis procedures - as they will be described in the following - the problem is mostly reduced to the two measurements of amplitude and phase curve. The only difference is that all the frequencies in which we are interested are given to the input at the same time.

If you do not have computer-aided modern processes available you should note the following tricks and tips:

(d) Always trigger the input signal and do not change the amplitude of u_{in} (if possible, select $\hat{U}_{in} = 1V$) during the entire measurement series.

(e) Turn the frequency range to be examined manually on the function generator and note the area where the amplitude of u_{in} changes most strongly. Make most of the measurements in this area.

(f) Select for the measurement of the amplitudes ($\hat{U}_{out}$ depending on the frequency) such a large time base that the sinusoidal alternating current appears as a "bar" on the screen. The amplitude can be most easily determined in this way. (See Illustration 96, top).

(g) To measure the phase difference $\Delta\varphi$ set exactly a half period of the input signal u_{in} by means of the uncalibrated time base regulator, which can be set as desired. T/2 is for example 10 cm on the screen scale and corresponds to an angle of π rad. Now read the phase difference $\Delta\varphi$ (or time shift) between the zero crossings of u_{out} and u_{in} and you obtain (initially) x cm. Finally, by means of the *rule of three* using a pocket calculator determine the phase difference $\Delta\varphi$ in rad for every x value. If u_{out} is *trailing* as in Illustration 96 (bottom) $\Delta\varphi$ is *positive*, otherwise it is negative.

For complete and careful measurement with evaluation you require roughly 2 hours. To awaken your enthusiasm for modern computer-based processing - for the same measurement and evaluation with a much greater degree of precision you need only a fraction of a second!

Sweep

It is possible to get a quicker overview of the frequency-dependent behaviour of output amplitude $\hat{U}_{out}$ by means of the sweep signal (Illustration 96). Here the idea is as follows. Instead of setting the frequency range continuously by hand from the lower starting frequency f_{start} to the upper stop frequency f_{stop} this is carried out in the instrument by means of a Voltage Controlled Oscillator VCO. The sweep range and the sweep speed are determined by a corresponding sawtooth voltage. If the saw tooth voltage rises linearly the frequency of the sine also changes linearly but if the sawtooth voltage changes logarithmically, this also applies to the frequency of the sinusoidal output voltage. The amplitude $\hat{U}_{in}$ remains constant during this sweep process.

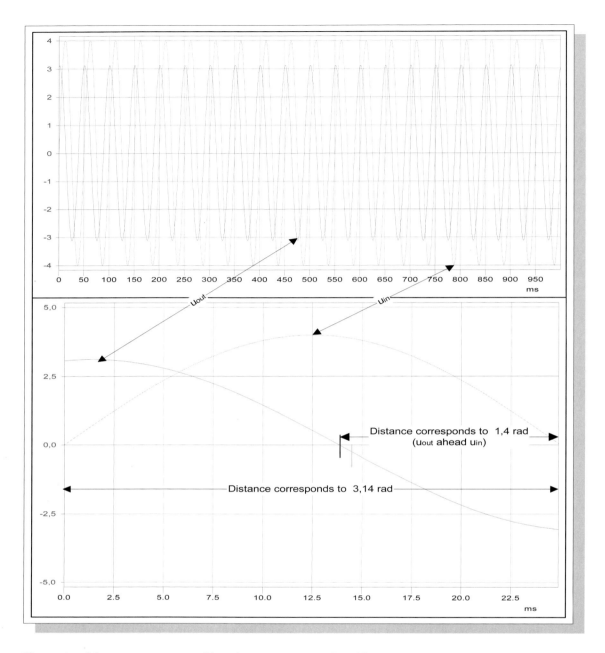

Illustration 96: **Function generator and oscilloscope**

By means of a traditional function generator and oscilloscope it is possible - with time-consuming measurement, reporting, evaluation (using a pocket calculator) and a graph of the curves - to determine the frequency response according to the amplitude ($\hat{U}_{out}/\hat{U}_{in}$) and phase ($\Delta\varphi = \varphi_{out} - \varphi_{in}$). Computer-aided processes deal with this in fractions of a second.

The Illustrations betray two tricks for establishing the frequency response. The top representation suggests that the amplitude curve depending on the frequency can be registered very quickly without constantly switching the time base of the oscilloscope if the time base chosen is big enough. The (sinusoidal) input signal and the output signal appear as "bars" the level of which can be easily read. The precision of the phase curve can be maximised whereby for each measured frequency by means of the time base regulator - usually a knob on or next to the time base switch - exactly one period half of u_{in} is presented over the entire scale. By means of the calculation using the rule of three the phase shift can easily be calculated. As a result of $\Delta\varphi = \varphi_{out} - \varphi_{in}$ and $\varphi_{in} = 0$ in the above situation this results in $\Delta\varphi = 1.4$ rad or 81° (u_{out} rushes ahead).

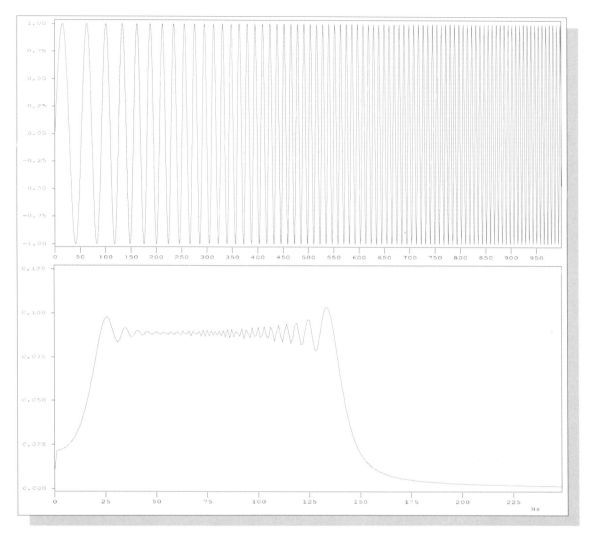

Illustration 97: The sweep signal as „frequency-range-scanner"

The sweep signal was the first step towards an automatized frequency response recording. "Little by little" all sine-wave oscillations or frequencies (with $\hat{U}$ = constant) from a start frequency f_{start} to a stop frequency f_{stop} are connected with the input of the test circuit. At the output $\hat{U}_{out}$ depends on the frequency characteristics of the test circuit. The time axis of the sweeping-signal indirectly represents a frequency axis from f_{start} to f_{stopp}.

*The problem of this procedure is its inaccuracy: Due to the **UP** there cannot be an "instantaneous frequency" as the sine-wave oscillation of a specific frequency must last for a long time so that $\Delta f \rightarrow 0$. If the duration of these "instantaneous frequencies" is too short, the system can only react with a distortion or not at all (transient response error). This is also shown by the FT of the sweep signal (bottom). In Actual fact an rectangular frequency window from f_{start} to f_{stopp} should result. Contravention of the **UP** results in a wavy, blurred frequency window.*

Thus, the circuit is gradually tested with all the frequencies of the frequency range with which we are concerned here. In Illustration 97 the whole sweep signal is represented on the screen, on the left the starting frequency and on the right the stop frequency. In this way the curve of uout on the screen shows not only the time curve but also indirectly the frequency response of the system under examination.

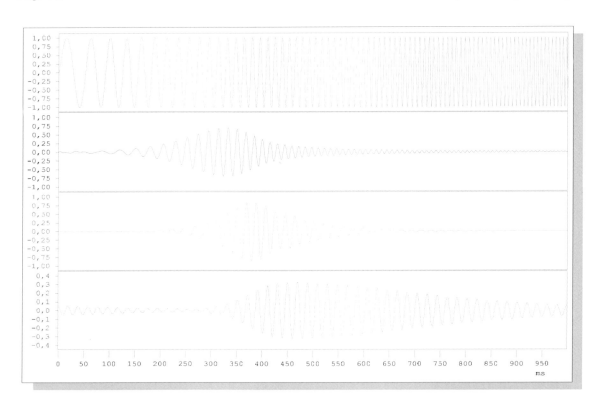

Illustration 98: ***Frequency-dependent reaction of a bandpass filter to the sweep signal***

The upper series shows the sweep signal which is connected in the same way with the input of three bandpass filters. They all have the same mid-frequency (!), but are of different quality (2nd series Q = 3; 3rd series Q = 6 and 4th series Q = 10).

The sweep response u_{out} of the three bandpass filters does not tell us very much. In the second series - of BP with Q = 3 - it is apparently possible to to see exactly at what temporary frequency the bandpass filter reacts most strongly, the amplitude curve seems to correspond to the frequency response of the filter.

In the third series the maximum is further to the right although the bandpass filter has not changed its mid-frequency. The frequency seems unlike the sweep signal not to change from left to right. Finally, in the lower series - of the BP with Q = 10) - the sweep response u_{out} clearly has the same instantaneous frequency over the whole period. The sweep response certainly does not reproduce the frequency response.

But careful! Never forget the Uncertainty Principle **UP**. The FOURIER transform of this sweep signal shows the consequences of beginning a signal suddenly, changing rapidly and ending abruptly. The sweep signal ought really to result in a clearly rectangular frequency response. Hopefully, you now know that this is not possible. The faster the sweep signal changes, the shorter the "*instantaneous frequency*" and the more inaccurately it is measured.

In Illustration 98 the sweep measurement is carried out using bandpass filters with different bandwidths or qualities Q. The quality Q of a simple bandpass filter is a measurement of the abililty to filter out in as narrow a bandwidth as possible or a yardstick for measuring the steepness of the sides of the filter. First a "poor quality" i.e broad width bandpass filter without steep sides (Q=3) is swept. The frequency response can be recognised here clearly via the amplitude curve. As a result of the large bandwidth B = Δf all the instantaneous frequencies were sufficiently long so that $\Delta f * \Delta t \geq 1$ was fulfilled (**UP**).

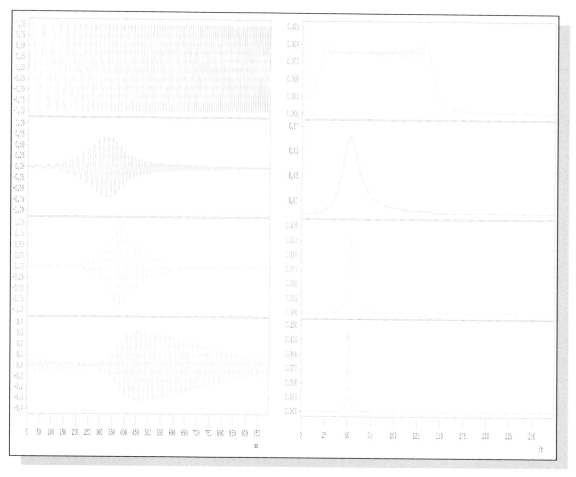

Illustration 99: ***Sweep - a method of measuring with pre-programmed errors***

Here we see the three sweep signal responses from Illustration 98 once again. In the time domain they are very different from each other; above all in the lower series "mysterious" effects occur.

*By contrast the FT of these "spoilt" signals shows the frequency response very precisely; accordingly the correct information on the frequency response must be present in u_{out} (see next section: Transients). The FT seems, therefore, to be the only suitable means of saying anything precise about the frequency behaviour of a circuit/process. Now the signal in the lower series is easy to explain. As the bandpass filter is extremely narrow here ($\Delta f \rightarrow 0$) it can practically only let one frequency through - see bottom left - and as a result of **UP** u_{out} must also last longer.*

*As a result of **UP** the sweep method contains too many errors - unless measuring is carried out very slowly - because the envelope possibly does not reflect the true amplitude curve. As can be seen on the right all three bandpass filters have the same mid- frequency 50Hz.*

Think of a curve which connects the upper maximum values (i.e. the amplitudes of the "instantaneous frequencies") with each other. This curve is intended to represent the frequency response (of the amplitude) of the filter.

If the bandwidth of the bandpass filter is set narrower and narrower by means of the quality Q - z.B. Q = 6 and Q = 10 - and the steepness of the sides increases, the sweep curve no longer shows the frequency response apparently because the **UP** has been contravened. The output signal u_{out} rather reflects the diffuse oscillations as a reaction to the sweep signal but not the amplitude frequency response.

We can escape from this dead end thanks to a modicum of intelligence and better methods (computer-assisted!). Apparently the bandpass signal of a higher quality Q has already signalled by its behaviour in the time domain what transfer properties it has in the frequency domain. It has to some extent shown by its behaviour in its transient response (in the time domain) what its behaviour in the frequency domain will be.

This should be stated more precisely. By means of the sweep signal gradually (altogether t = 1s) all the frequencies - sinusoidal signals - of the frequency range to be examined with a constant amplitude are directed to the input. The output signal must contain all the frequencies with a certain uncertainty, which have passed the bandpass filter with a certain strength (amplitude) and phase shift.

$\hat{U}_{out}$ should therefore be subjected to an FT by means of the computer. The result is shown in Illustration 99. The "erroneous" sweep curve "distorted" by transient responses in the frequency domain apparently shows the frequency response of the narrow bandpass represented as an amplitude spectrum. According to this the information is contained in the output signal u_{out} - but this information can only be recognised via an FT.

Preliminary conclusions:

- The transmission features of circuits/components/systems can only be deduced inaccurately and in a very time-consuming way with traditional measuring instruments - analog function generator and analog oscilloscope.

- Above all the possibility to use FT and IFT (inverse FT) is lacking. This is however easily possible with the digital signal processing system (DSP) by means of the computer.

- *There is only one correct path from the time domain to the frequency domain and vice versa - FT and IFT!*

The future of modern signal generation and processing and the analysis of signals and systems lies therefore in computer-aided digital signal processing (DSP). Analog signals can be stored digitally and evaluated according to the desired criteria and represented graphically by means of a straightforward sound card and the *DASYLab* S-version.

Modern test signals

In this age of computer-based signal processing other test signals are gaining in practical significance on account of their theoretical importance because each theoretical mathematical process can be implemented in a real sense via a certain progam algorithm.

Other important test signals are

δ-pulse	GAUSSian pulse
Step function	GAUSSian oscillation pulse
Burst pulse	Si-function
Noise	Si-oscillation pulse

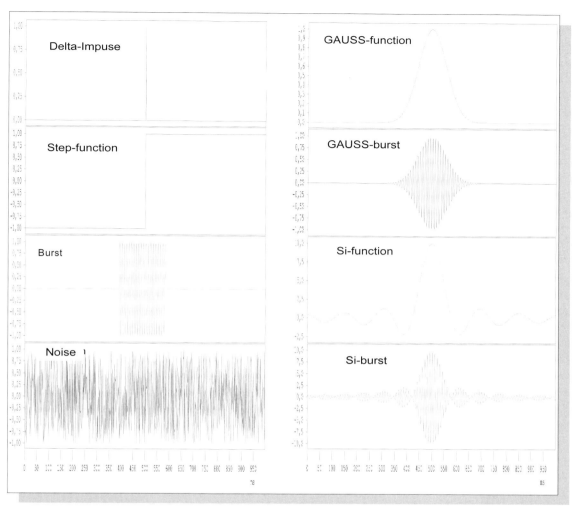

Illustration 100: ***Important forms of test signals for computer-based methods***

The signals presented here have so far been almost exclusively of theoretical importance. As a result of the computer, theory and practice are becoming more and more of a unit because theory is concerned with mathematical models of processes and the computer can easily calculate the results of these mathematical processes (formulae) and represent them graphically.

All these test signals are generated on the basis of formulae by means of a computer and not by means of special analog circuits.

Each of these test signals has certain advantages and disadvantages which will be outlined briefly here.

The δ-pulse

How can the oscillation properties of a car - the mass of the car, the springs, shock absorbers form a strongly damped mechanical bandpass - be measured in a straightforward way? Easy - by driving over a pothole in the road at speed! If the car continues vibrating for a longer time - because of the **UP** it is a narrow band system - the shock absorbers are not in order, that is, the car as a mechanical oscillation system is not sufficiently damped.

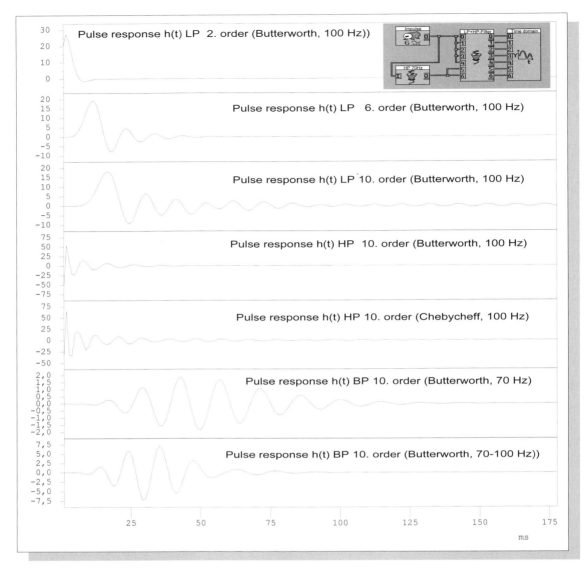

Illustration 101: **From the ugly duckling….**

Although we ought by now to be familiar with the properties of a δ-pulse - it contains all the frequencies and sinusoidal oscillations with the same amplitude, i.e. the system is tested simultaneously with all frequencies - the result of this measurement analysis never fails to surprise. The pulse responses at first sight look so insignificant but do tell the expert a great deal about the frequency-related behaviour of the filter.

*In the case of higher order filters the edge steepness is greater and consequently the conducting state region of the filter is smaller. As a result of this the pulse response h(t) lasts longer on account of the **UP**. This can be seen quite clearly in the case of the narrow bandpass filter (70Hz). And, the greater the edge steepness the greater the delay in the beginning of the pulse response.*

A highpass filter is always broadband, its pulse response starts with a step and builds up to its cutoff frequency. As a result of the high degree of edge steepness of the Chebycheff highpass filter, h(t) lasts much longer than in the Butterworth high pass of the same order.

It seems like a miracle that the FT of these insignificant-looking pulse responses results exactly in the transfer function H(f) according to the absolute value and phase (see Illustration 102). The phase spectrum is however only given correctly if the δ–pulse is positioned at the reference point of time t = 0. In contrast to the traditional methods of measuring using an oscilloscope and a function generator the computer-based measurement and evaluation only takes fractions of a second.

The electrical equivalent to the pothole in the road is the δ-pulse. The reaction of a system to this spontaneous, extremely short-lived deflection is the so-called δ-pulse response at the output of the system (see Illustration 101). If it is extended in time the frequency domain is according to **UP** strongly restricted, i.e. a kind of oscillating circuit occurs. Any strongly damped oscillating circuit (intact shock absorbers or ohmic resistance) is broadband, i.e. it therefore produces a brief pulse response. This is the only way the physical behaviour of filters (including digital filters) can be understood.

The δ-pulse response (generally called the "*pulse response*" *h(t)*) provides via the **UP** qualitative information on the frequency properties of the system being tested. But it is only the FT that provides precise information on the frequency domain. It alone shows what frequencies (and their amplitude and phases) the pulse response contains.

If a system is tested by a δ-pulse it is - unlike the sweep signal - tested simultaneously with all frequencies (sinusoidal oscillations) of the same amplitude. For example at the output of a high pass filter the low frequencies below the cutoff frequency are almost completely lacking. The sum of the (high) frequencies allowed to pass from the pulse response h(t). In order to obtain the frequency response the pulse response h(t) must simply undergo an FT.

> *The importance of the δ–pulse as a test signal is based on the fact that the FOURIER transform FT of the pulse response h(t) already represents the transfer function/frequency response of the system tested.*

Definition of the *transfer function* H(f):

For every frequency, amplitudes and phase displacements of u_{out} and u_{in} are compared with each other.

$$H(f) = (\hat{U}_{out}/\hat{U}_{in}) \quad (0 < f < \infty) \qquad \Delta\varphi = (\varphi_{out} - \varphi_{in}) \quad (0 < f < \infty)$$

Conclusions and notes:

- Unlike the sweep signal in the case of the δ–pulse the system is measured simultaneously with all frequencies ($\hat{U}$ = constant).

- As in the case of the δ-pulse $\hat{U}_{in}$ = constant for all frequencies, the amplitude spectrum of the pulse response forms the curve of the absolute value H(f) of the transfer function. On account of this relatedness the pulse response is represented internationally as h(t).

- Thus, in shorthand we can write: |FT(h(t))| = |H(f)| ("the absolute value of the FT of h(t) is equal to the absolute value of the transfer function H(f)"). In addition, the following holds: IFT(H(f)) = h(t) (" the inverse FT of the transfer function H(f) is the pulse response h(t)").

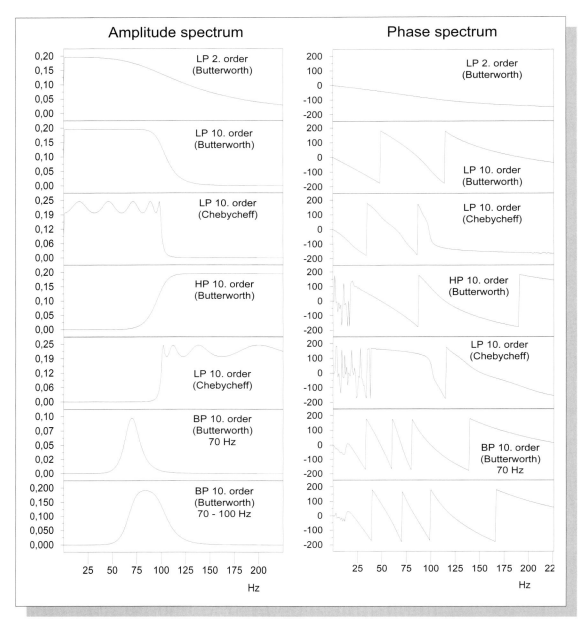

Illustration 102: **...... to the beautiful swan**

The FT of the insignificant-looking pulse responses h(t) from Illustration 101 produces as if by magic - although the explanation is quite clear - the transfer function according to absolute value and phase.

In the case of the phase spectra you see "strange steps". The reason is that the phase spectrum is only represented between $-\pi$ (or -180 degrees) and π (or 180 degrees). If the latter value is exceeded the curve "jumps" downwards as both angles are identical. The irregular, "noise-like" phase curve arises through inaccuracies in calculation.

It can be seen quite clearly what the price of edge steepness is. The Chebycheff type has with the same order steeper edges in the conducting state region but also greater ripple content.

The preceding considerations show the particular (theoretical) importance of δ-pulses as test signals. In practical application they would have to be very high in order to have sufficient energy. In this case they represent something like a "spark", i.e. like extremely brief high voltage. Hence, they represent a danger for all circuits in microelectronics. Does theory provide an alternative, i.e. a test signal with sufficient energy, which is easy to generate and which represents no danger for microelectronics?

- The phase curve of the FT of the pulse response is in general not exactly identical with the phase curve of the transfer function H(f). The phase spectrum of the δ-pulse is also included and that depends on the position of the δ-pulse with regard to t = 0 s. If the δ-pulse lies exactly at t = 0 s - which as a result of the physically finite width Δt of the δ-pulse is never exactly possible - both phase curves agree exactly with each other.

- The computer can calculate this "adjustment" and then give the phase curve of the transfer function.

The disadvantage of δ-pulses as test signals is their low energy as a result of their brief duration. This can only be increased by increasing the level (e.g. 100V). Such a high voltage pulse could destroy the input microelectronics. For acoustic, that is electromechanical systems such as loudspeakers the δ-pulse is completely unsuitable, even dangerous as a test signal. If the signal were strong enough the loudspeaker membrane might be destroyed. It would have the same effect as a brief blow with a hammer on the membrane or like the pothole in the road that was just that bit too deep.

The step function

The test signal generally used in automatic control engineering is the step function (see Illustration 100). This is based on the following philosophy. The system is exposed to one extremely brief change of state (in the case of the δ-pulse there are two extreme changes of state following each other whereby the first change of state is immediately reversed). This change of state in connection with the step function causes a certain reaction in the system. This then reveals everything about its own physical behaviour with regard to oscillations within the system.

Whereas the pothole driven over at speed corresponds as a test signal to the δ-pulse, the properties of the step function can be compared with driving a car on to or down from the kerb (provided the kerb isn't two high!). The briefly vibrating car tells us whether the shock absorbers are OK. From an electrical point of view the step function is equivalent to switching on or off of a direct voltage, i.e. it is very easy to produce this test signal.

The step function has the advantage compared with the δ-pulse that it has more energy. This can also be understood by considering the car as a "mechanical oscillating circuit". A pothole in the road - equivalent to the δ-pulse - may not be noticed by the driver if it is "absorbed". Only from a certain size will it make itself felt. The "jump" of a car from the curb will always be noticeable however fast the car drives off it.

As every oscillating system will react to such a "step" or any oscillating circuit is set in motion with its own particular frequency (natural frequency f_E) the one-off step function must contain all the frequencies.

This property will be exploited in an experiment presented in Illustration 103 to obtain information on the curve of the amplitude spectrum of the step function.

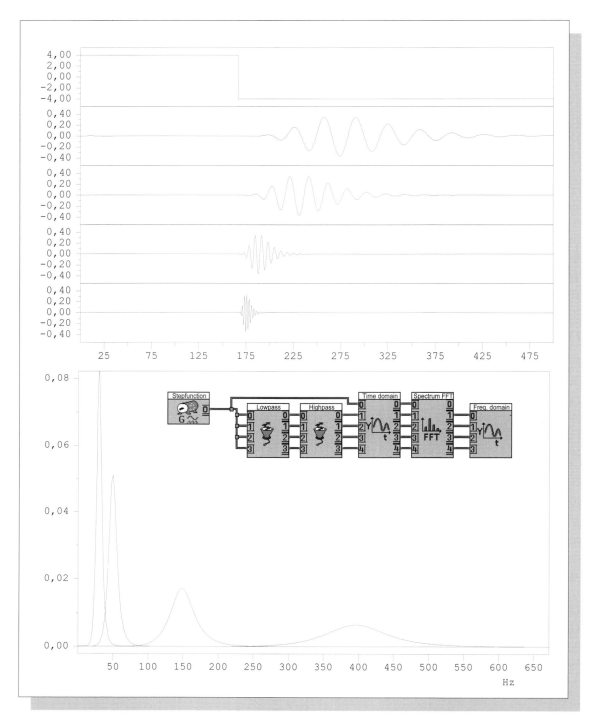

Illustration 103: **Does the step function contain all the frequencies?**

A step function is connected with an extremely narrow bandpass filter (formed from LP and HP) which only lets through one frequency. Hence the step responses are „brief sinusoidal signals". Only the center frequency is varied without changing the sensitivity (Quality Q) of the filter. The spectrum - the FT of the step responses - shows that the low frequencies are much more strongly represented in the spectrum than the high ones. Put more precisely - if the frequency is doubled, the height of the amplitude spectrum of the step response is halved. This suggests: $U \sim 1/f$.

Chapter 6: System analysis — Page 155

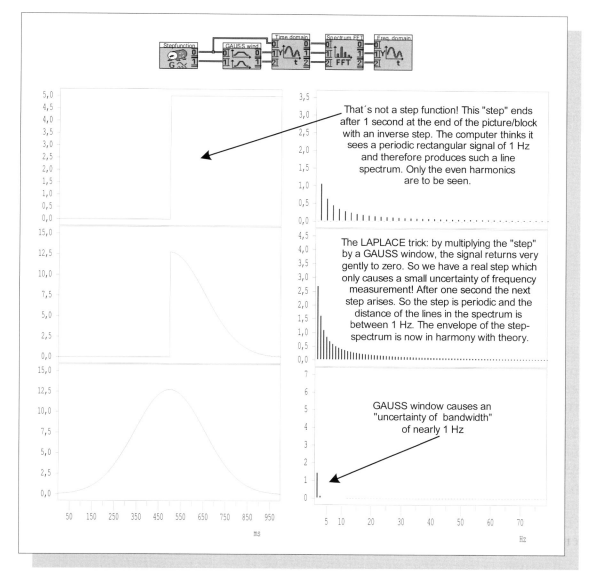

Illustration 104: **The trick with the Laplace transformation**

The step function contains mathematical and measuring technology problems because it is unclear what happens after the steps. The end of the measuring procedure implies the end or the "jumping back" of the step function. Thus, (in retrospect) we have not used the step function as a test signal but rather a rectangular of the width τ. This problem is got rid of by means of a trick. The step function is ended as gently as possible as it fades exponentially or according to a GAUSSian function. The more slowly this takes place, the more this test signal has the features of the (theoretical) step function. This trick in connection with the subsequent FT is called Laplace transformation.

Carrying out the step function of an FT is not easy without a trick because it is neither periodic nor is its duration defined. (What happens after the "step" - a leap backwards?).

Thinking about this takes us a step further - the step function is a kind of rectangle in Illustration 33 - a measure for the zero positions of the rectangular amplitude spectrum. The greater τ, the smaller the distance between the zero positions of the spectrum. The second "step" (jump back) is practically where the measurement procedure is ended.

Depending on the duration of "recording" Δt, τ has a quite specific value. Considerable difficulties in a mathematical and measurement technology sense are involved here.

Mathematicians have recourse to their bag of tricks and this trick is called *Laplace transformation*. This consists in limiting the duration of the step function artificially thus making it possible to evaluate it via an FT. Generally an e-function or or the GAUSSian function is used here as shown in Illustration 104. It forms the transition so "gently" that it is hardly noticeable in terms of frequency. The use of such "gentle" transitions ("windowing") in the FT analysis of non-periodic signals has already been explained (Illustration 52-Illustration 54).

> *Note:*
> Only periodic signals - although they are unlimited in time - are precisely analysable as they are always repeated in the same way. The length of the analysis must be exactly an integer multiple of the period length T.
>
> Otherwise signals can only be analysed with a frequency resolution Δf which on account of the **UP** is equivalent to the inverse value of the duration Δt of the signal length.

This limitation thus takes place "unnoticeably" whereby the step value (e.g. 1 V) after the step gradually and exponentially approaches zero. For the duration Δt of this "approach to zero", Δt >1/Δf should apply on account of the **UP** whereby Δf is the desired frequency resolution.

The theory provides the most intelligent method of combining the advantages of the step function - sufficient energy - with the advantage of the δ-pulse the FT of the pulse response directly gives the transfer function and the frequency response. A δ-pulse can be produced from a step function by differentiation. Differentiation is one of the most important mathematical operations used in natural sciences and technology. Chapter 7 (Linear and non-linear processes) will deal with this in greater detail. At this point it is enough to know what effect it has on a signal.

Differentiation makes it possible to establish how rapidly a signal is changing at a given point in time. The faster the signal increases at a given moment the greater the immediate value of the differentiated signal. If the signal decreases at a given moment the immediate value of the differentiated signal is negative. Simply look at Illustration 105 and you will know what the effect of the differentiation of a signal is.

By means of the sequence of the operations in Illustration 105 the method is to a certain extent optimised. As you can see from the caption text the test signal step function has enough energy in spite of the slight height of the step - it doesn't destroy the microelectronics - but the pulse response h(t) is evaluated, the FT of which directly provides the transfer function H(f).

The step function is also of great importance as a real test signal. It has sufficient energy and is extremely easy to generate. By means of computerised processing of the step response - first differentiation, then the FT - we obtain the transfer function H(f), i.e. the complete information on the frequency domain of the system tested. All the mathematical and physical difficulties which we usually have with the step function are now avoided.

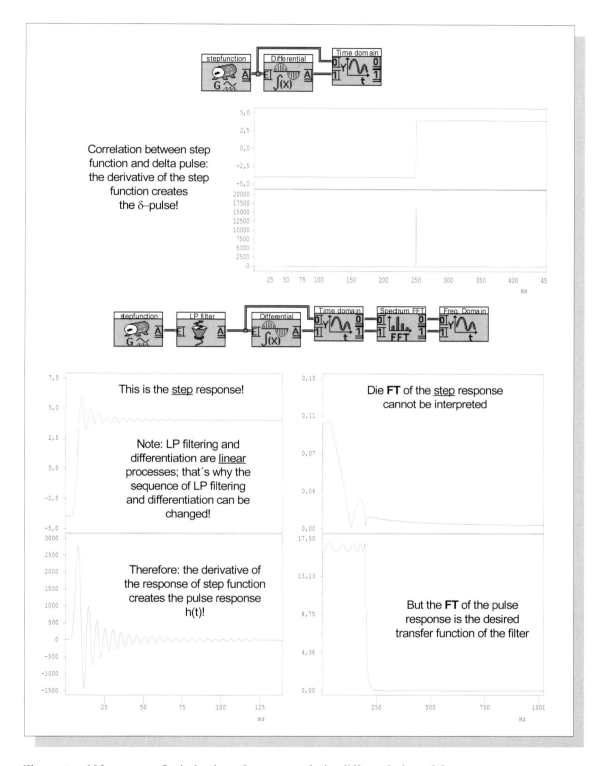

Illustration 105: ***Optimisation of system analysis: differentiation of the step response***

*If the differentiated step function produces a δ-pulse, the differentiated step response also produces the pulse response h(t). This follows from the fact that both the low pass and differentiation represent **linear** processes. Therfore the order of the sequence can be changed. The step function which has sufficient energy is directed at the lowpass circuit or the system. The step response is differentiated, the pulse response h(t) is obtained. The system was thus indirectly tested by a δ-pulse. This process cannot be understood without the theoretical background.*

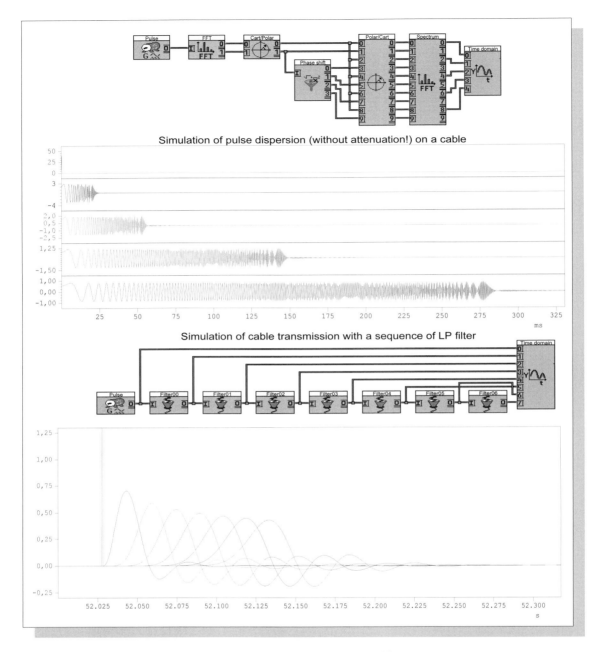

Illustration 106: **_Dispersion of pulses on cables_**

As far as their form is concerned needle pulses would at first sight be ideal for measuring transit time in cables or for establishing the speed of propagation in cables. However, physical phenomena occur in homogeneous cables which alter the pulse form: as a result of frequency-dependent dispersion - the speed of propagation depends on the frequency- the phase spectrum changes; as a result of the frequency-dependent absorption (damping) the amplitude spectrum of the signal changes. A different signal appears at the output from the input.

Here the "pure case" of the dispersion of a δ–pulse is represented. The cable was simulated by several serially connected allpasses. They are purely dispersive, i.e. they do not dampen depending on frequency but only change the phase spectrum. The δ–pulse at l = 0 km has practically the same amplitude spectrum as at l = 16 km. Interestingly, the δ–pulse disperses to become an oscillation pulse which resembles a sweep signal.

A real cable can be easily simulated by means of a series of lowpass filters because a lowpass has frequency-dependent damping and phase displacement (absorption and dispersion, see below).

The GAUSSian pulse

In this age of DSP (Digital Signal Processing) pulses play an extremely important role as any binary pattern, i.e. any binary information from a physical point of view consists of an arrangement of pulses (pulse pattern).

Rectangular pulses have a very wide frequency band as a result of their "steps". This wide frequency band makes itself inconveniently felt in transmission technology. Thus, every cable represents a "dispersive" medium which makes the transmission of pulses more difficult.

> *A medium is dispersive, if the propagation speed c in it is not constant but depends on a given frequency.*

As a result the phase spectrum changes along the cable because the individual sinusoidal oscillations spread at different speeds and at the end they are in a different position in relation to each other, i.e. the phase spectrum of the signals have changed. As a result of this effect the individual pulses disperse along the cable; they become flatter and wider until they overlap. The information is thus lost.

For this reason it is hardly possible to measure the propagation speed c along a stretch of cable by means of a rectangular pulse over the transit time τ. The pulse appearing at the end of the cable has become so dispersed in the case of a longer cable that the beginning and end of the pulse can no longer be established.

For transmission technology - and also for pulse measurement technology referred to here - a pulse is required which for a given duration τ and height U has as narrow a frequency band as possible. Such pulses hardly disperse - they are merely damped - because as a result of the narrow frequency band the frequencies hardly spread at different velocities.

As the GAUSSian pulse begins very gently, changes gently and ends gently it exhibits according to the knowledge we so far have (**UP!**) the desired behaviour.

> *For this reason the GAUSSian pulse form is suitable not only for the rapid transmission of binary data but also for measuring transit times in systems.*

There are other signal forms which begin and end gently and which at first sight hardly differ from the GAUSSian pulse. All these signals play an important role as "windows" in determining and analysing of non-periodic signals.

An important property of the GAUSSian pulse should be pointed out here: the FT of a GAUSSian function is also a GAUSSian function, i.e. the spectrum - including negative frequencies - has the same form as the signal in the time domain. This is true of no other function.

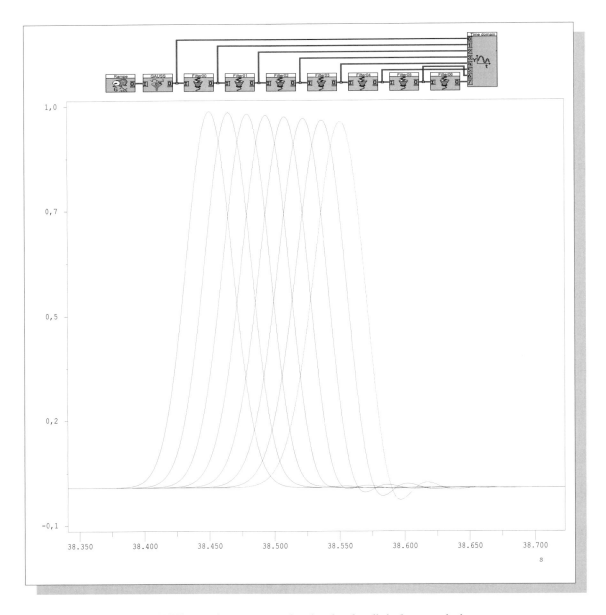

Illustration 107: **GAUSSian pulses as an optimal pulse for digital transmission.**

Instead of δ-pulses GAUSSian pulses are here exposed to dispersion. The result is unambiguous: the pulse form does not change considerably. Thus, GAUSSian pulses are the optimal pulses for transmission via cables. This is not limited to transit time measurements but also applies to digital transmission processes in which generally information in the form of binary patterns is transmitted. The GAUSSian pulse is suitable as a pulse type for these binary patterns although other modulation methods are state of the art here.

Because it begins and ends gently, i.e. has no rapid transitions its bandwidth is minimal, the frequencies which hardly differ spread at almost the same velocity. In this way dispersion does not have much of a chance.

The GAUSSian oscillation pulse

While the GAUSSian pulse has a narrow spectrum symmetrical to the frequency f = 0 Hz this spectrum can be moved to any position of the frequency domain by means of a trick. This results in the GAUSSian oscillation pulse in the time domain.

For this purpose the GAUSSian pulse is simply multiplied by a sine of the frequency f_C (C as "carrier frequency") by which the spectrum is to be displaced. The result is then a sine in the time domain which begins and ends gently (see Illustration 100). In the frequency domain this causes a displacement of the (symmetrical) spectrum from $f = 0$Hz to $f = f_C$ Hz (and $f = -f_C$). The narrower this signal called a GAUSSian oscillation pulse in the time domain, the wider its spectrum in the frequency domain. (**UP**; see also Illustration 46). This important trick using multiplication is dealt with in more detail in Chapters 7 and 8).

The GAUSSian oscillation pulse is - like the GAUSSian pulse - suitable for determining in a straightforward way the *group velocity* v_{gr} in cables or in entire systems and - in addition - for comparing it with so-called *phase velocity* v_{ph}.

The phase velocity v_{ph} is the velocity of a sinusoidal wave. In a disperse medium - e.g. along a cable - v_{ph} is not constant but depends on the frequency f and the wavelength λ. Group velocity v_{gr} is understood to mean the velocity of a group of waves, i.e a group of waves limited in time and place (e.g. GAUSSian oscillation pulse).

> Note:
> Energy and information propagate themselves at the group velocity v_{gr}. If the GAUSSian oscillation pulse is used as a group of waves the transit time t of the sinusoidal carrier signal is a yardstick for the phase velocity v_{ph}, the GAUSS shaped curve of the envelope on the other hand is a yardstick for the group velocity v_{gr}. If v_{ph} is constant $v_{gr} = v_{ph}$. An interesting physical property is the fact that the group velocity v_{gr} can never be greater than the speed of light in a vacuum c_0 ($c_0 = 300,000$ km/s). This is not necessarily true for the phase velocity v_{ph}. The maximum upper limit for the transportation of energy and information is thus the speed of light, or the speed of the electromagnetic energy of the medium involved. In cables it is between 100,000 and 300,000 km/s.

Preliminary conclusions:

> *The GAUSSian pulse and the GAUSSian oscillation pulse are less important as test signals for measuring the frequency response of circuits and systems. They can be used in the measurement of pulses to determine in a straightforward way the transit time, group and phase velocity.*

The Burst signal

The burst signal is - as we already know from Illustration 45 - a time-limited "sine". However it begins and ends abruptly and this has consequences for the bandwidth and the uncertainty of the frequency of the sine wave.

Thus a burst can be used to test the frequency selectivity of a circuit or system qualitatively in the sphere of the value of the sine frequency (mid-frequency). However, the spectrum has zero positions; these represent frequency gaps.

Transients of frequency selective circuits can be superbly demonstrated by means of a burst. See the next section on "Transients".

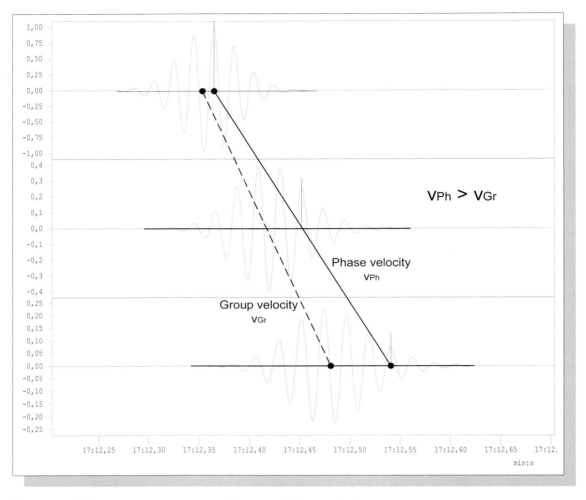

Illustration 108: ***Group and phase velocity***

This GAUSSian oscillation pulse is directed here to a series of the same lowpass filters according to Illustration 106. It is affected by damping and phase displacement.

The phase velocity is visualised here by means of the maximum value of the "short-lived" sine or the line drawn through the three maxima. The group velocity corresponds to the velocity of the enveloping GAUSSian function. Both velocities are not identical because neighbouring maximums do not keep the same level.

The Si-function and the Si-oscillation pulse

The disadvantages of the δ–pulse as a test signal were its low spectral energy and its "dangerous" pulse level. In addition the energy distributes itself equally over the entire frequency domain from 0 to ∞.

What might an ideal test signal look like? - a test signal that has all the advantages of the δ–pulse without its disadvantages. Let us try to describe a signal of this kind:

- The energy of the signal should be confined to the frequency range of the system to be tested

- In this range all the frequencies (as in the case of the δ–pulse) should have the same amplitude so that the FT directly produces the transfer function H(f).

Chapter 6: System analysis

- All the frequencies should be present simultaneously so that - unlike the sweep signal - the system to be tested can respond at the same time to all frequencies.

- The energy of the test signal should not be put in abruptly or in a very brief space of time, but continuously so that the system should not be destroyed. Please note in this connection that as a result of the **UP** a frequency band limited signal Δf must of necessity have a certain duration Δt.

- The maximum value of the test signal must not be so large that the microelectronics are destroyed.

There is a signal which does all these things and which we are already very familiar with. The Si-function and the Si-oscillation pulse. Why has it not been used much up to now. Simple. It can only be generated i.e. calculated and evaluated using a computer. The use of this signal is therefore inseparably bound up with computer-based DSP (Digital Signal Processing).

Illustration 48 already showed:

The Si-function (and the Si-oscillation pulse) is nothing but a segment cut out from the spectrum of a δ–pulse.

The Si-function (in the time domain) results as a frequency segment from the δ–pulse from $f = 0$ to $f = f_{Si}$ Hz with the frequency response of a virtually perfect lowpass filter LP. Put more precisely, the spectrum of the Si-function ranges from $f = -f_{Si}$ to $f = +f_{Si}$. See also Illustration 84.

The Si-oscillation pulse (in the time domain) results as a frequency segment of the δ–pulse from f_{bottom} to f_{top} with the frequency response of a virtually ideal bandpass filter BP.

There are two methods of generating the Si-function and the Si-oscillation function. First purely by means of a formula (e.g. by using the module "formula interpreter" from DASY*Lab*). You would then have to do a lot of experimenting until the Si-function or the Si-oscillation pulse have exactly the right bandwidth.

The method already suggested on this page of simply cutting the desired segment out of the spectrum of a δ–pulse is much more elegant. This is the method which was used in Illustration 88 and is now used in Illustration 109. Only at this stage does it become clear how elegant the method of transforming "back and forwards" can be in practice.

Note:

- The wider the frequency band Δf of the Si-function or the Si-oscillation function the more it resembles the δ–pulse. In this case the tip of the Si-pulse can possibly endanger the microelectronics.

- The longer the Si-function lasts altogether - i.e. the more the Si-curve tends towards zero - the more the frequency response resembles the ("ideal") rectangular curve (**UP!**).

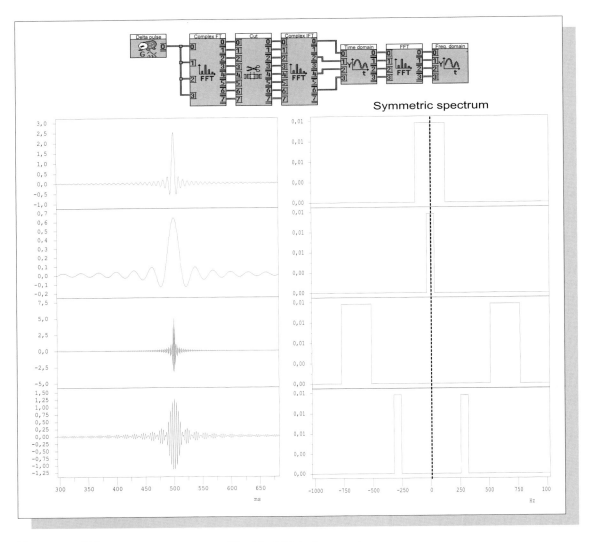

Illustration 109: **The ideal Si-test generator**

This circuit which in principle was already used in Illustration 88 makes use of the fact that the Si-function and the Si-oscillation pulse result as frequency segments cut out from the δ–pulse.

The δ–pulse is first transformed in the frequency domain (FT). There the real and imaginary section are both cut out in the same way. If a lowpass signal is required the frequencies from 0 to the cutoff frequency f_{Si} are cut out. In the case of a bandpass signal only the relevant section etc. The rest of the frequency band is then transformed back into the time domain (IFT). It is now available as a test signal with a precisely defined frequency range.

Note that the upper lowpass has double the bandwidth from $-f_{Si}$ to $+f_{Si}$. The third signal (bandpass) has exactly the same bandwidth as the first (lowpass). The equivalent is true of the second and fourth signal. This can be seen in the time domain. The Si-oscillation pulses shown have precisely the Si-signals described as envelopes.

Noise

The most "exotic" test signal is without doubt pure stochastic noise. As shown at the end of Chapter 2 it results from random processes in nature. Such processes can be simulated as desired by the computer with the result that noise can also be generated by means of the computer.

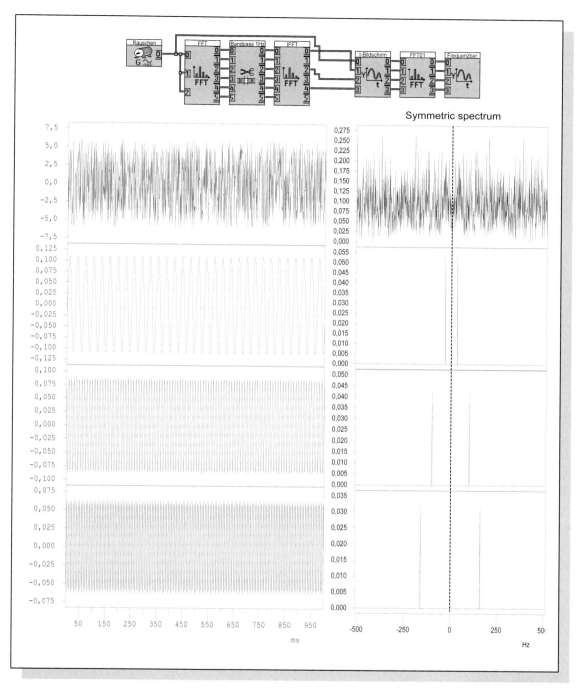

Illustration 110: **_Does noise really contain all the frequencies?_**

This can be checked easily by using our "supercircuit". Any four frequencies are filtered out in the frequency domain - in this illustration 32, 128, 512 and 800 Hz from top to bottom.

That sounds simple but at the same time quite amazing. Where do filters exist - bandpasses - whose band width only lets a **single** frequency pass, or the steepness of whose sides tends "towards infinity". Our filter bank is only limited by the **UP**: in the case of a signal duration of 1 s the bandwidth or the frequency uncertainty must be at least 1 Hz. You will have noticed that the frequencies filtered out or the sinusoidal oscillations have different amplitudes. If you construct and operate the circuit yourself you will note that the amplitude and phase position fluctuate (magnifying glass). It is not possible to see any regularity as noise is a stochastic or random signal. When, do you think, will these fluctuations be greater - in the case of a short noise signal or in the case of a very, very long one?

If we have defined information as an arranged meaningful pattern (see Chapter 2) stochastic noise appears to be the only signal which according to our definition does not possess any information. A pattern in the signal gives it a tendency to preserve itself. This means that as the transmission of a pattern requires a certain period of time in the case of two time segments A and B which directly follow each other something in segment B must remind one of segment A. To put it in another way: there must be a certain similarity or relation between time segments A and B.

Similarity or relation are complex concepts. In the case of signals they may refer to three areas:

> *Signal patterns - may resemble each other in the time and/or frequency domain, the relation/similarity may also refer to* ***statistical*** *information.*

An example: radioactive decay occurs in a random manner. Pronounced radioactive decay - made audible or visible via a detector - makes itself felt as noise. The time intervals between two processes of decay which follow each other or between two clicks statistically subject to an "exponential distribution". Brief intervals between two clicks occur very frequently, long intervals occur rarely. Accordingly two signals which have come into being as a result of radioactive processes of decay may have a certain (statistical) relationship with each other.

Such considerations unavoidably take us to *information theory*. We cannot leave out the theoretical aspect in the case of test signals. While a noise-signal u_{in} does not contain any information about the features of the system, the response of the system, i.e. u_{out}, should contain and make transparent all the information about the system. The test signals dealt with up to now have a certain tendency towards preservation, that is, a certain regularity in the time and frequency domain. This is not the case with the noise-signal - apart from statistical features.

> *All information which the noise response u_{out} contains derives originally from the system, all the information says something about the system. Unfortunately this information about the system is not directly recognizable either in the time or frequency domain as the noise response still has a random component.*

In order to obtain the features of the system in a pure form from noise response we must have recourse to statistics. For example, it would be possible to add a few noise responses and calculate the average. The FT of this mean noise response is a clear improvement.

The average value plays a decisive role in statistics, partly because the mean value of a series of stochastic measurement readings must be equal to zero. If this mean value were not zero certain quantities would occur more frequently than others. The measurement readings or the signal would than not be purely random.

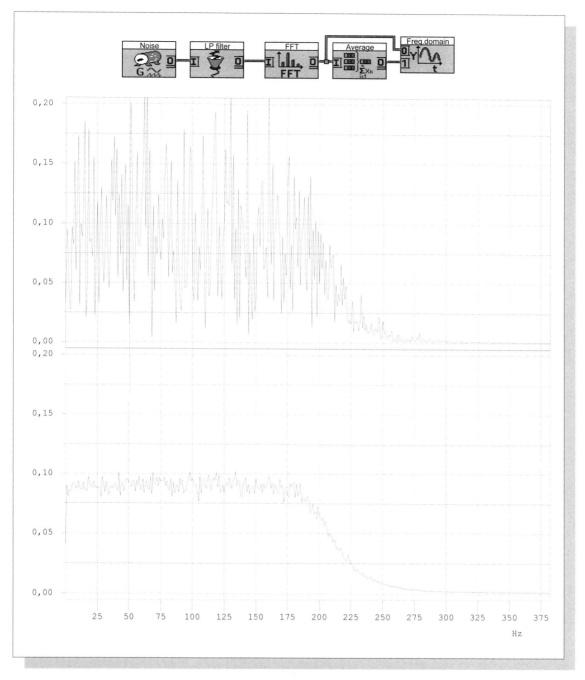

Illustration 111: ***Noise as a test signal***

The longer a noise signal lasts the less it is random with regard to the amplitudes of the frequencies contained in the noise. This is proved by this measurement method.

Instead of a long noise signal the mean of the results of many noise signals of 1s in length is calculated in the illustration (top). At the bottom you see the spectrum of a lowpass filtered noise signal of 1s duration.

The noise is generated here by means of an algorithm (computational method) and strictly speaking no longer corresponds to the ideal case. This is shown here because even after extremely lengthy block averaging the filter curve is not "smooth". In some places the peaks continue to exist.

Transients in systems

In the case of some test signals - such as δ–pulses or step functions - we expose a circuit or a system to a sudden change and observe the reaction (u_{out}). The pulse response h(t) and the step response reproduce this reaction. Frequency selective circuits exhibit a certain time delay as a result of the **UP**. Illustration 101 shows this for all filter types.

It is possible to estimate the bandwidth Δf of the system as a result of $\Delta f * \Delta t \geq 1$ (see Illustration 102.

> *The transient response reflects the reaction of the system to a sudden change in the original condition. It lasts as long as the system is attuned to the change or has reached its so-called stationary state.*

Any change imposed from outside to a (linear) system capable of oscillation - this is always a frequency selective system - is not abrupt or sudden but requires a transition phase. This transition phase it called a "transient". The steady state follows which is usually referred to as the stationary state.

The pulse response h(t) describes a typical transient. As the FT results in the transfer function H(f) this example makes the following clear: see Illustration 101.

> *The transient of a system betrays its oscillatory features in the time and hence in the frequency domain. For this reason this can be subjected to a System analysis.*
>
> *During the transient the system provides information **about itself** and then largely conceals the information on the actual signal to be transported from the input to the output. The actual transmission of a signal is restricted essentially to the sphere of the stationary (steady state) condition.*
>
> *In order to achieve a considerable flow of information throughout the system the transient time Δt should be as short as possible. According to the **UP** the bandwidth Δf of the system should be as large as possible.*

In Illustration 112 a special experiment is intended to show what actually happens during the transient in a frequency selective system.

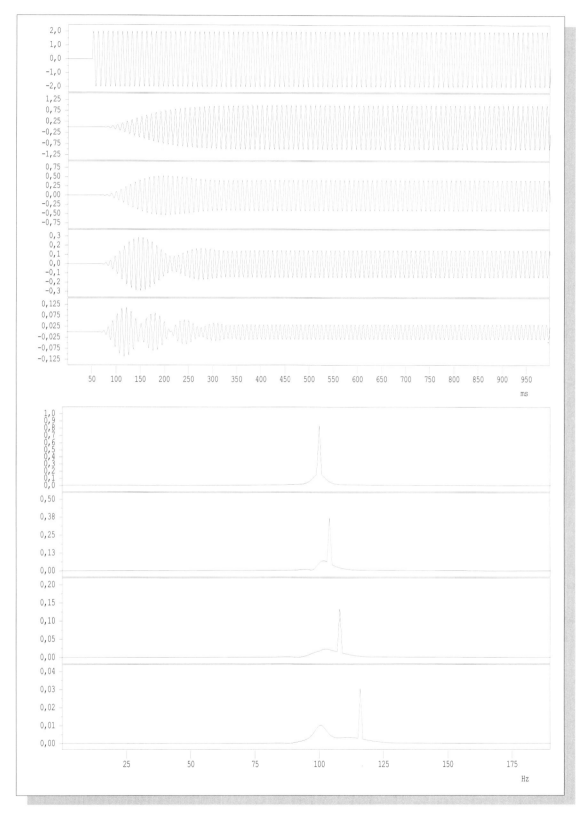

Illustration 112: **Transient: the system says something about itself**

This experiment has results which are of great importance for information technology. It shows and **explains** *why a narrowband system can only transmit a small amount of information per unit of time. You will find a more detailed explanation in the main body of the text.*

We select an extremely narrow bandpass filter - an oscillation circuit - because the transient period τ is great enough to make it possible to see and interpret the processes more easily. Let the conducting state frequency (natural frequency, resonance frequency) equal 100 Hz

Definitions:

- Natural frequency f_E is understood to mean the frequency with which a resonant circuit oscillates after it has been started - for example by a δ–pulse and then left to itself.

- The resonance frequency f_R indicates the frequency at which a resonant circuit started by a sine and forced to oscillate shows its *maximum reaction*.

- The natural frequency and the resonance frequency have the same value in resonant circuits of high quality. In the case of high damping of the resonant circuit the resonance frequency is slightly higher than the natural frequency: $f_R \geq f_E$

Select a burst pulse of which the beginning but not the end can be seen in Illustration 112 as a "test signal". The center-frequency f_M of the burst signal is now slightly varied. At first it is 100 Hz, then 104, 108 and finally 116 Hz. The four bottom rows in Illustration 112 show the transient - that is the output signal - in the sequence given.

First of all the time is noticeable which passes after the start-up process (top series) until the reaction appears at the output of the bandpass/resonant circuit. The time is 20 ms. After this the transient begins which ends at roughly t = 370 ms. After that the stationary state begins.

As a result of the four different "center-frequencies" of the burst signal the transients are all different. Note also the different scalings on the vertical axes. Thus the voltage level of the lowest transient (at 116 Hz) is only roughly 1/10 of the upper transient at 100 Hz.

An experienced person immediately sees the *beat*-like appearance of the three lower transients. What is meant by this?

> *A beat arises as a result of the addition of two sinusoidal oscillations of roughly the same frequency (see Illustration 113). It expresses itself in a periodic amplification and attenuation with the beat frequency $f_b = |f_1 - f_2|/2$.*
>
> *The beat is a typical interference phenomenon in which the maximum amplification or damping arises as a result of the same phase position or a position displaced by p of the two sinusoidal oscillations.*

There *must* be two frequencies involved although we started the system using only one (center-) frequency. Where does the *second* one come from?

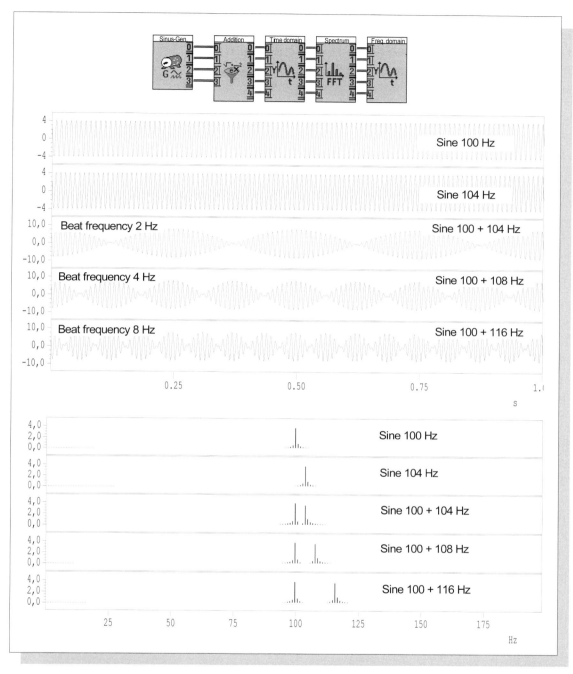

Illustration 113: **Beat effect and beat frequency**

A beat arises by the addition of two sinusoidal signals of virtually the same frequency. The small difference in frequency has at times the effect of a phase displacement.

In the upper two series you see two sinusoidal oscillations and in the third the addition (sum). While to the left and the right the two sinusoidal signals appear to have the same phases, in the middle the "phase displacement" is practically 180° or π rad. On the left and right they amplify each other to reach twice the amplitude, but in the middle they extinguish each other.

Acoustically a beat is characterised by a rise and fall in volume. The rhythm with which this takes place is called the beat frequency. It can be recognised visually by the envelope of the beat and is calculated by the formula $f_B = |(f_1 - f_2)|/2$

Our bandpass filter is - in a physical sense - a simple resonant circuit which can practically only resonate with one frequency. Like a children's swing. Once it has been set in motion it swings with only one frequency, its natural frequency f_E. If it is pushed regularly - periodically - it will only reach the maximum swing when this occurs precisely in the rhythm of its natural frequency. The second series in Illustration 112 shows this case.

If the rhythm of the energy supply gets somewhat bigger or smaller, cause and effect are in certain moments in phase, then less and less until they are contra-phase. In this case the swing would be given a push when it approaches the person pushing and slowed down here. "In phase" is equivalent to the case in which the swing is pushed in the direction of motion. The deflection is then temporarily greatest.

In the three bottom cases in Illustration 112 the "swing" bandpass filter is given a push at times in the right or at least almost right rhythm and then in the wrong rhythm. The deflection increases or decreases in a certain rhythm. Experts call this the beat frequency f_B. This is reproduced by the envelope.

The following result should be noted: even when a sinusoidal voltage is switched on the reaction of the resonant circuit reveals the process which takes place physically within it. It tries to oscillate with its natural frequency. At the beginning it swings like a swing with its natural frequency if the activating frequency differs slightly. The more the activating frequency differs *temporarily* from the natural frequency the more weakly the resonant circuit resonates in the stationary state.

The bottom part of Illustration 112 shows this very clearly. It shows the four transients in the frequency domain. The peaks in the spectrum represent the *input signals* activating the bandpass (100, 104, 108, 116 Hz). The remaining spectrum derives from the bandpass/ resonant circuit and occurs mainly at 100Hz (*natural frequency*).

Exercises on Chapter 6

Exercise 1

(a) Select a sweep generator from the library of DASY*Lab*-circuits and try varying the start and stop frequency and the sweep speed systematically.

(b) What „trick" makes it possible to increase the linear time sweep frequency?

Exercise 2

Discuss the concept of *instantaneous frequency*. Why does this word contain an inconsistency?

Exercise 3

In order to measure the frequency response you plan to "sweep" the system. What do you have to take into consideration in order not to measure nonsense?

Exercise 4   Modern test signals

(a) Why does the δ–pulse appear to be an ideal test signal from a theoretical point of view but not from a practical point of view?

(b) You wish to measure the entire *transfer function* of a filter according to the absolute value and phase. What do you have to take into account?

(c) Describe a method of measuring the *normed* transfer function according to absolute value and phase. Select the appropriate circuit from the library of the DASY*Lab* circuits and analyse the method.

(d) Complete this process and represent the transfer function as a local curve in the GAUSSian plane.

(e) You are to construct the hardware of a test signal generator. In the case of which signal is this easiest?

(f) What problems are there with the step function as a test signal and how can they be avoided?

(g) How can test signals be generated which have a constant amplitude curve in any specifically defined frequency range?

(h) What advantages and disadvantages do the last mentioned test signals have?

(i) What are the possible fields of application for burst signals as test signals?

(j) What makes the noise signal so interesting as a test signal? And what makes it so difficult?

(k) Measure the correlation factor between two different noise signals and one and the same noise signal. What correlation factors do you anticipate and which do you measure? How does the correlation factor depend on the length of the noise signal? What does the correlation factor represent in terms of *content*?

Exercise 5

Where can GAUSSian pulses or GAUSSian oscillation pulses be used meaningfully in measurement technology? What physically interesting properties do they have?

Exercise 6

(a) What do transients tell us about the features of a system?

(b) A system - for instance a filter - needs more time to respond. What can you deduce from this?

(c) Does the pulse response h(t) describe the transient?

(d) Does the step function g(t) describe the transient?

(e) Under what conditions does the "burst response" b(t) describe the transient?

(f) Explain the terms "natural frequency" and "resonance frequency" in the context of transients.

Make the necessary experiments with DASY*Lab* in order to be able to answer the questions confidently.

Chapter 7

Linear and non-linear processes

In communications engineering there is theoretically - as I have already pointed out - an infinite number of signal processes. This is another good example that clearly demonstrates the discrepancy between theory and practice because only roughly two or three dozen of these processes are usable and of relevance for practical work. This number roughly corresponds to the number of letters in our alphabet. As we have learned to read and write by using letters and words in a meaningful way we should not have any problems doing the same with the relevant processes in signalling.

System analysis and system synthesis

Some chips nowadays contain whole systems consisting of millions of transistors and electronic components. It is thus impossible to deal with each individual transistor. Chip manufacturers therefore explain the chips signalling and system properties by means of block diagrams. We can interpret these systems by analysing step by step the modules (processes!) contained in the block diagrams. And on the other hand, we can design such systems by combining adequate processes (modules!) to create a block diagram. This procedure is called synthesis.

It leads to a better overall view if the different relevant signalling processes/components can be categorised in accordance with the characteristics they have in common. They can actually be categorised and allocated to two (for the time being) groups. One group is linear, the other group non-linear processes/components.

> *All theoretically possible signalling processes, including the practically usable ones, can be distinguished by their linear and non-linear properties.*

Measuring a process to reveal whether it is linear or non-linear

I am not going to beat about the bush, but tell you straight away what the distinguishing feature of these two groups is: how do you find out if a process is linear or non-linear when measuring it?

> *If a sinusoidal signal of any frequency is entered into the input(s) of a module and if only a sinusoidal signal of the same frequency appears at the output, the process is linear. Otherwise it is non-linear.*

Please note:
The sinusoidal oscillations of the output signal may vary with regard to amplitude and phase. The amplitude can increase (amplification) or decrease dramatically (extreme damping). Remember that phase displacement is just a time modification of the output sinusoidal oscillation compared with the input sinusoidal oscillation. After all, any process takes some time before it is completed.

Line and space

One of the most important examples of a linear system in communications is the line. At the end of a line there will always be a sinusoidal oscillation of the same frequency as the one which was fed into the line at its other end, no matter whether the line measures only a few yards or several miles. The line dampens the signal with increasing length, i.e. the longer the line the lower the amplitude. Transmission time also increases with length - the transmission speed of lines is about 100,000 and 200,000 km/s. This becomes noticeable as phase displacement on the oscilloscope.

It would be disastrous if a line had non-linear properties! Fixed-line telephony for example would hardly be possible because the voice-frequency band at the other end of the line would be in a completely different, possibly inaudible, frequency range, and the length of the line would play a role as well.

The most important „linear system" is space. Otherwise the signals of a radio station would reach its listeners in a different frequency, or several different frequencies for that matter, from the one intended by the station. And again, just as with lines, there would be a dependency on the distance between transmitter and receiver. This example shows the importance of linear processes more than any other. Later in this book we will take a look behind the scenes to see why it is that lines and space have linear properties and the physical reasons behind it.

Inter-disciplinary significance

Linearity and non-linearity play an extremely important role in mathematics, physics, technology, and in natural sciences generally speaking. Linear equations in mathematics for instance can be solved quite easily, whereas non-linear equations can rarely, if ever, be solved. Quadratic equations are examples of non-linear equations and generations of pupils have been tortured with them. It is really hard to believe: mathematics (these days) mostly fails when it comes to solving non-linear problems. As theoretical physics finds its most important support in mathematics, non-linear physics is rather underdeveloped and, as it were, still in its infancy.

The behaviour of linear systems is generally speaking easier to understand than that of non-linear systems. The latter can even lead to chaotic, i.e. principally non-predictable, behaviour or create so-called fractal structures which - if represented graphically - can be of great aesthetic beauty. Furthermore they possess a „universal" property which seems to be becoming more and more important: that of being fractal. If you take a closer look they basically display the same structures whether you enlarge or diminish the scale.

What is interesting is that these highly complex structures can be created by very simple non-linear processes. So it is obviously wrong to assume that highly complex systems are the result of highly complex causes. Some of these fractal objects created by mathematics bear a stunning resemblance to certain plants so that similarly simple laws are assumed to be behind certain biological processes. We therefore know that non-linear processes must be responsible for the great variety in natural phenomena and research in this field is quite brisk at the moment.

Computers meanwhile have become the most important medium with which to create, map and examine non-linear structures. This has given new momentum to modern mathematics. Even at our level it is possible in many cases to illustrate non-linear interrelations and characteristics. We will take a closer look at very simple non-linear processes and examine them using the computer later on in this book.

Mirroring and projection

But enough of mysterious advance remarks. Let me now try and explain the terms linearity and non-linearity in signal processing in plain words without using any mathematics. For the time being flat and curved mirrors will suffice to provide an explanation.

At least once a day - in the morning - we look at our reflection in the (flat) mirror of our bathroom. In the mirror we see a lifelike mirror image of our face (however, the whole thing is the wrong way round). We expect a mirror image to be a true mapping of the original image in terms of proportions. This is an example of a linear depiction caused by linear mirroring. Another example is a photo. The object depicted is usually much smaller than the original, but the proportions are identical. Enlargement and reduction - or multiplication by a constant to use a mathematical expression - is therefore a linear operation.

You may have been to a hall of mirrors at a fun fair looking at your reflection in distorting mirrors. The surface of these mirrors is non-linear, i.e. not flat, but uneven and bulging. These are examples of non-linear mirrors. What does the reflected image look like? Your body is grotesquely distorted. One mirror shows a tiny head on an enormous torso with short leg stumps attached to it. Another one enlarges the head and reduces the rest of the body to a shrivelled sausage.

Our bodies are distorted in a non-linear way. The mirror image does not reflect the proper shape because the proportions are distorted.

Changing a signal by means of signal processing can be described in a similar way. A characteristic curve - which is the equivalent of the mirror - describes the relation between the input signal u_{in} and the output signal u_{out} of the component or the process.

Illustration 114 shows that the curve of the signal is projected vertically upwards onto the characteristic curve and from there horizontally to the right. The projection is the equivalent of the mirroring process.

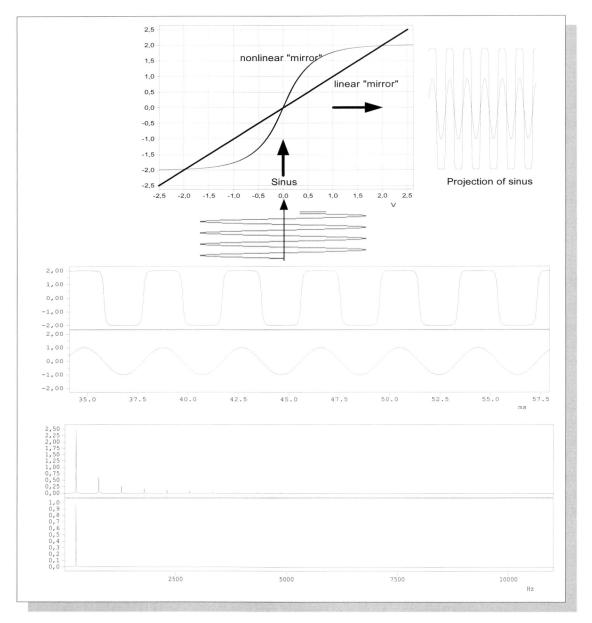

Illustration 114: **Non-linear distortion of a sinusoidal oscillation**

A new signal is created by projecting a sinusoidal oscillation at a non-linear characteristic curve. The sinusoidal oscillation contains merely one frequency whereas the new signal contains several frequencies in the spectrum (bottom spectrum).

The angle of incidence of each beam, however, is not equal to its angle of reflection. If the characteristic curve is linear the proportions are maintained in the projection. If it is non-linear the proportions are distorted. Illustration 114, bottom, shows a sinusoidal oscillation which is transformed into an obtuse rectangular pulse.

While the sine contains just one frequency, the analysis of the obtuse rectangle shows that there are several frequencies. It follows from that that non-linear characteristic curves create distortions which in the frequency range also create additional frequencies, i.e. non-linear distortions.

A complex component: the transistor

The most important component in microelectronics is no doubt the transistor. Unfortunately, transistors have in general non-linear characteristic curves for physical reasons. It is therefore rather difficult to develop circuits with transistors which work very precisely, even though the message conveyed in classes and lectures is a different one. It requires experts to build a linear transistorized amplifier that amplifies in a way which „maintains the proportions". Such an amplifier can to a certain extent be linearized by means of numerous circuit technology tricks - the best trick is called *negative feedback*. But for the above reasons no amplifier is completely linear. The so called *distortion factor* is a measure of its non-linearity.

First consequences:

The above illustrations and explanations show in what cases linear and non-linear processes are generally applied:

> *If the frequency range of a signal is to be - or can be - changed, e.g. by relocating it to a different range, it can only be achieved by using non-linear processes.*
>
> *If the frequencies contained in a signal are not to be changed or new frequencies added this can only be achieved by using linear processes.*

This means that linearity and non-linearity are about whether or not new frequencies, e.g. sinusoidal oscillations, are created in a process.

Unlike mathematics, communications technology deals with linearity and non-linearity only in the context of sinusoidal oscillations. This is understandable enough as according to the underlying FOURIER Principle any signal can be regarded as composed of sinusoidal oscillations.

Almost every signal processing causes a change in the time and frequency domains. The link between these two changes is (of course) the FOURIER transformation, which is the only way of getting from the time into the frequency domain and vice versa. All the results of the signal processing described below will therefore be described in the time and in the frequency domains.

There are only few linear processes

The situation is quite clear. There is just a total of five or six linear processes and most of them appear ridiculously simple at first sight. But they are still enormously important and appear in numerous applications. In contrast to the indefinite variety and complexity of non-linear processes, we know almost everything about them and there are hardly any surprises, even when several linear processes are combined to form a linear system.

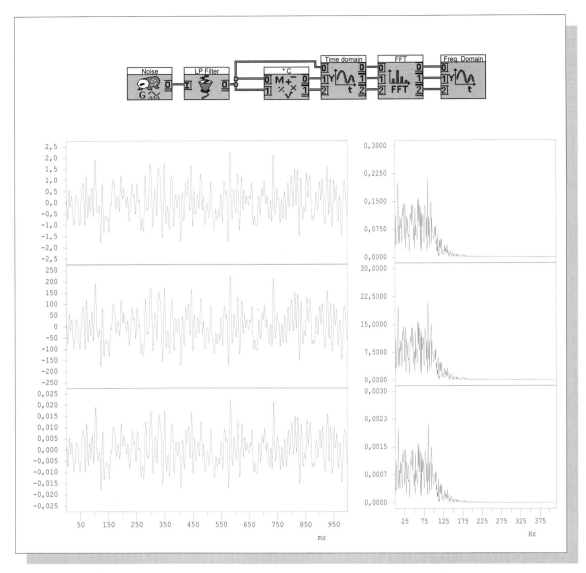

Illustration 115: **Multiplication by a constant: amplification or damping**

The top series shows the original signal - filtered noise - in the time and frequency domains. In the centre: the signal after amplification by one hundred, and finally at the bottom, the signal after damping by one hundred (multiplied by 0.01).
Amplification and damping are given in dB (decibels), a logarithmic scaling. The factor one hundred here equals 40 dB.

It may come as a surprise that there are mostly extremely simple mathematical operations behind these processes which, at first glance, have little to do with communications engineering. Besides, our aim was to leave mathematics out of it.

Multiplication of a signal by a constant

This sounds and is very easy. But behind it there are important concepts such as amplification and damping. A hundredfold amplification means the multiplication of all instantaneous values of variables by the factor 100, and in the frequency domain stretching the amplitude response by one hundred.

Addition of two or more signals

Strictly speaking, the **FP** can be seen as the addition of (an infinite number of) sinusoidal oscillations of different frequencies, amplitudes and phases: *any signal can be seen as composed of different sinusoidal oscillations (addition).*

Why is the process of addition linear? If two sinusoidal oscillations of e.g. 50 Hz and 100 Hz are added the output signal will contain just those two frequencies. No new frequencies have been added. Thus the process is linear.

The addition of sinusoidal oscillations is not necessarily identical with the simple addition of figures. As the relevant sections in Illustration 113 show, the addition of two sinusoidal oscillations of the same amplitude and frequency may equal zero if they are phase-shifted by π. If they are not phase-shifted the result is a sinusoidal oscillation with twice the amplitude.

It follows that sinusoidal oscillations are only added in a correct way if the relevant instantaneous values of variables are added. In other words: when sinusoidal oscillations are added their individual phase positions in relation to each other are automatically taken into account. For this reason sinusoidal oscillations can only be added in a correct way if - from the point of view of the frequency domain - amplitude and phase spectrum are known.

Illustration 35 show that, if seen section by section, the sum of (an infinite number of) sinusoidal oscillations of different frequencies can equal zero. In the case of the δ–pulse the sum of the „infinite number" of sinusoidal oscillations always equals zero in the time domain, except for the point where the δ–pulse appears. In mathematics such a point is called singularity.

To conclude the section on addition we should take (another) look at a particularly interesting and important special case: two successive δ–pulses (Illustration 116). Each of the two pulses initially contains all frequencies of the same amplitude. The amplitude spectrum is thus a constant function. Accordingly, two pulses ought to contain all frequencies of twice the amplitude!? No, they don't! The reason is that one of the δ–pulses is time-staggered compared with the other one and therefore has a different phase spectrum. As Illustration 81 and Illustration 82 show the spectrum is sine-shaped. At the zero position of the spectrum the sinusoidal oscillations of the same frequency and size are added in a counterphase way, i.e. they cancel each other out and the result equals zero.

All these examples show how important a simple process like addition is in the field of analysis, synthesis and processing of signals.

Delay

If a sine is fed into a longish cable and input and output signal are compared on the screen of a two-channel oscilloscope, the output signal appears phase-displaced, i.e. time-delayed compared with the input signal.

> *A time delay means that the phase spectrum in the frequency domain has changed.*

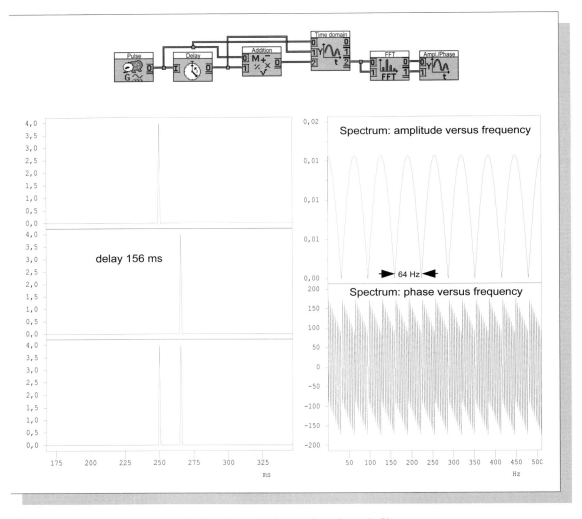

Illustration 116: **Delay plus addition = digital comb filter**

This Illustration shows a simple way of doubling a δ–pulse: the input pulse is delayed (in this case by 1/64 s) and fed into an adder. On the right hand side you can see - similar to Illustration 81 and Illustration 82 - the spectrum of the two bottom δ-pulses according to absolute value and phase.

The amplitude spectrum is cosine-shaped with a zero position interval of 64 Hz. The first zero position is at 32 Hz (the first zero position in the negative frequency range is at -32 Hz).

All digitalised signals consist of lines of numbers which can be represented as „weighted" δ–pulses (see Illustration 37 and Illustration 116 bottom). The envelope reproduces the original analog signal. If you added to each of the δ–pulses - here after 1/64 s - a pulse of the same level, all the odd multiples of 32 Hz, i.e. 32 Hz, 96 Hz, 160 Hz etc, would be filtered out of the spectrum.

A filter, which, like a comb, has gaps at regular intervals, is called a comb filter. In the case of this digital comb filter the gaps come at intervals of 64 Hz. Would you have known that it is that easy to design a filter?

It is hardly possible to delay a given signal precisely by a planned value by means of analog technology, but there is no problem whatsoever if digital signal processing DSP is used.

Together with addition and multiplication by a constant, delay forms a triad of elemental signal processing. If you take a closer look, many extremely complex (linear) processes merely consist of a combination of these three basic processes. A good example are digital filters. Chapter 10 will demonstrate exactly how they work and how they are designed.

Differentiation

Differential and integral calculus are where „higher mathematics" is generally thought to begin. Both types of calculus are important because the most important natural laws and technological relationships can only be modelled by using this.

But do not worry - we will not use formal mathematics. This forces us to describe the substance of a problem and not simply point casually to „trivial" mathematics.

You should therefore simply recognise by means of Illustration 117 what differentiation as a technical signal process in the time domain actually means. Try to answer the following questions:

- At what points of time of u_{in} do local maxima and minima occur in the case of the differentiated signal u_{out}
- What property does the input signal u_{in} have at these points of time?
- At what points of time of u_{in} is the differentiated signal u_{out} equivalent to zero?
- What property does the input signal u_{in} have at these points of time?
- In what way does u_{in} differ at the places where u_{out} has a positive local maximum and a „negative local maximum" (equal to a local minimum)?

You should arrive at the following result:

> *The differentiated signal u_{out} indicates how rapidly the input signal u_{in} is **changing**.*

By way of example two of the most important laws of electrical engineering will be mentioned in this context.

> *Law of induction:*
>
> *The faster current changes in a coil the greater the induced voltage.*
>
> *Law of capacity:*
>
> *The faster the voltage at the capacitor changes the greater the current that flows in or out (charging current or discharge current of a capacitor).*

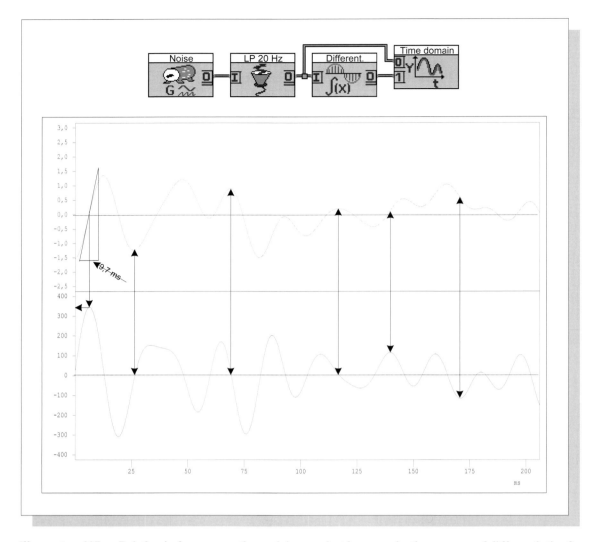

Illustration 117: **Brief quiz for non-mathematicians: what happens in the process of differentiation?**

In order to make statements which are generally valid a random signal curve is selected. Here it is lowpass filtered noise. Compare carefully the input signal u_{in} (top) with the differentiated output signal u_{out} (bottom).

A number of lines have been entered to help. What behaviour does the input signal u_{in} exhibit at the points of the local maximum as marked in the case of the differentiated signal u_{out}? And its behaviour at the local minimum as marked? How rapidly does the input signal u_{in} change instantaneously if the differentiated signal equals zero?

At the top left a „gradient triangle" has been entered. First the tangent at the point of the curve which is of interest here, then the horizontal and vertical component are entered. The vertical section divided by the horizontal section (Kathetes) ought to give exactly the value of the differentiated function.

*Note: a **local maximum** is understood to mean the highest point in any defined environment. Thus there are smaller values immediately to the left and right of this local maximum. The same is true of the **local minimum**.*

In the case of differentiation the „*speed of change*" is measured. From a mathematical point of view this corresponds to the momentary gradient of the signal curve. Differential calculus is inseparably bound up with the concepts of upward gradient and downward gradient. For example, a sphere accelerates on an increasing downward gradient and is slowed down on an upward gradient.

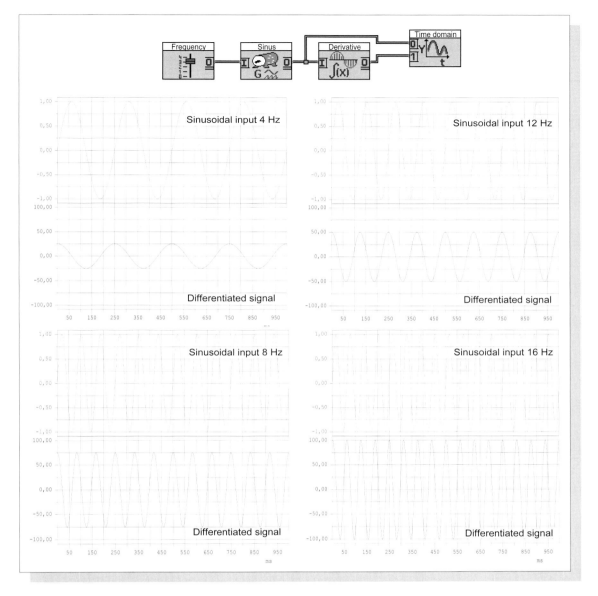

Illustration 118: ***Differentiation of sine voltages of different frequencies***

First the somewhat confusing module system for differentiation should be explained. What is actually to be seen is the integral sign which would go better with integration - the next linear operation. Differentiation and integration are linked with each other, which in the case of integration, has still to be shown experimentally. The word „derivative" is an alternative expression used for differentiation in mathematics.

The grid was shown so that you can see the strict proportionality between the frequency of the input signal and amplitude of the (differentiated) output signal. This opens up a range of extremely accurate applications in measurement technology.

> *The signal process of differentiation can therefore also be used as a signalling device whether something is changing too quickly or too slowly or as a measurement device showing how quickly something is changing.*

We should pay particular attention to the way in which the differentiation process affects sinusoidal signals of *different* frequencies.

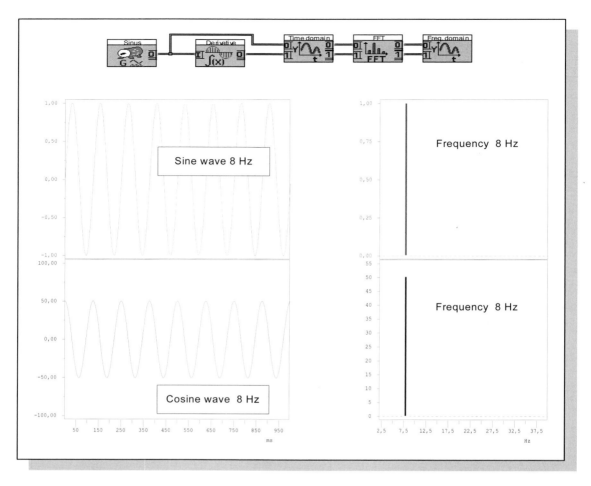

Illustration 119: ***Experimental evidence: Differentiation of a sine shows a cosine***

At first sight the differentiated signal of a sine has a cosine- and/or sine wave oscillation. But is it really precisely a cosine? If it was not the case the differentiation would be non-linear.

In the frequency domain we have a reliable procedure to prove this. As the differentiated signal in the frequency domain - like the sine at the input - has a single frequency of 4 Hz it can only be a linear process. It causes a phase shift of π/2 (Cosine!). Its amplitude is strictly proportionally the frequency ... and the amplitude of the input signal (why ?).

These results are not surprising and result directly from the general property of differentiation which can be seen in Illustration 117.

- A differentiated sine produces a cosine, that is it displaces the phase by π/2 rad. This could not be otherwise because the gradient is greatest at the zero crossing; hence the differentiated signal is also greatest at this point. That this is really surprisingly a cosine is proved by the frequency domain (Illustration 119): a line of the same frequency is still present there.

- The higher the frequency the faster the sinusoidal signal changes (also at the zero crossing). The amplitude of the differentiated voltage is also strictly proportional to the frequency.

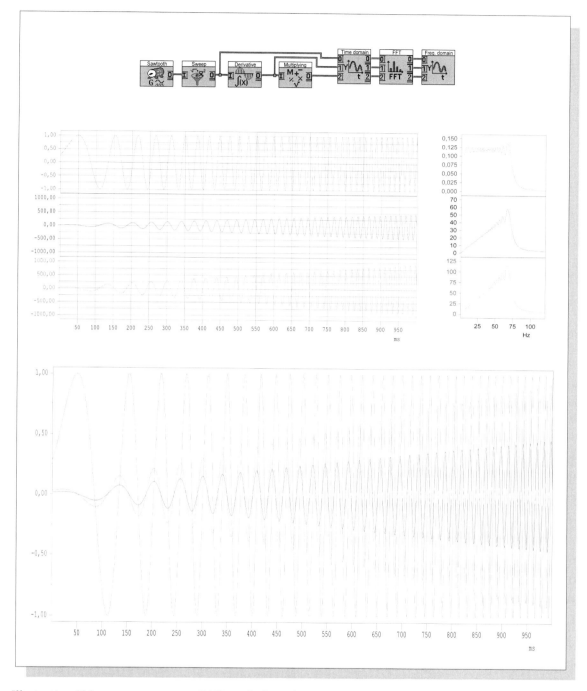

Illustration 120: **Differentiation of a sweep signal**

The sweep signal was set in such a way that the instantaneous frequency changes in a completely linear way in time. This can be proved by means of the following differentiation. The amplitude increases completely) linearly in time, i.e. the envelope produces a straight line as you can easily check using a ruler.

By means of a following amplifier - multiplication by a constant - the gradient of the envelope and thus the sensitivity of the circuit towards changes in frequency can be set virtually at will.

This results in important practical applications which exploit the linearity of cable and space as transfer media.

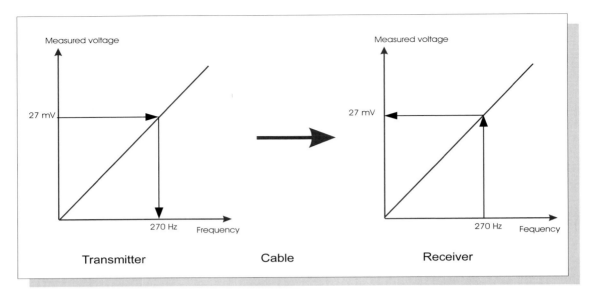

Illustration 121: ***Principle of a telemetric system (with frequency coding and decoding)***

In practice many sensors only provide tiny measurement voltages which it is difficult to transmit without distortion. The telemetric methods presented here make it possible in principle to transmit such (tiny) measurement readings with a high degree of accuracy and freedom from distortion.

Note: the new version of DASYLab S uses a sound card for the input and output of analog signals. As a result of the input filter it is not able to register signals which change very slowly. This would be possible, however, by using a VCO.

The situation can be illustrated beautifully with a sweep signal (Illustration 120) the frequency of which changes linearly over time. Accordingly, the differentiated signal exhibits a linear increase in amplitude.

This makes possible a sensational technical application: by means of a differentiator a frequency-voltage converter with a perfectly linear characteristic can be created (Illustration 121). By means of a following amplifier - multiplication by a constant - the steepness of the characteristic can be set virtually at will. This is equivalent to the sensitivity of the frequency-voltage-converter towards changes in frequency.

In telemetrics the important thing is to transmit measurements as accurately as possible via cables or cable-free. But in a transmission of this kind everything can be distorted: the entire form of the signal or even in the case of a sine the amplitude and phase. One thing can, however, not change because both media - cables and space - are *linear* and that is the *frequency*.

In order to make absolutely sure, the measurements should be frequency-coded: at the point of reception a frequency gauge is adequate in practical terms to receive the measurement with a high degree of accuracy. At the point of transmission, however, a highly accurate voltage frequency converter (VCO: Voltage Controlled Oscillator) is required. This is shown in Illustration 121. In principle we now have a highly accurate telemetric system which exploits the linearity of the medium of transmission. VCOs with a high degree of accuracy are available on the market. In differentiation we have a virtually ideal process for transforming any frequency completely linearly into a voltage of a corresponding level.

We have thus taken the first step in the direction of frequency modulation and demodulation (FM), a particularly distortion-prone method of transmission which is used in the VHF range and also in television (sound). Details will be given in the next chapter.

Now we shall summarize the effects of differentiation in the frequency domain:

The differentiation of a sweep signal shows quite clearly the linear increase of the amplitudes with the frequency. Differentiation therefore has highpass properties i.e. the higher the frequency the better the conductivity.

Illustration 118 makes possible a precise definition of the mathematical context (determining the proportionality constants and the gradient constants). The amplitude of all imput signals u_{in} is 1 V.

Frequency f (Hz)	Ûout (V)
4	25
8	50
12	75
16	100

Thus the amplitude increases linearly with the frequency. The proportionality factor or gradient constant is:

$$25 \text{ V}/4 \text{ Hz} = 50 \text{ V}/8 \text{ Hz} = 75 \text{ V}/12 \text{ Hz} = 100 \text{ V}/16 \text{ Hz} = 6.28 \ldots = 2\pi \text{ /Hz}$$

Hence, it follows that:

$$\hat{U}_{out} = 2\pi f \; \hat{U}_{in} = \omega \; \hat{U}_{in}$$

> *A differentiation in the time domain therefore corresponds to a multiplication by $\omega = 2\pi f$ in the frequency domain.*

As it should be, in conclusion we determine the frequency response or the transfer function of the differentiator. As a test signal a δ-pulse at the position $t = 0$s is selected.

Surprisingly the spectrum at first rises linearly but then increasingly loses steepness. Is there something wrong with the formula above? The needle pulse is what is wrong. It is not an ideal δ-pulse but a pulse of infinite width (here 1/1024 s). The spectrum of the needle pulse in Illustration 122 (top) is constant, i.e. all the frequencies have the same amplitude. The spectrum of the differentiated needle pulse reveals the truth: an ideal δ-pulse does not exist.

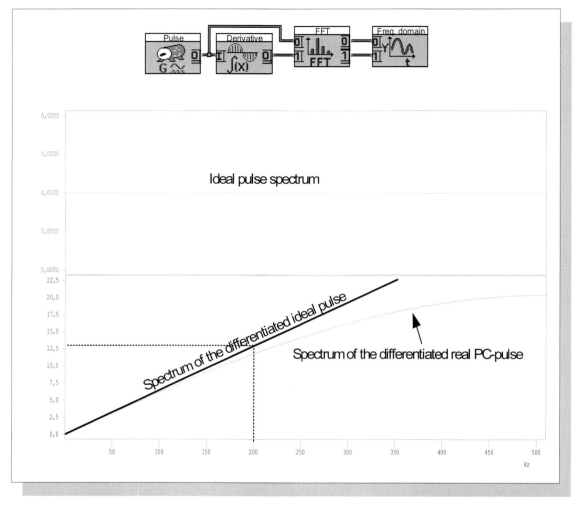

*Illustration 122: **Transfer function of a differentiator***

In accordance with the insights described the transfer function of the module „derivative" would have to have a perfectly linear ramp. The measurement (bottom) shows this only for the lower frequency range; after that the ramp is constantly reduced.

This is an example of the limitations of precise computer-aided measurement techniques and signal processing. We are working conceptually with a δ-pulse which practically cannot exist. Our needle pulse must have a finite duration (width) which results from the block length and scanning rate. This is nowhere revealed so clearly as in differentiation.

The expected linear ramp is entered as a tangent at the zero point. The proportionality factor is (initially) $2\pi f = \omega$. In this connection note in the drawing (top) that the amplitude for all frequencies is only 0.01 V.

Integration

The properties of the integration process are to be determined experimentally. Integration must be related in a certain way to differentiation - the two processes would otherwise not be included in one and the same DASY*Lab* module. We first select a random signal curve - lowpass filtered noise - and carry out an integration (Illustration 123).

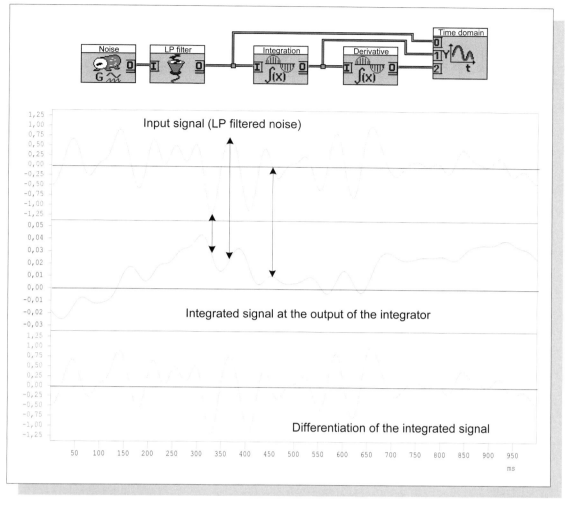

*Illustration 123: **Visualisation of the relationship between integration and differentiation***

Closer examination leads one to suspect that the middle signal after differentiation produces the upper one. The appropriate experiment shows the agreement: the differentiation of the integrated signal produces the original signal.

As we are focussing on differentiation it ought to strike an attentive observer that the middle (integrated) signal when differentiated ought to give the input signal. The local maxima and minima of the input signal are again to be found at the steepest points of the input signal.

Closer examination bears out this surmise. Thus, differentiation appears to be the reversal of integration. Is integration the reversal of differentiation? An experiment presented in Illustration 123 bears this out too. The fact that in both instances random signals were selected practically ensures that this is generally valid

> *Differentiation can be seen as the reversal of integration and integration as the reversal of differentiation.*

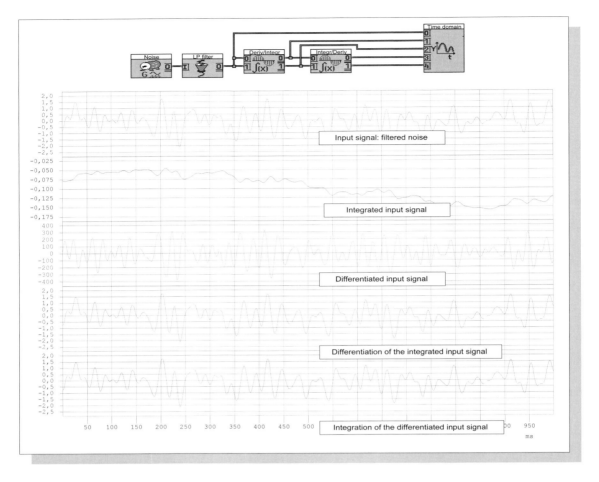

Illustration 124: ***Changing the order is possible***

As was proved experimentally the order of the processes of differentiation (derivative) and integration can be changed round. „First differentiation and then integration" leads back to the original signal exactly as „first integration and then differentiation" does.

On closer examination of the signals two further aspects in the time domain may be surmised:

- The integrated signal has far less edge steepness than the input signal. This would imply the suppression of higher frequencies in the frequency domain, i.e. a lowpass characteristic

- The curve of the integrated signal looks as if the mean value were formed successively from the input signal. The curve decreases as soon as the input signal is negative and increases once it is positive.

In order to test the first assumption we select a periodic rectangular signal. It has the greatest conceivable edge steepness at the step points. The integrated signal ought not to exhibit „vertical" sides. The spectrum of the integrated signal ought to have a much smaller proportion of higher frequencies. At the same time the curve of the rectangular signal is constant - apart from the step points. This makes possible an answer to the question as to what an integrator does with a constant function.

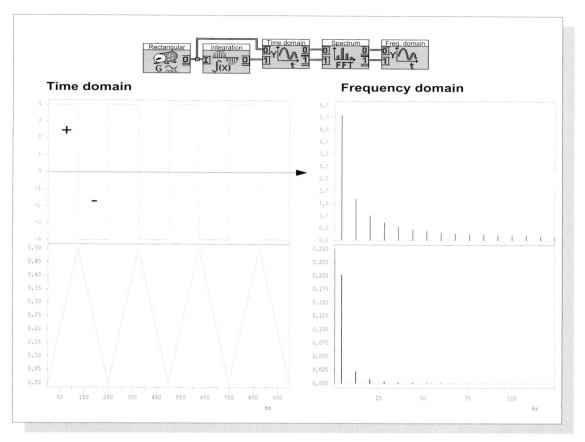

Illustration 125: **Integration of a periodic rectangular signal.**

We select a very straightforward signal curve in the case of the periodic rectangular signal. This reveals the property of integration particularly clearly.

Constant positive segments lead to a rising straight line and constant negative segments lead to a falling straight line. Apparently, seen from a geometrical point of view, the integral successively measures the **area** *between the curve of the signal and the horizontal time axis. The „area" of the first rectangular signal is 4 V∗125 ms = 0.5 Vs and it is precisely this value that the integration curve gives after 125 ms. We must continue to differentiate between positive and negative „areas" as the next - in terms of the absolute value equally large - area is deducted, so that after 250 ms the value is again zero.*

The frequency domain (right) provides unambiguous qualitative and quantitative pointers. Initially, the integration quite clearly has a lowpass characteristic. Both signals - the rectangular and the triangular signals - only contain the odd multiples of the base frequency (see Chapter 2). The rule for the decrease in amplitudes is quite straightforward and can be measured using the cursor. For the sawtooth $\hat{U}_n = \hat{U}_1/n$ (e.g. $\hat{U}_3 = \hat{U}_1/3$ etc) holds. For the spectrum of the triangular curve $\hat{U}_n = \hat{U}_1/n^2$ (e.g. $\hat{U}_3 = \hat{U}_1/9$) applies.

If the same frequencies of the spectra are compared with each other the result is $\hat{U}_{triangle} = \hat{U}_{sawtooth}/2\pi f = \hat{U}_{sawtooth}/\omega$. Integration as the reversal of differentiation also in the frequency domain is thus fully borne out here.

The first supposition is fully borne out by the measurement shown in Illustration 125:

> *Integration exhibits lowpass behaviour. To put it accurately, the frequency domain of the input signal is divided by $2\pi f = \omega$ in integration. The amplitudes of the higher frequencies are thus disproportionately reduced in size.*

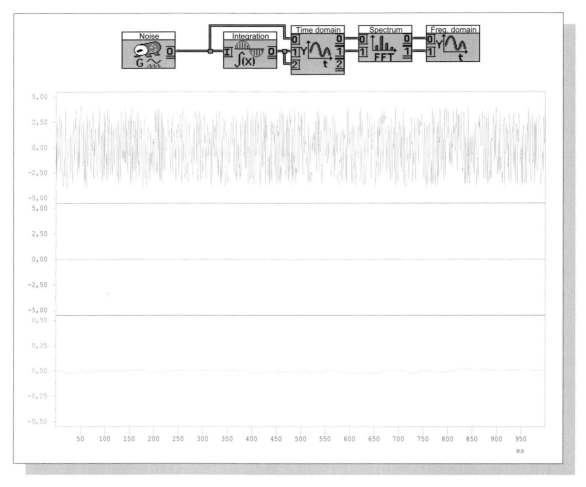

Illustration 126: ***Forming mean value by integration***

In this case a noise signal is integrated. The result is practically zero with small fluctuations (bottom higher resolution. The process of integration adds the positive and negative „areas" together. Because everything is random in the case of a noise signal an equal distribution of positive and negative „areas" must be present. In the average they must all be zero.

Accordingly, integration can be used to form the mean value, an extremely important aspect for measuring technology. In addition, forming a mean value apparently eliminates noise better than a normal filter could.

It is striking how quickly the mean value zero is reached in the case of noise. As noise consists of a „stochastic sequence of individual pulses", i.e. practically of a sequence of weighted δ–pulses, this is easy to understand. Positive and negative „areas" which have an important effect only arise when a state is maintained „for not inconsiderable periods of time".

In the examination of the integration process as a generator of mean values Illustration 126 arrives at the following result:

- As a result it is clear that noise has the mean value of zero. The integration of a noise signal practically gives this value, i.e. the mean value can be determined by means of integration.

- Apparently, noise elements can be removed from a signal by means of integration, better than this is possible with normal filters.

Chapter 7: Linear and nonlinear processes

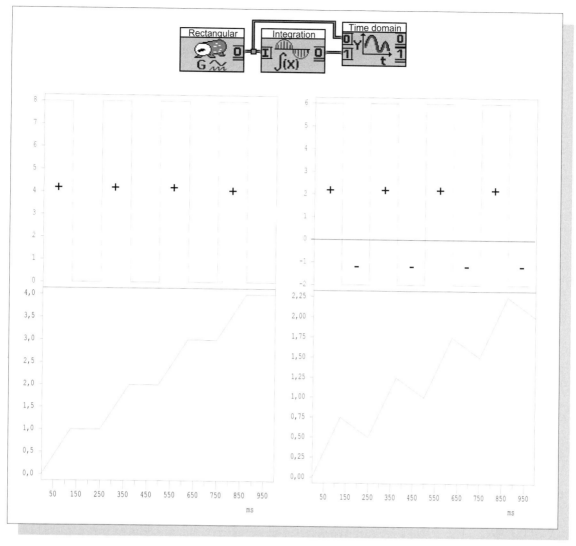

Illustration 127: ***Closer examination of the „measurement of areas" and the forming of the mean value.***

On the left a rectangular signal is integrated which lies entirely in the positive sphere. In accordance with Illustration 125 the area increases linearly from 0 to 125 ms. After that it remains constant to 250 ms because the rectangular curve is zero at this point. On the right however the area decreases linearly because from 125 ms the curve of the signal lies in the negative sphere.

If one continued to carry out the integration of the periodic rectangular signal, the curve of the signal would increase more and more in both cases and tend towards infinity. Where does the mean value lie?

The mean value of the input signal top left lies at four - as can be seen without difficulty - that of the input signal top right lies at two. The curve of integration shows precisely these values after 1 s.

The results in Illustration 127 show:

> *If the curve of the signal lies predominantly in the positive (or negative sphere), the integrated signal rises more and more (or falls more and more). In the case of signals of this kind in pure integration the correct mean value is indicated exactly after 1 s!*

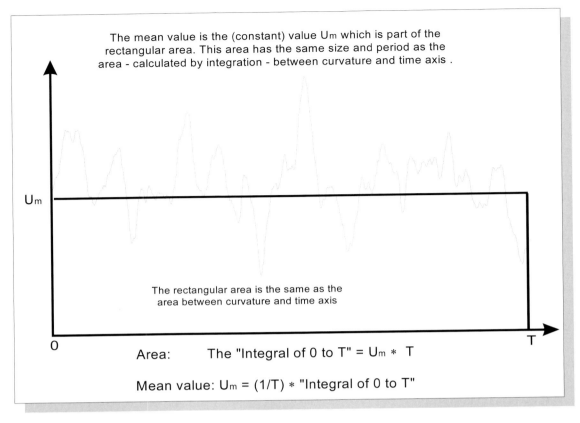

Illustration 128: **Precise definition of the (arithmetical) mean value**

As is clear from the above illustration and the notes and definitions given there the precise procedure for determining the (arithmetic) mean value is as follows: determine the area from 0 to T between the curve and the time axis by means of integration. Transform this area into a rectangular area of the same size with the same time duration T. The height of this rectangle represents the (arithmetic) mean value U_m.

The subject „mean value" is actually part of measuring technology. For this reason we shall not go into all the possible forms of mean value here. The (arithmetic) mean value is simply the average value from a sequence of numbers. If the sequence of numbers is, for example, (21; 4; 7; -12) the mean value is $U_m = (21 + 4 + 7 + (-12))/4 = 5$. Here it is simply presented as the possible outcome of integration and its relevance as a technical signal process described.

We can now see the following from Illustration 126 and Illustration 127:

- In the process of integration the mean value U_m is only indicated precisely after 1s. $T = 1$ s applies here

- The mean value U_m for any other duration T can be determined by dividing the integral value from 0 to T by the duration T

- In the context of signal technology the forming of the mean value describes a process of inertia which is not able to recognise rapid changes (see Illustration 126) but only an averaged value. This is similar to the behaviour of a lowpass filter. This calculates a kind of „gliding mean value" in which - similar to Illustration 52 and Illustration 61 - the mean value of overlapping „windows" is shown continuously.

Chapter 7: Linear and nonlinear processes

Note: Our eyes do this, for instance, with television. Instead of 50 (half-) pictures a second we see a continuous „averaged" sequence of images.

As a result of integration the curve of a signal is changed in a certain, mathematically very important way. The simplest of all functions, the constant function u(t) = K becomes by integration a linear function u(t) = K * t, as shown by Illustration 127. Is there a rule which occurs in the case of repeated integration and which can be reversed by differentiation? Illustration 129 shows the relationships involved:

- A constant function of the type $\qquad$ u(t) = K

  becomes by integration a linear function of the type $\qquad$ u(t) = K * t

  This becomes by integration a „quadratic" function of the type $\quad$ u(t) = $K_1 * t^2$
  (whereby K_1 = K/2)

  By integration this becomes a „cubic" function of the type $\qquad$ u(t) = $K_2 * t^3$
  (whereby K_2 = K_1/3) etc. This is only precisely true for the
  integration carried out (blockwise) with signal curves („specific integral").

- By (multiple) differentiation it is possible to reverse this (multiple) integration as shown in Illustration 124.

Integration is in practice such an important operation - which everyone ought to be aware of - that we almost forgot the most important question: what happens in the case of integration to a sinusoidal signal?

This is shown by Illustration 130. It shows the behaviour expected if we regard integration as the reversal of differentiation, i.e. differentiation of the integrated signal must result in the original signal.

Malicious functions or signal curves

Not just for the sake of completeness we ought at this point to mention certain functions or signal curves which can be integrated but not differentiated.

There is for instance the „chaotic" noise signal". If you differentiate this signal you will obtain a result. But noise has a property which normal signals do not have - the successive random values change in „steps". The computer insists on measuring the gradient from the difference between two neighbouring random results and the (constant) time interval. The result is a different noise signal. This also shows that a computer-generated noise signal is a simplified copy of natural noise processes. Natural noise does not produce „clicks" at constant time intervals.

In order to satisfy your curiosity here is another tip: try differentiating a periodic rectangular signal or sawtooth signal. At what points are there problems or extreme values?

These are the „step" points! This is why mathematicians have created a graphic term to designate the precondition for *differentiability* - the *continuity* of a function. This can be graphically defined as follows:

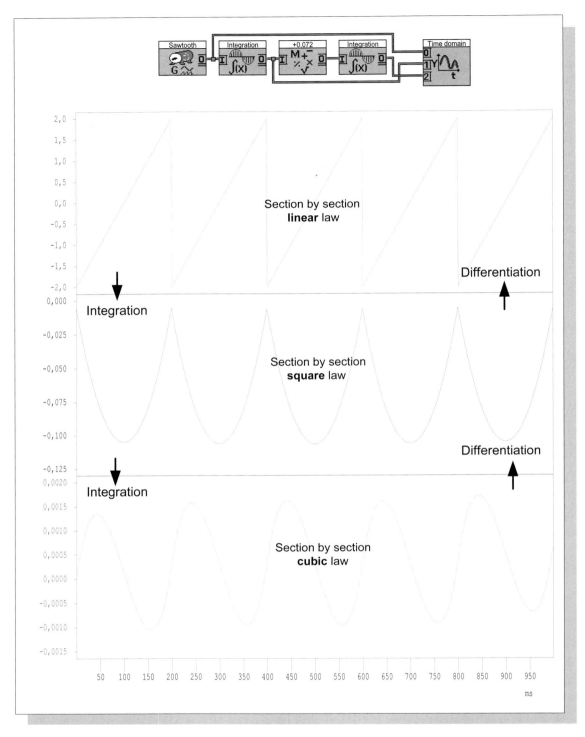

*Illustration 129: **Integration: from the constant to the linear, quadratic, cubic function and back via differentiation***

We cannot completely do without a basic knowledge of function theory. However, everything is visualised here, described in simple words without a rigorous mathematical approach. It is important to follow this up experimentally using the above circuit. What do you observe after a time? What does the curve of the signal look like if you do not add the constant to it?

Note that the lower function does not have a sinusoidal curve but is in segments „cubic", i.e. proportional t^3.

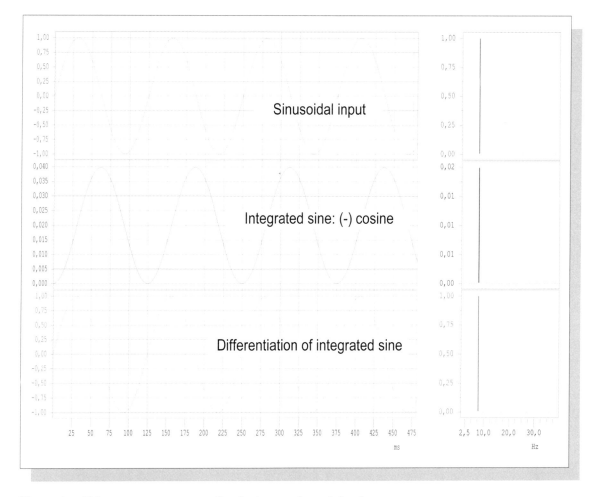

Illustration 130: **On the integration of the sine wave**

The sine (top) which has been integrated results in a (-)cosine-shaped curve (centre). This can be reproduced via the differentiation of the integrated signal, i.e. the reversal of the integration. The „gradient curve" of the integrated signal corresponds to the original sine (see top and bottom).

> *A function is continuous if the function can be drawn in a mathematically correct way without putting the „pencil" down.*
>
> *Only continuous functions can be differentiated because the steepness at „step" points is (theoretically) infinite.*

Filters

When linear processes are listed filters are often forgotten although they are indispensable in information technology and signal processing. The linear processes mentioned so far referred to the time domain as far as their names are concerned. Thus the process differentiation of a signal means that the differentiation is carried out in the time domain. This is equivalent, as already explained, to multiplication in the frequency domain (amplitude spectrum) by $\omega = 2\pi f$, and to a shift of the phase spectrum by $\pi/2$ (the differentiation of a sine results in a cosine; this is the equivalent of this phase shift!)

The behaviour of filters refers to a process in the frequency domain. Thus a „lowpass" filter allows the low frequencies to pass, the high ones are largely blocked out etc. The type of filter always indicates what range is filtered out.

Note:

The word „filter" is reserved in signal processing for the frequency domain.

The equivalent process in the time domain is the „window".

Radio and television are at present (still) technologies which are inseparably linked with filtering. Every station operates within a specific frequency band. Above and below this frequency band there are other stations. The aerial receives all the broadcasters, the task of the „tuner" is above all to filter out this station accurately. Accordingly, a tuner functions in principle like a *tracking filter*.

The history of (analog) filter technology is an example of the basically futile attempt to develop filters using analog components - coils, capacitors which come as near as possible to the ideal of a rectangular filter. In Chapter 3 - „The Uncertainty Principle" - it was shown that rectangular filters are basically impossible to realise because they would contradict the laws of nature.

Digital filters represent a quantum leap in filter technology. They come close to the rectangular ideal although they can never achieve it. As will be shown in Chapter 10 - see also Illustration 116 - they can without exception be realised by means of three extremely straightforward linear processes: delay, addition and multiplication of a signal by a constant. All this in the time domain.

First, a number of traditional filter types used in analog technology will be described briefly and examined from a measurement technology point of view using DASY*Lab*. We will confine ourselves to three types which can be realised as lowpass, highpass, bandpass filters and band elimination filters. Bandpass and band elimination filters can in principle be put together from lowpass and highpass filters.

These three types were named after scientists who created the mathematical basis for calculating the circuits:

   1. Bessel filter

   2. Butterworth filter

   3. Chebycheff filter

These filter types are important within the framework of analog filter technology. They can, however, also be realised digitally. However, there are much better filters available in digital signal processing.

In Illustration 131 these three filter types are contrasted using the example of the lowpass filter. Their behaviour in the context of transmission technology demonstrates their different advantages and disadvantages and points to possible areas of application.

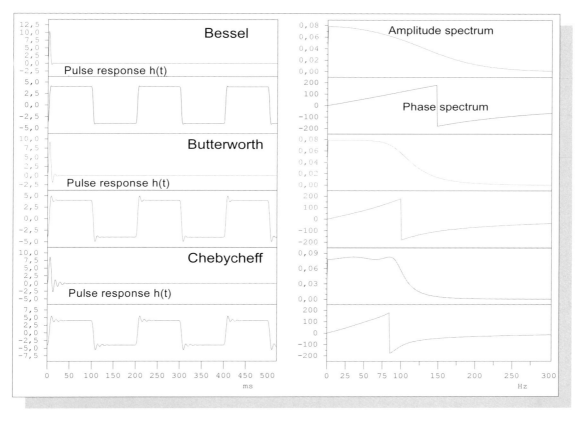

Illustration 131: ***Traditional analog filter types***

These three types are used in analog technology - in particular in the low frequency domain - depending on the area of application. In each series top left you see the pulse response h(t) of each filter type, among them the linear distortion of a sequence of rectangular pulses at the output of this filter. Top right, the amplitude curve and below this the phase curve.

How important a linear phase curve can be in the conducting state region can be seen in the case of the Bessel filter using the sequence of periodic rectangular pulses. Only then is the symmetry of the signal and its „form conservation" optimally preserved. Only in this case are all the frequencies (sinusoidal oscillations) contained in the signal delayed by exactly the same value.

In the case of the Chebycheff filter the non-linear phase curve - careful! filtering is a linear process - leads to the overshoot of the pulse form. On the other hand the steepness of the sides is very good. The ripple content of h(t) or of the sequence of periodic rectangular pulses is equivalent to the cutoff frequency of the conducting state region. The Butterworth filter is a frequently used compromise between the other two types.

All these filter types are available in different qualities (orders). Filters of a higher order generally require greater circuit complexity and/or components with very small tolerances. The filter types presented here are of the 4th order.

In the case of the Bessel filter great store is set by a linear phase response, on the other hand the filter steepness between the conducting state region and the blocking state region is poor, the same is true of the amplitude curve in the conducting state region.

In the case of the Chebycheff filter great importance is attached to edge steepness. On the other hand, the phase response is very non-linear and the amplitude curve in the conducting state region is extremely „ripply".

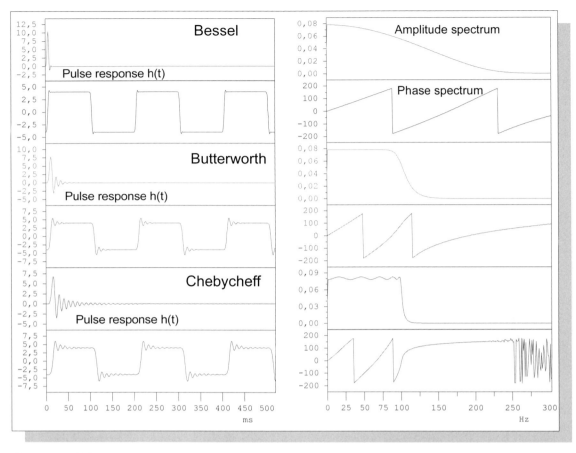

Illustration 132: **Traditional analog lowpass filters of the 10th order**

The arrangement of the signals or their amplitude and phase curve corresponds exactly to to Illustration 131. However, all the lowpass filters here are 10th order, roughly the maximum of what can be created in analog terms with a reasonable expenditure of effort.

In the distortions of the rectangular signal you see the weaknesses and virtues of the different filter types. Note that the pulse response h(t) lasts longer the steeper the edges of the filter. The trained person can give very precise information on the curve of the transfer function H(f) from the curve of h(t).

The steps in the phase curve should not confuse you. They always go from π to -π or from 180 degrees to -180 degrees. Both angles are identical. For this reason it is usual to enter the phase curve only between these two cutoff values. An angle of 210 degrees is plotted at -180 + 30 = -150 degrees.

Bottom right you see a phase curve with „random steps". It results from the numerical calculation and has nothing to do with real curve. Precisely speaking, the computer has difficulty with the division of values which are „close to zero".

The Butterworth filter represents an important compromise between these two types. It has a reasonably linear amplitude curve in the conducting state region and tolerable edge steepness in the transition from the conducting state region to the blocking state region.

In the case of the traditional analog filters there is therefore only the possibility of obtaining a tolerable filter characteristic at the expense of other filter values. There are two straightforward reasons for this:

- In order to construct filters of a higher quality, highly accurate - for instance, accurate to 10 decimal points - analog components (coils, capacitors and resistances) would have to be available. This is however not possible because tolerances in manufacture and temperature-related fluctuations make this impossible.

- All the high-quality digital filters are based - as will be shown - on the process of exact *delay*. Precise delay is not possible in analog technology.

The quality of the individual filter type can be enhanced by technically more complex circuits whereby, for example, two filters of the same type (decoupled) are series-connected one behind the other. The „order" is a mathematical/physical yardstick for this quality. Analog filters of the types described can be realised to the 10th order with reasonable effort.

> Note: Analog filters of a high quality can be constructed with considerable effort by exploiting other physical principles. Quartz-crystal filters and surface wave filters should be mentioned here. Their application is mainly in the high frequency field. They are not dealt with here.

You have already come across a special and easily understood type of *digital* filter several times. Look closely at Illustration 88, Illustration 109, and Illustration 116. This type is virtually ideal as its phase response is linear, its edge steepness limited only by the **UP** and its amplitude curve in the state conducting region is nearly constant.

How is this achieved? Easy - using the computer! First the signal segment (block length) is subjected to a FOURIER transformation (FFT). Thus, this signal segment is available in the frequency domain. Using a cutting device (Module „Cut out" the frequency range is defined that is to be allowed to pass. Then comes an *inverse* FOURIER transformation (IFT) by means of which the „rest" of the signal is „beamed" back into the time domain. Within the frequency domain you see two signal paths which must be set identically. The amplitude and phase spectrum must be blocked for the same area, so to speak.

In Illustration 109 a practically ideal bandpass filter has thus been realised which has a bandwidth of only 1 Hz. Unfortunately this filter does not operate in real time, i.e. it cannot filter a continuous signal of some length without loss of information. It can only filter a signal limited in time, e.g. a signal block of 1024 measurement readings with a sampling rate set at 1024 measurements per second.

The following can be said in general terms about filters:

> *The more the transfer function of a filter approaches the rectangular ideal, the greater the complexity of circuits in the case of analog filters and the greater the amount of calculation in the case of digital filters.*

> *From a physical point of view the following applies: As a δ–pulse appears „smudged" in time at the output of the filter, the filter must contain a chain of interconnected energy stores. The better the filter the more complex this „retardation system".*

Non-linear processes

The four linear operations dealt with up to now -

> *Multiplication by a constant*
> *Addition and substraction*
> *Differentiation*
> *Integration*

- are the four *classical* linear mathematical operations as they are known from the *"Theory of linear differential equations"*. They were dealt with here mainly from a signal technology perspective.

They are of very great importance because they are sufficient to describe in mathematical terms the most important natural laws - electromagnetism and quantum physics. This is why space and cables exhibit linear behaviour in the propagation of signals.

In the specialist literature on the *Theory of Signal Processing Systems* non-linear processes or systems are hardly mentioned. This gives the impression that non-linear processes/systems with the exception of one or two „exotic" processes were of hardly any practical importance. This is quite wrong.

Remember Chapter 1: here the *Theory of Signal Processing Systems* was understood as the mathematical modelling of signal processing on the basis of physical phenomena. It is true that in the case of non-linear processes and systems the mathematical modelling is successful in the form of equations but with few exceptions these equations are insoluble.

> *Mathematics fails almost completely in the solution of non-linear equations.*

For this reason non-linear processes are hardly mentioned in these books on theory. It is only with the help of computers that such non-linear processes can be investigated and the results at least visualised. For a number of reasons this seems increasingly important for research:

- While there are roughly half a dozen different linear processes the number of different non-linear processes is infinitely greater.

- The importance of non-linear processes for all the really interesting „systems" is becoming more and more evident. This appears to apply above all to biological processes. At last we wish to understand, for instance, according to what laws the growth of plants and animals occurs. But in inanimate nature, too, non-linearity rules supreme. Turbulences in the atmosphere are an example of this, as is any water droplet or wave. The popular term for research into the non-linear is „chaos theory". In this connection there is a range of gripping popular scientific literature which draws people's attention to what will be the main field of scientific research in the next few decades.

Given the infinite diversity of non-linear processes we have to be selective. At this point we shall deal with a few fundamental processes. As necessary, others will be added - for instance, in measuring technology. It is important to clarify what common features non-linear processes (in the time domain) have in the frequency domain.

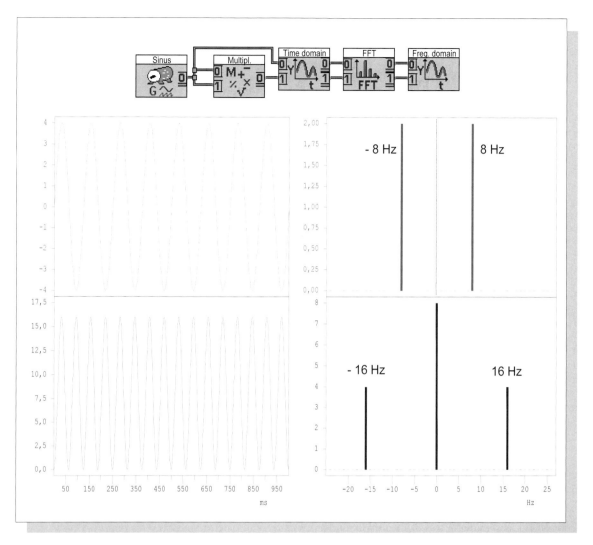

*Illustration 133:   **Multiplication of identical sinusoidal signals***

The multiplication of a sinusoidal signal by itself may be used as a technique for doubling the frequency. The output signal contains a different frequency from the input signal(s). It is thus a non-linear process.

The output signal is always positive. The frequency domain shows that it consists of a sinusoidal signal (16 Hz) and an offset („zero frequency" 0 Hz).

Multiplication of two signals

What happens when two signals are multiplied with each other? Let us proceed systematically. As representative of all signals as a result of **FP** we select two sinusoidal signals, initially of the same frequency, amplitude and phase size.

Illustration 133 shows the astonishing result in the time and frequency domain. In the time domain we see a sinusoidal signal of twice the frequency, which has a direct voltage superimposed on it. Further experiments show the following: the level of the direct voltage depends on the phase displacement of the two sinusoidal signals in relation to each other. In the frequency domain we see a line in the doubled frequency - thus it is really a sinusoidal signal.

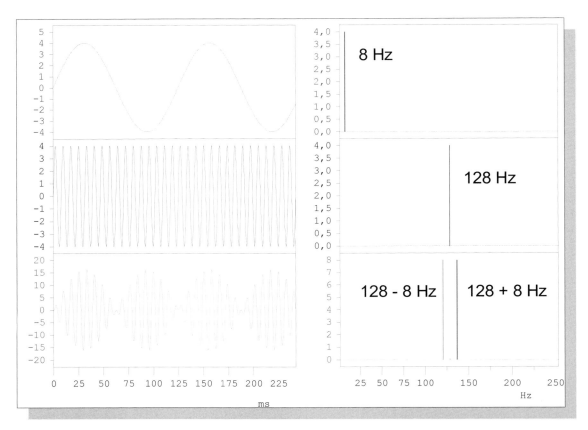

Illustration 134: **_Multiplication of two sinusoidal signals of clearly different frequencies_**

In this case the principle emerges clearly for the first time which characterises the frequency spectrum of most non-linear processes - non-linear linking of two signals. The literature talks of sum and difference frequencies, i.e. all the frequencies contained in this spectrum result from the sum and difference of all the frequencies contained in the two spectra of the imput signals. In the language of symmetrical spectra - see Chapter 5 - according to which the mirror-image spectrum exists in the negative frequency domain, in the case of these non-linear processes only the sums of all negative and positive frequencies are formed.

> *The multiplication of a sinusoidal signal by itself can be used technically to double the frequency.*
>
> *The multiplication of two sinusoidal signals is obviously a non-linear process as the output signal contains different frequencies from the input signals.*

Now we select two sinusoidal signals of a clearly different frequency as shown in Illustration 134. There is a signal in the time domain which is equivalent to a beat (see Illustration 113). This is also shown in the frequency domain. There are two neighbouring frequencies contained here.

Appropriate experiments with DASY*Lab* usually give

$$f_1 \pm f_2$$

To put it precisely, f_1 and f_2 are the - symmetrical positive and negative - frequencies of a sinusoidal signal (see Chapter 5). The following holds:

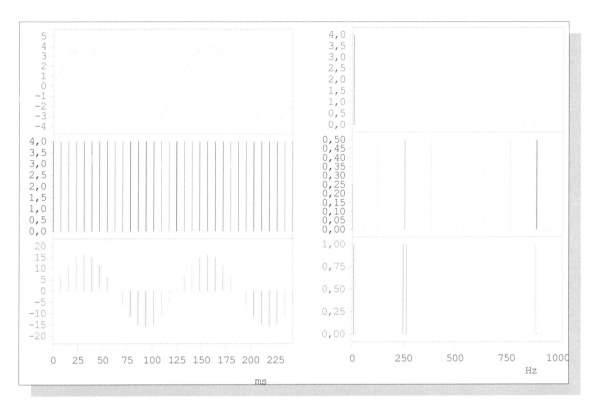

Illustration 135: **Sampling as the multiplication of a signal by a sequence of δ-pulses**

Using the example of a sinusoidal oscillation the multiplication of a signal by a (higher frequency) sequence of δ-pulses is shown. The sequence of δ-pulses has an infinite number of sinusoidal oscillations of the same amplitude, always at the same interval of 128Hz.

In the lower spectrum you will only find the „sum and difference frequencies" of any two frequencies of the two upper spectra.

> *The smaller one of the two frequencies is, the nearer the sum and difference frequency are to each other.*
>
> *A beat can thus be generated by multiplication of a sinusoidal oscillation of a low frequency by a sinusoidal oscillation of a higher frequency. The beat frequencies are in a mirror-image symmetry to the higher frequency.*
>
> *Note that the envelope of the beat corresponds to the sinusoidal curve of half the difference frequency.*

What other multiplication could be of technical importance? Intuitively, it would seem sensible to include the second most important signal - the δ-pulse in the multiplication.

The most important practical application is the multiplication of a (band limited) signal with a sequence of δ-pulses. As was clear from the Illustration 37 and Illustration 85 (bottom) this is equivalent to a sampling process by which samples of the signal are taken at regular intervals. Normally, the frequency of the sequence of δ-pulses is much greater than the (highest) signal frequency. This sampling is always the first step in the transformation of an analog signal into a digital signal.

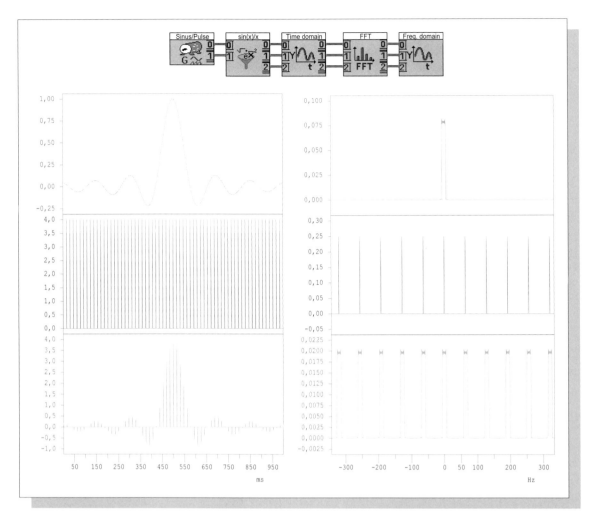

Illustration 136: **Convolution in the frequency domain as the result of a multiplication in the time domain**

A simple explanation for the strange behaviour of the frequency domain in the case of the multiplication of two signals in the time domain was already given at the end of Chapter 5 „The Symmetry Principle" - periodic signals in the time domain have line spectra of equidistant frequencies. Thus equidistant lines in the time domain must result in periodic spectra for reasons of symmetry. The symmetrical mirroring of the LF-spectrum at every frequency of the δ–pulse is called **convolution**.

> *The first step in the conversion of an analog signal to a digital signal is always multiplication of the signal by a periodic sequence of δ–pulses, that is a non-linear process. Thus a digital signal above all in the frequency domain has properties which the original analog signal did not have. Most of the problems in the later retrieval of the original information of the analog signal arise from this fact.*

These frequency-related phenomena are to be demonstrated from a measurement-technology point of view using the example of two forms of signal. The first selected is a sine wave, the simplest form of signal. After that an Si-shaped curve (see Illustration 109) as an almost ideal band-limited signal.

The following are the results of these investigations (Illustration 135 and Illustration 136):

- In the experiment with a sine two lines appear quite clearly in a mirror-image relationship to each frequency of the periodic sequence of δ–pulses in the intervals between the frequency of the sine. For every frequency f_n of the sequence of δ–pulses there is therefore a sum and a difference frequency of the form $f_n \pm f_{sine}$.
- As for reasons of symmetry (Chapter 5) we have to attribute a positive and negative frequency to a sine, these two frequencies are to some extent reflected in every frequency f_n of the periodic sequence of δ–pulses. We must imagine the whole spectrum as mirrored in the negative frequency region.

The situation is presented even more clearly and above all in a way more relevant in practice in the sampling of an Si-function:

- As shown in Chapter 5 „Symmetry Principle" the full bandwidth of the Si-function, that is the originally positive and negative frequency regions, are convoluted at each frequency f_n of the sequence of periodic δ–pulses.
- The (frequency-related) information on the original signal is therefore contained (theoretically) an infinite number of times in the spectrum of the sampled signal.

> *Multiplication in the time domain results in a convolution in the frequency domain.*
>
> *For reasons of symmetry the following must apply: a convolution in the time domain results in multiplication in the frequency domain.*

Formation of the absolute value

The formation of the absolute value is a particularly straightforward non-linear process. The rule is:

> *In the case of the formation of the absolute value the minus sign is deleted in front of all the negative values and is replaced by a plus sign; the originally positive values remained unchanged.*

In a sense all the signs „rectified". The (full-wave) rectification of currents or signals in electrical engineering is the best known example of the abstract term formation of the absolute value. In Illustration 137 the examples in the time domain show what effect the formation of absolute value will probably have in the frequency domain: the periodicity of the base oscillation doubles compared with the periodicity of the input signal.

This is true of the first two signals - sine and triangle - in Illustration 137. As a consequence of the formation of absolute value the period length is halved, i.e. the base frequency is doubled. The first three signals - sine, triangle and sawtooth - have the symmetrical position in relation to the zero line in common. Therefore on would probably think that the absolute value of the sawtooth would be doubled in its base frequency. That is, however, not the case; it becomes a periodic triangular oscillation of the same frequency.

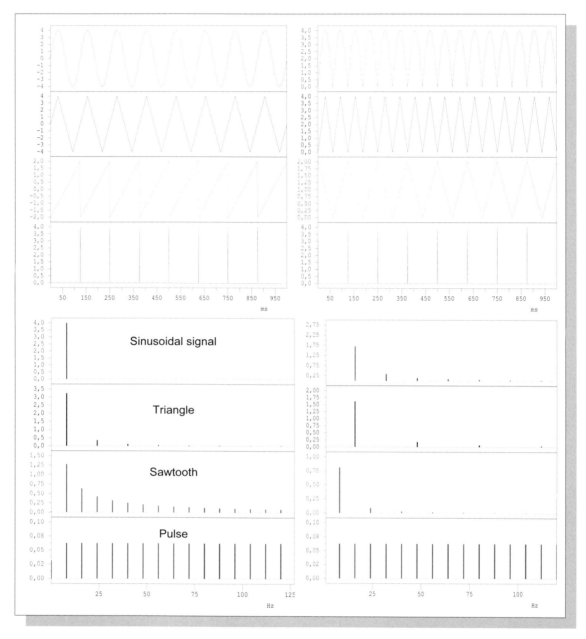

Illustration 137: ***The absolute value response of different signals***

Top left you see the input signals in the time domain, below this their frequency spectra; top right you see the response signals in the time domain and below this their frequency spectra. In the case of the two upper signals - sine and triangle - a generally valid principle for the formation of absolute value appears to have been discovered in the doubling of frequency; unfortunately this does not apply to the periodic sawtooth which also lies in a symmetrical relationship to the zero line and which becomes a periodic triangle of the same base frequency. But have frequencies - the even multiples - disappeared?
Finally, the sequence of δ–pulses remains unchanged because it lies in the positive region only on the zero line. A linear process?

Has the sawtooth been distorted in a non-linear way or have new frequencies been added? In a sense, yes, as the even multiples of the base frequency are now lacking. They must have been deleted by the new frequencies. These would then have a phase position displaced by π compared with the frequencies already present. Or is it a linear process after all?

Finally the lower periodic sequence of δ–pulses lie without exception at zero in the positive region. The formation of absolute value also leads to an identical signal here. This would correspond to a linear process.

This simple process of „formation of absolute value" demonstrates clearly how dependent the results are on the given input signal.

> *Certain signalling processes may show non-linear behaviour depending on the form of signal and linear behaviour in the case of other signal forms.*
>
> *Any (practically) linear amplifier shows non-linear behaviour, for example, if it is overdriven).*

Quantization

Finally, we shall deal with quantization, an extremely important non-linear process which always occurs in the conversion of analog into digital signals.

An analog signal is continuous both in time and value, i.e. it has a certain value at a given point of time and passes through all (an infinite number of) intermediate values between its maximum and minimum value.

A digital signal, on the other hand, is discrete in time and value. Before quantization the signal is only regularly sampled at quite definite discrete points of time. Afterwards, this measurement is converted into a (binary) number. However, the stock of numbers is limited and is not infinitely large. A measuring instrument with three positions can only show 999 measurement values. Illustration 138 shows the difference between these forms of signal. The analog signal is continuous in time and value, the sampled signal is discrete in time and continuous in value and finally the quantized signal - the difference between two neighbouring values is here 0.25 - is discrete in time and value. On the right you see the quantized signal as a string of numbers.

A very important method particularly in the case of A/D converters is sampling. The sampled value is retained until the next „sample" is taken. The sampling curve is in the first instance a „stepped" curve with steps of differing height. In the subsequent quantization (Illustration 139) the height of the steps becomes uniform. Every step corresponds to a discrete value.

Accordingly, there is a difference between the original signal and the quantized signal. In the final analysis there is a distortion of the signal. This difference is represented accurately in the middle. As the spectrum of the difference shows it is a kind of noise. The expression „quantization noise" is used.

In order, for example in HiFi technology, to keep this difference as small as possible so that this quantization noise is inaudible, the step curve must be made smaller until there is no visible or audible difference between the two signals.

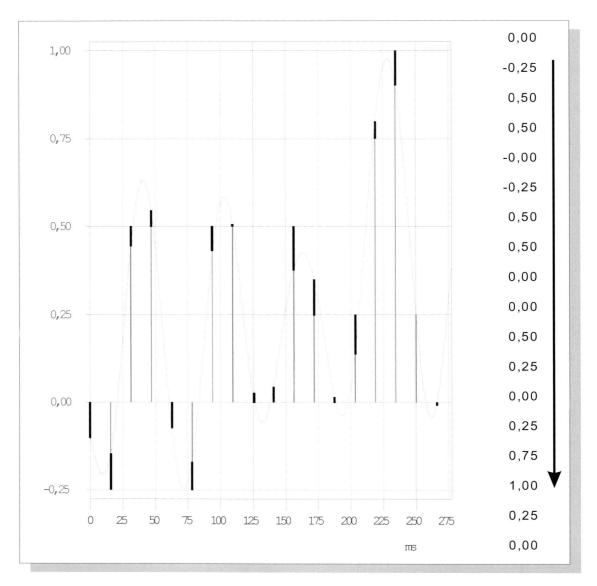

Illustration 138:   ***Illustration of the quantization process***

In order to represent the situation more precisely the three signals have been entered above each other. There is first of all the segment from the analog signal continuous in time and value. The sampled signal is also to be seen. It is a time discrete signal still continuous in value which extends from the zero line exactly as far as the analog signal. These are therefore exact samples of the analog signal. The difference between the sampling signal and the quantized signal is represented by the broad bars. These broad bars thus represent the area of error which is inherent in the digital signal.

The quantized signal samples begin at the zero line and end on the allowed values which are shown here by the hatching. The quantized signal is represented on the right as a string of numbers.

Both representations are so roughly quantized that they would not be acceptable in a technical and acoustic sense.

In HiFi technology 16 bit A/D converters are used. They allow $2^{16} = 65536$ different discrete numbers within the value range. The technology of A/D converters will be dealt with in a later chapter. As a result of the properties of our ears the quantization noise then falls below the audible level.

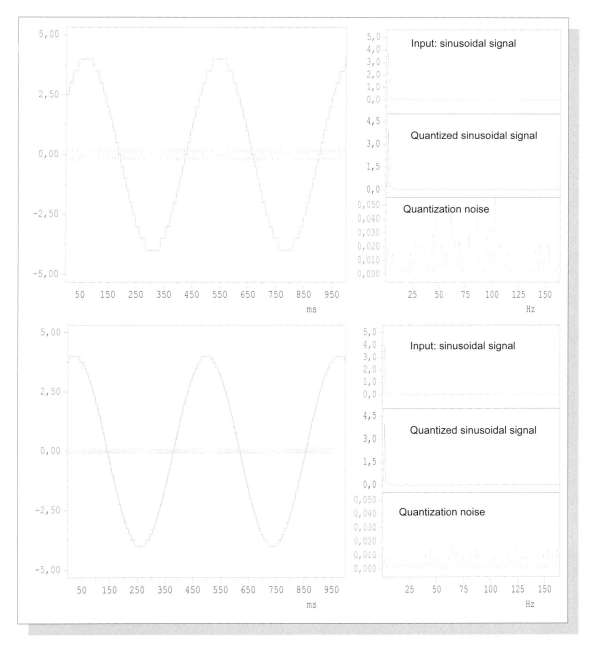

Illustration 139: ***Quantization of a sampled signal***

There are three signal curves to be seen in the time domain: the sinusoidal analog signal continuous in time and value, the time and value discrete quantized sample signal (step curve) and the time and value continuous difference signal in the centre of the illustration.

At the top you see an extremely inaccurately quantized signal. Try to determine the number of allowed quantized values within the measured region of -5V to +5V. The spectra of the three signals are shown on the right. The spectrum of the sine shows a single line to the far left on the vertical axis. Below this is the spectrum of the quantized signal. Additional small irregular lines are to be seen clearly within the frequency domain. The difference signal incorporates the quantization noise. The amplitudes reach 0.05 V. By increasing the number of possible quantization steps (how many are there here?) the step curve of the quantized signal appears to be much nearer the analog signal: As a result the difference voltage is much smaller. The quantization noise is less than 0.02V.

Both representations are so roughly quantized that they would not be acceptable in a technical acoustic sense.

Sampling and quantization are always the two initial steps in the conversion of analog signals into digital signals. Both processes are fundamentally non-linear.

In sampling periodic spectra arise as a result of convolution. The information is contained in them (theoretically) an infinite number of times, the bandwidth of the sample spectrum is therefore infinitely great.

The consequence of quantization is quantization noise, which in HiFi-technology must be reduced to an inaudible level. This is done by increasing the quantization steps accordingly.

A digitalised signal necessarily differs from the original analog signal. This results in the requirement of making this difference as small as possible, so that it is either inaudible or scarcely audible.

Windowing

Longer, non-periodic signals - for instance, audio-signals - must be processed in segments as soon as the frequency domain is involved. As already explained in detail in Chapter 3 (see Illustration 51 to Illustration 54) and Chapter 4 (see Illustration 61), the source signal is multiplied by suitable „windows", which should overlap so that there is no loss of information compared with the source signal.

Illustration 51 shows the consequences if the „cutting out" is done with a rectangular window. These segments contain frequencies which were not contained in the source signal. „Windowing" is thus basically a non-linear process. The so-called „optimal windows" such as the GAUSSian window attempt to minimise these additional frequencies. As a result of the non-linear functions of these windows even the original spectrum of the source signal is distorted. In the time domain the central segment of the windowed signal is given greater weight than the two edges which begin and end gently.

Preliminary conclusions:

In the conversion of analog signals into digital signals the signals are always sampled, quantized (and coded). All these processes are non-linear and distort the source signal. The important thing is to minimise these errors to such an extent that they are below audibility level.

The digitalised signal differs fundamentally from the analog source signal, above all in the frequency domain. Digital signals have a periodic spectrum as a result of sampling.

Exercises on Chapter 7

Exercise 1

Using DASYLab design a test circuit which is as simple as possible by means of which you can demonstrate and measure the linearity or non-linearity of a system or process.

Exercise 2

What would be the possible consequences of the non-linear behaviour of the transfer media „open space" and „cable"?

Exercise 3

If a sequence of rectangular pulses are connected to the input of a long cable a distorted signal (a kind of „melted rectangle") appears at the distant end of the cable.

a) What are the causes of these distortions?

b) Are non-linear distortions or linear distortions involved?

> Note: You can simulate a cable as a chain of simple lowpass filters using DASYLab (see Illustration 106)

Exercise 4

How does the sum of two sinusoidal oscillations of the same frequency and amplitude change depending on the mutual phase displacement? Design a small test position using DASYLab.

Exercise 5

Explain the way the amplitude spectrum of two consecutive δ–pulses in Illustration 116) arises. How do the zero positions of the amplitude spectrum depend on the time interval between these δ–pulses?

Exercise 6

Using DASYLab design a voltage frequency converter VCO according to Illustration 121 which with an input voltage of 27mV produces a sinusoidal output signal of 270 Hz. The characteristic should be linear.

You will need a special setting of the „Generator" module and an input circuit to control this generator.

Exercise 7

What advantages does frequency coding of measurement values by means of VCO have in telemetrics?

Exercise 8

Check the following thesis using DASYLab: the sequence of linear processes in a linear system can be changed round without changing the system.

Exercise 9

Explain why processing differentiation exhibits highpass filter behaviour and processing integration exhibits lowpass filter behaviour?

Exercise 10

How could the system illustrated be simplified - for instance by cutting the number of modules - without changing the properties of the system. Check whether your solution is right using DASY*La*b.

Note: The filters must be absolutely identical

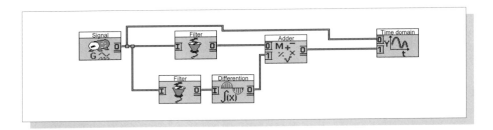

Illustration 140:   ***Optimizing a linear system***

Exercise 11

Develop a process by means of the module „formula interpreter" with which a sinusoidal input voltage with the amplitude 1 V of any frequency f always produces a sinusoidal output voltage with the fourfold frequency 4∗f. The output voltage should have an amplitude of 1 V and have no offset.

Exercise 12

What role do sampling and quantization play in modern signal processing?

Exercise 13

Explain the term „convolution". By means of what operation could „convolution" be caused in the time domain? Hint: Symmetry Principle.

Exercise 14

Develop a simple circuit for pure sampling (time discrete and value continuous) by means of the module „Latch".

Exercise 15

Develop by means of the module „formula interpreter" a circuit for the quantization of an input signal according to Illustration 139. It should be possible to set the quantization anywhere from fine to „coarse".

Chapter 8

Classical modulation procedures

All the processes which prepare the source signal for the transmission path are subsumed under the term *modulation*.

Signals are modulated in order to

- exploit in an optimum way the physical properties of the medium (e.g. choice of frequency range),
- guarantee largely undistorted transmission,
- optimise the reliability of transmission,
- respect data protection in telecommunications
- exploit transmission channels several times over (frequency and time multiplex) and
- free signals from redundant information.

Note: Alongside the term *modulation* the term *(en-)coding* is often used particularly in modern digital transmission processes. The two terms cannot be precisely distinguished. In this connection see Chapters 11-13. There is also a distinction between source coding and channel coding, which for reasons of information theory should always be carried out separately. Source coding serves the concentration of information or data compression, i.e. the signal is freed from unnecessary redundant information. The task of the channel coder is to guarantee a reasonably reliable transfer of signals in spite of the signal-distorting disturbances occurring along the transmission path. This is achieved by means of coding processes which recognise and correct errors. Check elements are added to the signal as a result of which compression by source coding is again diminished to some extent.

Transmission media

A distinction is made between

- "wireless" transmission, (e.g. by satellite) and
- wire communications (e.g. via twin wires and coaxial cables).

An important new medium is the optical wave guide, glass fibre

Modulation with sinusoidal carriers

The classical modulation processes of analog technology use the continuous modification of a sinusoidal carrier. These processes are still standard in radio and television technology. Modern digital modulation processes are gaining ground here and will increasingly replace classical modulation processes in future.

> *In the case of a sine three quantities can be varied: amplitude, frequency and phase. Accordingly it is possible to superimpose information on a sinusoidal "carrier" by means of a change in the amplitude, frequency or phase or by a combination of these.*

Note:
From a physical point of view - see Uncertainty Principle - an amplitude, frequency or phase change always implies frequency uncertainty in the frequency domain; so the sudden switching of the frequency of a sine in the time domain in reality implies frequency uncertainty in the frequency domain which is expressed in a frequency band.

A modulated signal with a sinusoidal carrier thus only has something like an "instantaneous sine" or an "instantaneous frequency" in the time domain, in the spectrum we always see a bundle of frequencies. This bundle of frequencies is wider the shorter the existence of the "instantaneous frequency" ($\Delta f > 1/\Delta t$) in the time domain.

In classical modulation processes only one of the three quantities amplitude, frequency or phase is varied continuously "in the rhythm" of the source signal.

The classical modulation processes are thus

- Amplitude modulation AM
- Frequency modulation FM
- Phase modulation PM

In traditional transmission technology - e.g. radio and television - AM and FM are mainly used.

> *AM, FM and PM are used to displace or transfer the source signal into the desired frequency range. All modulation processes represent non-linear processes because the modulated signals occupy a different frequency range from the source signal.*

Modulation and demodulation in the traditional sense

Any modulated signal must be demodulated in the receiver, i.e. it must be returned as accurately as possible into the form of the original source signal.

In the case of a radio station the signal is modulated in the transmitter, i.e. only one modulator is required. But each of the many thousands of receivers need a demodulator. In the early days of radio technology it was only possible to use a modulation process which required a simple and cheap demodulator in receivers.

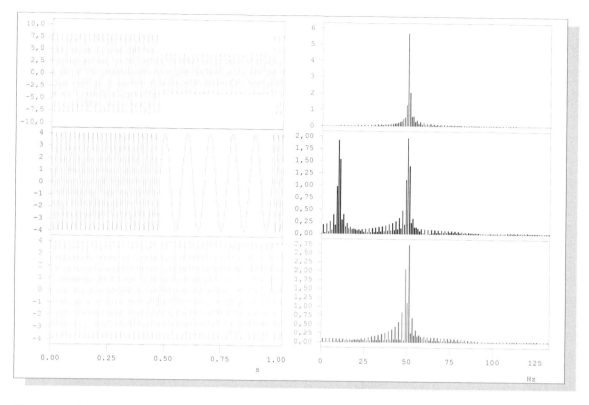

Illustration 141: ***Does "instantaneous frequency" exist?***

In the top series the amplitude, in the middle series the frequency and in the bottom series the phase of sinusoidal carrier oscillation is spontaneously changed. Precisely in the case of modulation types with a sinusoidal carrier in the time domain the term "instantaneous frequency" is often used.

A precise analysis in the frequency domain shows that because of the Uncertainty Principle this cannot exist. Any sinusoidal signal, after all, (theoretically) lasts an infinite period of time. In the centre there could otherwise only be two lines present at (20 and 50 Hz). However, we see a whole bundle of frequencies, a whole band of frequencies, so to speak. Any change in a quantity of the sinusoidal signal leads to a frequency uncertainty - as can be seen here. And in the bottom row the phase was only briefly displaced by π.

Amplitude modulation and demodulation AM

The history of early radio technology is at the same time a history of AM. From today's point of view there was the complicated attempt to carry out the most straightforward of signal processing - e.g. the multiplication of two signals - using unsuitable means and complex circuits. The problems arose because of the grossly inadequate properties of analog components (see Chapter 1, page 24). These attempts and the course of subsequent development will be recounted here.

As can be seen from Illustration 142 the information of the source signal - from the time domain point of view - lies in the envelope of the AM signal. What signal processing generates the AM signal?

If you look carefully at Illustration 134 will find the low frequency sinusoidal signal (top left) in the envelope of the multiplied signal (bottom left). However, with one difference: this envelope changes constantly from the positive to the negative region (and vice-versa). The multiplication appears in spite of this as a candidate for AM modulation. A small change to the source signal must first be made, however.

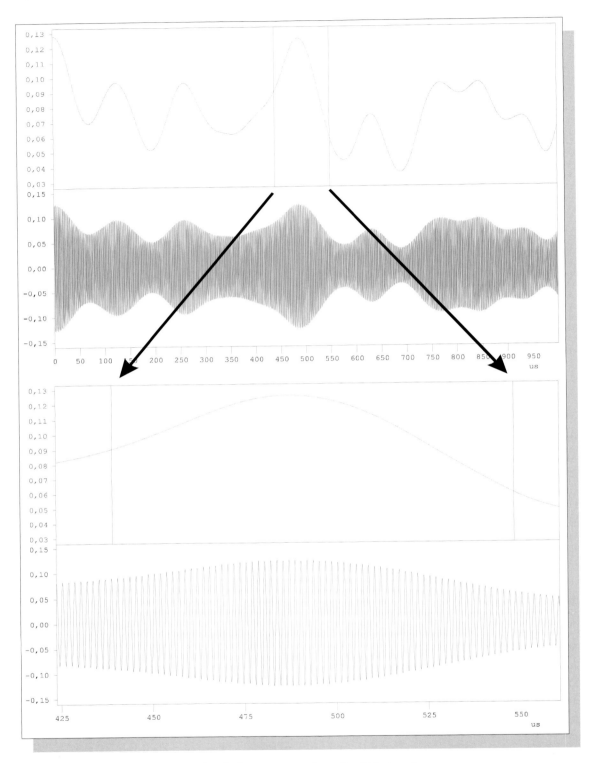

Illustration 142: **_Realistic representation of a AM-signal_**

Only by using a very rapid storage oscilloscope is it possible to represent a real AM signal in a qualitatively similar way as here. Here it was simulated by means of DASYLab. At the top you see a language-like LF signal - generated by filtered noise - below this the AM signal. The LF signal is contained in it as the envelope of the signal. Below this there is a small segment of the above signal in which the sinusoidal carrier is to be seen clearly. See also Illustration 143.

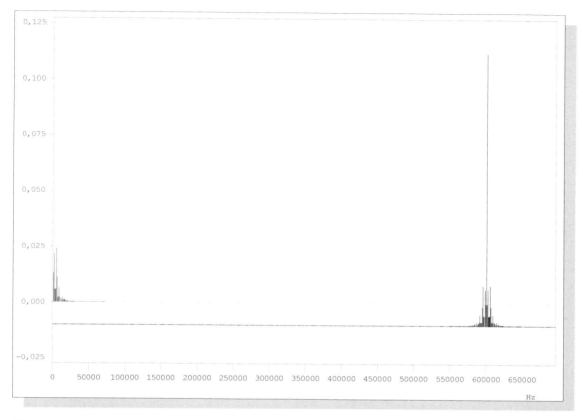

Illustration 143: **Frequency domain of an AM signal**

On the upper horizontal line you see on the left the frequency domain of the LF signal from Illustration 141, on the lower line at 600 kHz that of the AM signal. It is striking that the LF signal appears double at the carrier frequency, mirrored symmetrically. It is said that the LF spectrum is „convoluted" at the carrier frequency. This convolution is in the final analysis the result of a multiplication in the time domain.

Traditionally the right sideband is referred to as the „upper sideband" and the left sideband as the „lower sideband". From our knowledge of the Symmetry Principle this term is misleading as in the reality the LF signal is symmetrical to the frequency 0 Hz. The mirror-image half of the negative frequency domain is also part of the LF frequency domain (in this connection see Illustrations 83 and 84). The multiplication of a signal in the time domain by a sinusoidal carrier or the AM is the simplest method of displacing the frequency domain of a signal to any position desired.

This small change is shown in Illustrations 142 and 144. Here a (variable) direct voltage - an "offset" - is superimposed on the source signal until the source signal is completely in the positive region. If this signal is now multiplied by a sinusoidal carrier we obtain the form of an AM signal according to Illustration 142. The envelope lies exclusively in the positive region or inverted in the negative region.

Illustration 144 gives a rule of thumb for the offset. The direct voltage U must be greater than or at least equal to the amplitude of the low frequency sine or the (negative) maximum value of the source signal (see Illustration 144).

If the source signal is displaced completely into the positive (or negative) region by the addition of a direct voltage U, the envelope of the AM signal represents the original source signal after multiplication by the sinusoidal carrier. Then the so-called degree of modulation m is smaller than 1 or smaller than 100%.

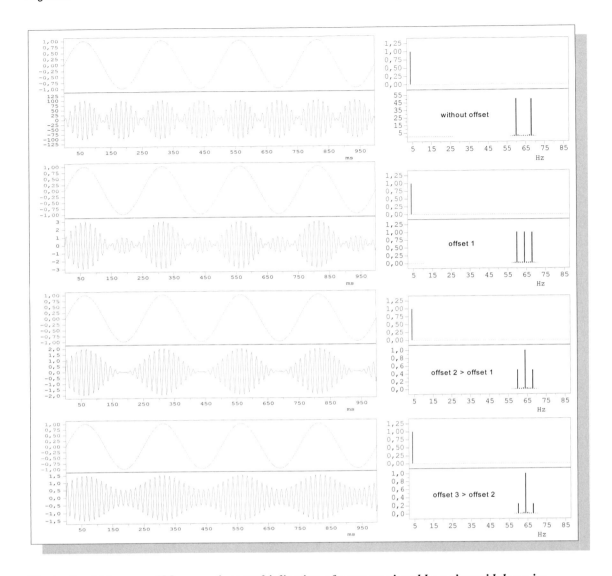

*Illustration 144: **AM generation: multiplication of a source signal by a sinusoidal carrier***

A low frequency sine is selected here as the most straightforward form of source signal. A direct voltage (offset) which becomes greater and greater is superimposed until the sine lies entirely in the positive region. When this is the case, the original source signal lies in the envelope of the AM signal after multiplication by the sinusoidal carrier.

In this form the AM signal with the above circuit consisting of diode D, resistance R and capacitor C can be demodulated, i.e this circuit makes it possible to retrieve the original source signal. There are, however, disadvantages involved in this type of amplitude modulation.

Note:
In the case of analog or computer- (digital) multiplication the frequently mentioned *degree of modulation m* often makes little sense. In theory the mathematical model of an AM signal is described by the formula

$$u_{AM}(t) = (1 + m \sin(\omega_{LF}t)) * \hat{U}_{carrier} \sin(\omega_{carrier} t)$$

The first bracket represents the sinusoidal source signal voltage with a direct voltage ("1") superimposed. The last term describes the sinusoidal carrier voltage.

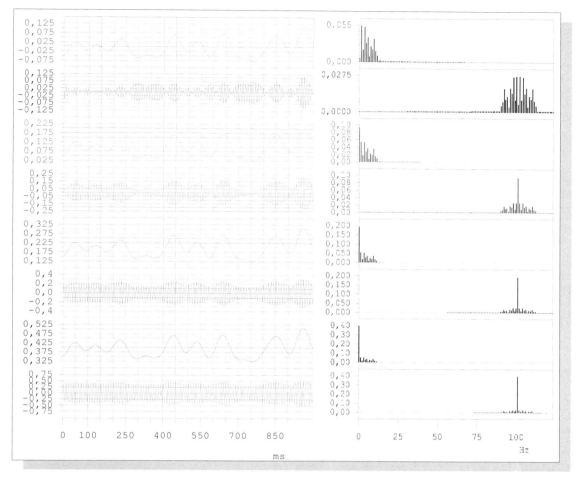

Illustration 145: ***AM of a low frequency signal curve***

In principle the same is represented here as in Illustration 144, only with a typical low frequency signal segment as the source signal. A direct voltage, which increases from top to bottom is also superimposed on the source signal.

In the frequency domain this offset corresponds to the sinusoidal carrier which from top to bottom becomes more and more dominant. The main part of the energy of the AM signal is accounted for by the carrier and not the information-bearing part of the AM signal.

*In the frequency domain the so-called **upper sideband** and **lower sideband** of the original source signal can be seen clearly to the right and left of the carrier, that is the information is present twice over. For this reason this kind of AM is often referred to as a "double sideband AM".*

If two voltages are multiplied by each other the unit (V∗V), i.e. (V^2) results. This is incorrect in a physical sense for at the output of a multiplier a voltage appears with the unit (V). We have recourse to a little trick - the term in the brackets is defined as a pure number without a unit. m is the *degree of modulation* and is defined as $m = \hat{U}_{LF}/\hat{U}_{carrier}$. (V) is now simplified and for $u_{AM}(t)$ the unit (V) results.

The term $(1 + m \sin(\omega_{LF} t))*\hat{U}_{carrier}$ can now be meaningfully interpreted as "time-dependent amplitude" which changes in the rhythm of the LF signal. If this time-dependent amplitude is always positive - the term in brackets must be greater than 0, i.e. positive - the time-dependent amplitude (envelope!) lies exclusively in the positive region or, inverted in the negative region.

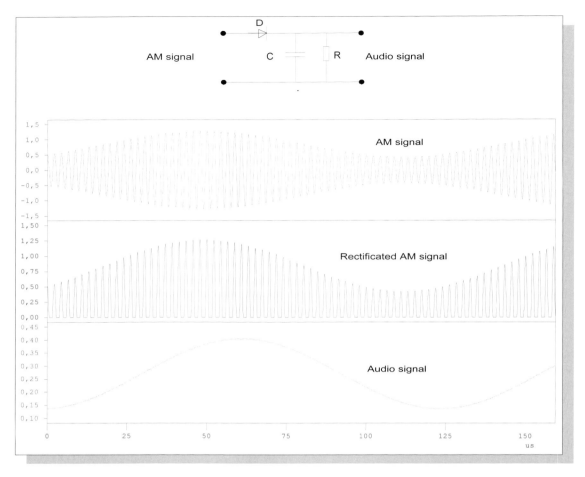

Illustration 146: **_Demodulation of two AM-Signals in the traditional manner_**

At the top you see the AM signal, the envelope of which represents the original signal. This signal is rectified by the diode, i.e. the negative part of the AM signal is cut off. The R-C part acts like a lowpass filter, i.e is so inert timewise that it does not notice "short-term" changes. It acts like a "floating averager".

If you look closely you will notice in the retrieved source signal slight steps at the bottom which derive from this (imperfect) floating averaging.

For this reason the requirement m < 1 applies for this traditional form of AM.

In addition, as a real signal is certainly not sinusoidal the *modulation degree m* cannot be meaningfully described as the relationship between two amplitudes of different sinusoidal voltages. Our above definition of the offset is simply more meaningful.

If the original source signal is not clearly recognizable in the envelope, it cannot be difficult to retrieve it by means of demodulation.

Since the beginnings of radio technology demodulation of the AM has been carried out by means of an extremely simple circuit consisting of a diode D, a resistance R and a capacitor C as shown in Illustration 146. The AM signal is first rectified by the diode, that is, the negative region is cut off.

Chapter 8: Classical modulation procedures

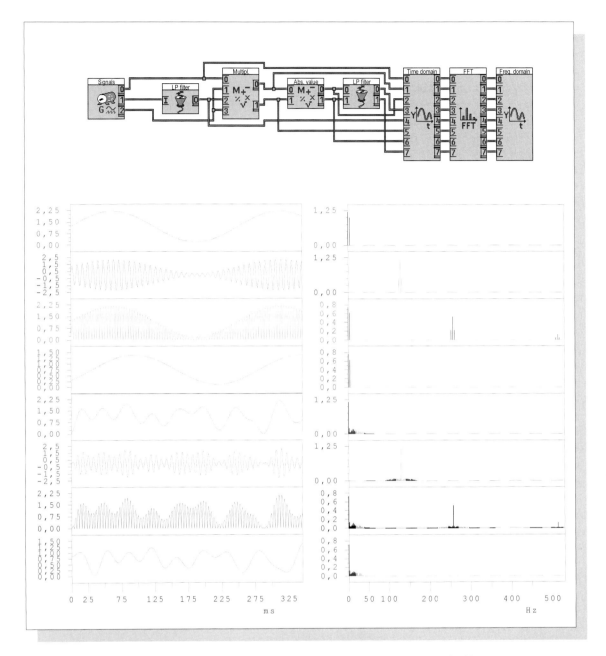

Illustration 147: **Demodulation of two AM signals in the traditional fashion**

As examples of source signals (top) a sine (4Hz) and bottom a realistic signal curve with an upper limiting frequency of 32 Hz have been selected here. The carrier thus lies at 128 Hz. In order that demodulation can be carried out in the traditional way, both source signals must lie completely in the positive region. As a result the envelope of the AM signals represents the source signal.

As a result of the absolute value process (rectification) and the subsequent lowpass filtering (R-C-link) the two source signals are retrieved. These signals are slightly time-displaced compared with the original signal because all signal processing takes time.

As explained in more detail in the text, the signal demodulation corresponds to a simple multiplication of the AM signal by a carrier of the same value (here 32 Hz). As precise analog multipliers did not exist, inventive scientists had to come up with something.

The following R-C circuit is something like a lowpass filter. In the language of the time domain the circuit is so sluggish that it cannot keep up with the rapid changes of the rectified sine. Given the correct time dimensions of the R-C circuit ($\tau = R*C$) something ressembling a sliding average value of the rectified signal appears at the output. This sliding average value is however nothing other than the source signal.

> Note:
> A television set provides an appropriate example. It shows 50 half-pictures or 25 full pictures a second. As a result of the sluggishness of our eye in time and the subsequent processing of the signal in the brain we do not perceive 25 individual images per second but a picture which changes continuously. The "sliding average value" of this sequence of pictures, so to speak.

In Illustration 147 instead of a diode the module "absolute value" was selected. The absolute value corresponds here to the so-called full wave rectification - the negative region of the sine is in addition represented positively - and represents a non-linear process which we should look at again more closely at this point.

According to Illustration 137 a kind of doubling of the frequency takes place in the frequency domain as a result of the "absolute value process" of a pure sine, just as in the multiplication of a sine by itself in Illustration 133. As analog multipliers did not exist 50 years ago *rectification* was used as a kind of substitute multiplication of the AM signal by a carrier signal of the same frequency.

Just as in Illustration 133 we obtain by means of this „absolute value process"- carried out in the time domain - an AM signal in the frequency domain with double the carrier frequency and one with the "carrier frequency" 0Hz, i.e. the original source signal. This cannot be seen in the time domain.

Now all that remains is to filter out the AM signal with double the carrier frequency by means of a lowpass filter and the demodulation is complete (see Illustration 147).

Wasting energy: double sideband AM with carrier

The classical process described has serious drawbacks which become apparent when the AM spectrum is examined. On closer examination the following results from Illustrations 144 and 145:

- By far the larger part of the energy of the AM signal is accounted for by the carrier, which does not contain any information. Please note that the electrical energy is proportional to the square of the amplitude, i.e. $\sim \hat{U}^2$.

- The information of the source signal seems to be present twice over, once in the upper sideband and once in the lower sideband. As a result the frequency domain is unnecessarily large.

We shall therefore first try to generate and demodulate a double sideband AM signal without a carrier in the spectrum. This is shown in Illustration 148. Demodulation is now carried out by a method based on the above remarks about multiplication by a sinusoidal carrier of the same frequency. As in Illustration 134 we expect two bands in the range of the sum and difference frequency, i.e. in the double carrier frequency and at 0 Hz.

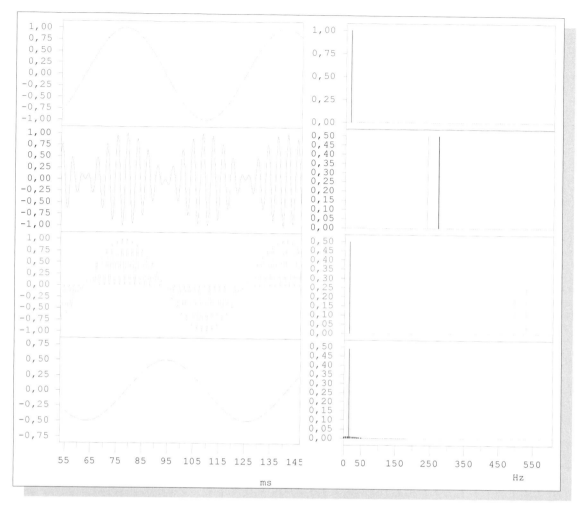

Illustration 148: **_Simple double sideband AM signal without a carrier_**

A sine of 16 Hz (first series) serves as a source signal. This source signal has no offset here and lies symmetrical to the zero axis. The multiplication of this sine by the carrier frequency 256 Hz results in the double side band AM signal without a carrier in the second series.

Demodulation takes place here (at the receiver) by multiplication of this carrier-less AM signal with a sine of 256 Hz. As a result of the multiplication we obtain the sum frequencies and the difference frequencies. The sum frequencies of 256 + 256 -16 Hz and 256 + 256 + 16 (right) are filtered out by the lowpass filter, the difference frequencies of 256 -(256+16) and 256 -(256-16) form the retrieved source signal.

This signal in the third series is practically the sum from the source signal (to be seen left at 0 Hz in the frequency domain) and the AM signal, which lies symmetrical to the double carrier frequency of 512 Hz. If this AM signal is filtered out by means of a lowpass filter, we obtain the retrieved, time displaced source signal in the lower series.

Single sideband modulation without a carrier

The public cable network of *Deutsche Telekom* is probably worth more than 250 bn euros. Laying cables is extremely expensive as roads have to be torn up, cable shafts installed etc. It would be a waste of money if there were no attempt to transmit as much information as possible per unit of time via these cables.

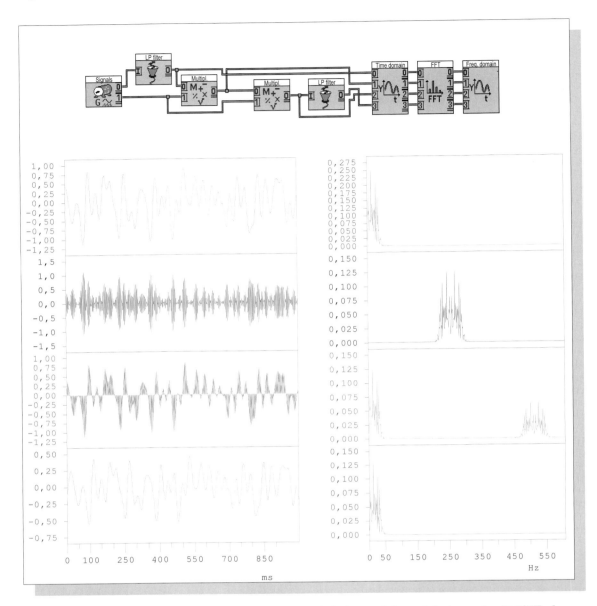

*Illustration 149: **AM signals without a carrier: modulation and demodulation using DASYLab***

The starting point here is a source signal which was generated by means of lowpass filtered noise (it contains all the information about the lowpass). In the second series you see the corresponding double-sideband AM signal without a carrier. The first step in demodulation consists in the multiplication of the latter signal by the "mid frequency" or the carrier of 256 Hz. In the third row there is in the frequency domain, on the left the source signal, and on the right the AM signal with a double mid or carrier frequency.

Finally, after filtering by the lowpass we get the original source signal back (bottom series).

For this reason - just as in the "wireless" sphere - it is extremely important to use frequency bands as effectively as possible.

Double sideband AM modulation is therefore uneconomical because all the information is contained in each of the two side bands. For a long time it has been technically possible to load thousands of telephone single sideband voice channels close together on a coaxial cable to exploit them better.

Chapter 8: Classical modulation procedures

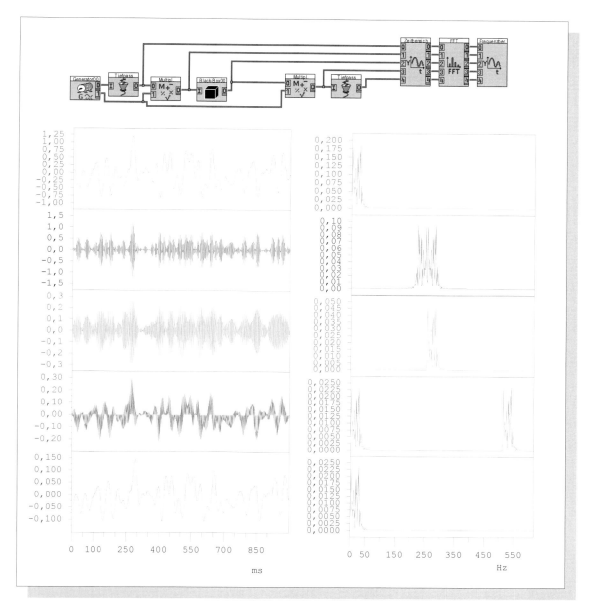

Illustration 150: **Modulation and demodulation in the case of single sideband AM**

First the source signal (1st series) is amplitude-modulated by simple multiplication by the carrier. As the source signal has no offset a double sideband AM signal without a carrier results. By means of a high precision bandpass signal in the "black box" (for the circuit see Illustration 88) the upper sideband - the regular position - is filtered out (3rd series). We thus obtain a one sideband AM signal.

This is multiplied by a carrier of 256 Hz for demodulation. As before we obtain the sum of two signals (see frequency domain 4th series), the source signal and a one sideband AM signal at the double carrier frequency or mid-frequency. Note that both bands have the same amplitude level. If the upper band is filtered with a lowpass filter, we obtain the retrieved source signal in the lower series.

This process will now be checked and precisely analyzed using DASY*Lab*. After a double sideband AM signal without a carrier was generated in Illustration 150 the upper sideband (regular position) is filtered out by means of a high precision bandpass filter. In the third series we thus obtain a single sideband signal.

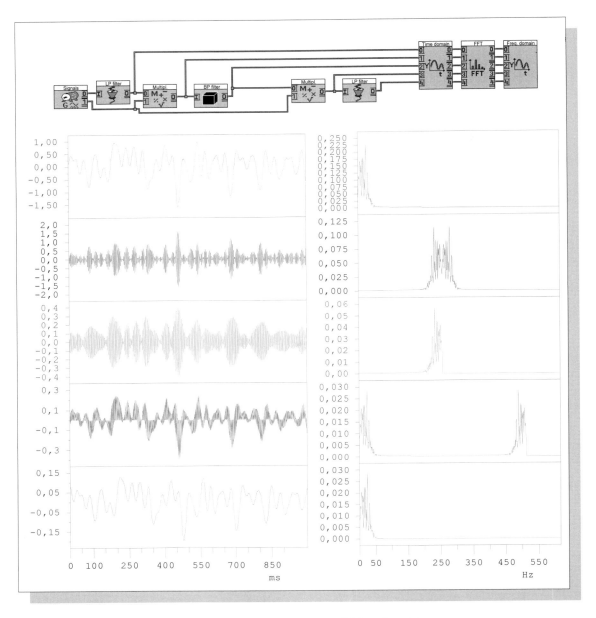

Illustration 151: **Single sideband in lower sideband position**

Unlike Illustration 149 here the lower sideband position is filtered out of the double sideband AM signal. The process which follows is completely identical and, although we have processed a completely different signal than in Illustration 149, we obtain the source signal in this way.

> *The single sideband modulation is the only (analog) modulation process in which the bandwidth is not greater than that of the source signal. The single sideband modulation is thus the most economical analog transmission process if the level of distortion on the transmission path is successfully kept to a minimum.*

The demodulation begins with the multiplication of this signal by a sinusoidal carrier (256 Hz) with which the AM signal was originally generated.

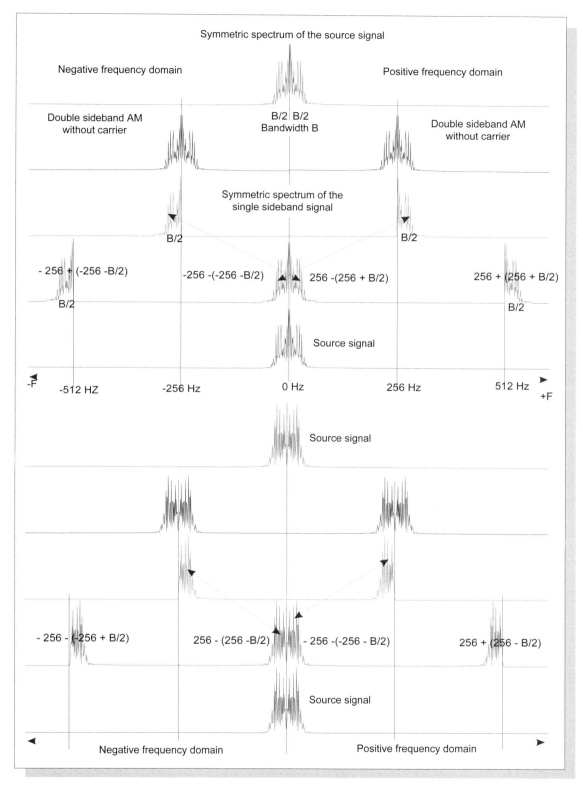

Illustration 152: ***Single sideband modulation: generation and demodulation***

It is only possible to understand the same result - retrieval of the same source signal - using the upper and lower sideband position when other conditions are constant - by knowledge of the Symmetry Principle. In the case of demodulation or reconversion of the lower sideband (lower part of diagram) in reality a sideband of the negative frequency domain is displaced into the positive frequency domain and vice versa.

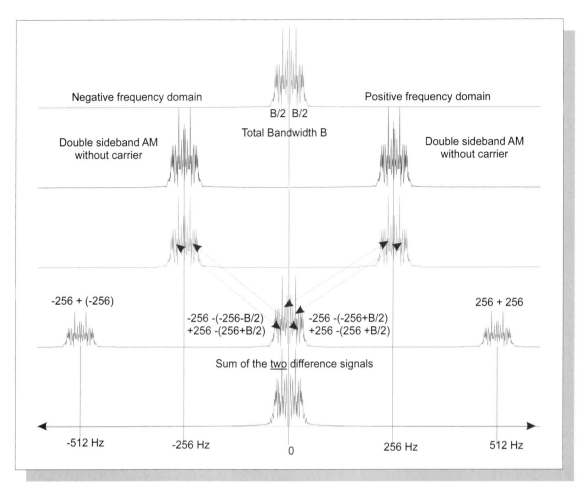

Illustration 153:   **Double sideband AM: modulation and retrieval of the source signal**

In Illustration 149 the spectrum of the source signal had the same level as every sum signal. Here however the retrieved signal has twice the level. This is not surprising in that both single sideband variants from Illustration 149 are simply added. The positive and negative frequency areas each make a contribution to each sideband of the retrieved source signal (bottom).

The spectrum in the fourth series results from the sum and difference formation of this carrier frequency and the single sideband frequency band. The single sideband sum band is twice as high as was originally the case, the difference band is the frequency band of the source signal.

This procedure is carried out again in Illustration 151 but here the lower sideband, the reverse position, is filtered out. Strangely, as our final result we obtain the source signal as a difference signal by proceeding in the same way - multiplication by a carrier of 256 Hz although the difference really ought to lie in the negative frequency domain.

In order to get a further clue, we examine Illustration 149 again, in which a double sideband AM signal (without a carrier) was demodulated. Here we obtain in contrast to the last two cases the spectrum of the source signal with twice the intensity and twice as high as the spectrum of the sum signal. How does this arise?

Chapter 8: Classical modulation procedures

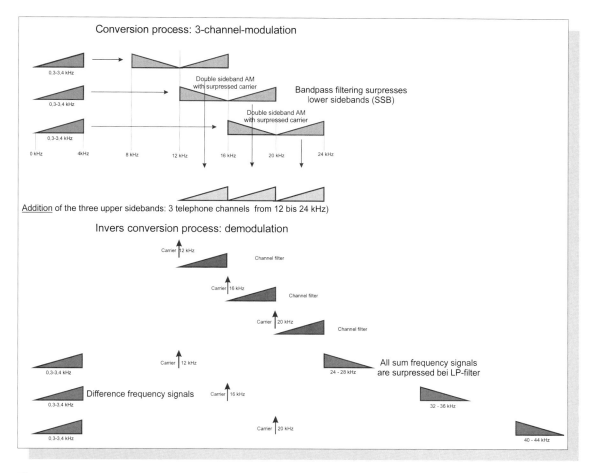

Illustration 154: ***Frequency multiplex illustrated by pregroup formation in telephone channels***

Over 10,000 telephone channels can be transmitted simultaneously by means of a coaxial cable by „traditional" analog communication technology. Small groups and bigger groups consisting of several smaller groups etc are formed by frequency staggering.

The smallest group is the so-called pregroup. Three telephone channels are grouped to form a pregroup which always occupies the range 12 to 24 kHz. This example illustrates the frequency conversion, processing and staggering of the channels.

The mirror-image negative frequency domain which together with the positive frequency domain forms a symmetrical spectrum is not shown here.

The Symmetry Principle provides an explanation. In reality frequency spectra have a positive and a negative domain which have complete mirror-image symmetry. Only if these facts are taken into account can we determine how the sum and difference response functions in the frequency domain as a result of multiplication in the time domain. This is shown in exact detail in Illustration 152. Whereas in Illustration 150 the generation of the source signal frequency band can be easily calculated, for Illustration 151 the result is that the spectrum of the source signal has arisen from the negative frequency domain.

Illustration 152 gives the numerical details. The width of the symmetrical spectrum of the source signal is used as bandwidth B of the source signal. The arrows indicate how the sum and difference frequencies arise.

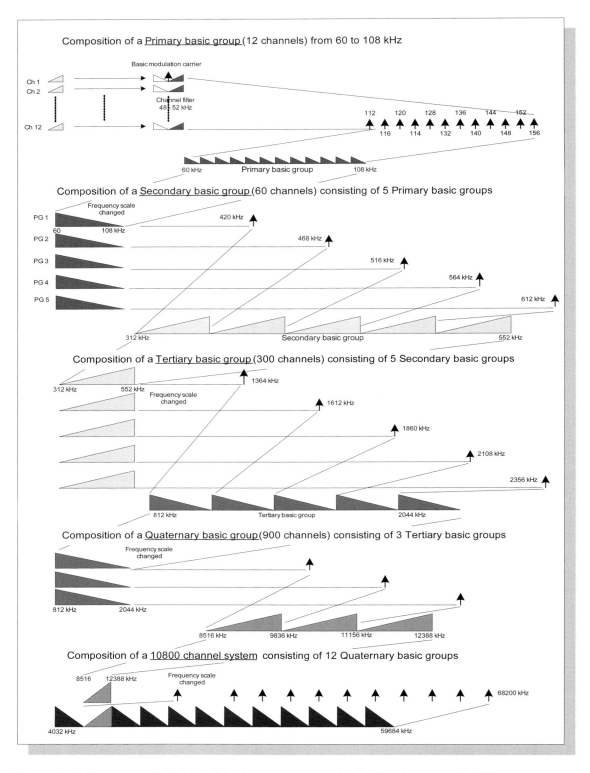

Illustration 155: **Principle of the frequency staggering by forming groups in TF systems**

This representation shows the structure of a V 10800 system. It makes it possible to transmit 10800 telephone channels simultaneously via a coaxial cable and at the same time represents the high point and conclusion of analog transmission technology. Future transmission systems will be founded purely on digital technology. They will exploit the fantastic possibilities of digital signal processing (DSP), which will be discussed in following chapters.

Both in the case of AM and single sideband modulation multiplication is the central signal processing. By means of this frequency bands can be moved around in the spectrum at will.

It should be noted that in the case of multiplication we have a sum and difference frequency, i.e. generally (at least) two frequency bands always arise which are accompanied by a "doubling of information".

> *The interpretation of the frequency bands which arise is only correct if a **symmetrical spectrum with a positive and negative frequency domain** is assumed so that in the difference (formation) the frequency bands can change these domains.*

The real problem with the single sideband modulation SSB are the high precision filters (bandpasses) which are required to be able to filter out one of the immediately adjacent side bands. In analog technology this was (and is) hardly possible using normal filter techniques (R-L-C-filters). Among others filters with mechanical resonance circuits or quartz-crystal filters were and are used.

In the high frequency range, surface wave filters offer excellent solutions. All these filter types exploit acoustic-mechanical physical effects and the fact that mechanical oscillations via the piezo electric effect can be easily transformed into electrical oscillations and vice versa.

Frequency multiplex

Many radio and TV transmitters reach our receiver at the same time via the aerial or cable. Up to now radio and TV have worked in frequency multiplex, i.e. all the stations are staggered in frequency or lie close together in the frequency band. For example, in the medium-wave radio range all the transmitters are double sideband modulated (AM with a carrier), frequency-modulated (FM) in the VHF range.

> *In frequency multiplex systems all the channels are transmitted simultaneously and staggered in frequency.*

The task of the tuner is to filter out the desired station and then to demodulate the filtered signal. As the aerial signal is very weak it has to be additionally amplified.

In the case of the telephony of Deutsche Telekom the telephone channels are staggered in frequency in a particularly economical way by means of the frequency multiplex process. Here single sideband modulation is used the bandwidth of which is exactly as large as that of the source signal. 10,800 telephone calls can be transmitted simultaneously (V-10800 system) via a single coaxial cable - as used as a connection between antenna and tuner.

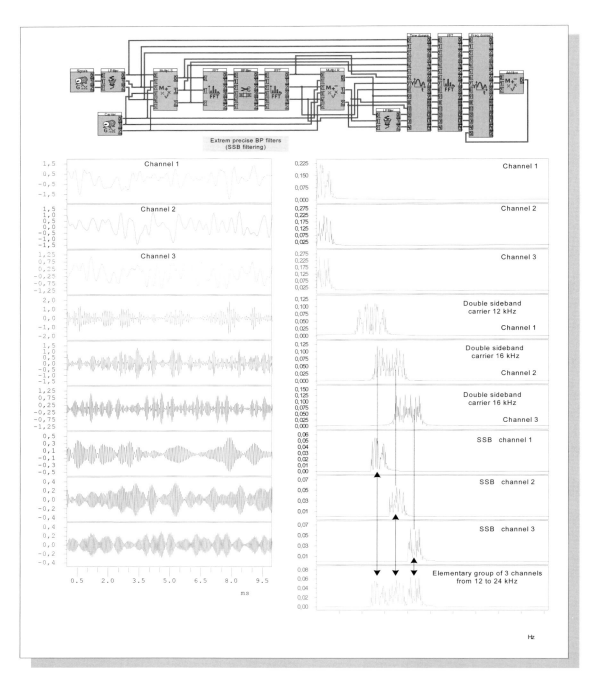

Illustration 156:   *Frequency multiplex: pre-group formation with three telephone channels.*

Here the pre-grouping of three telephone channels in the range 12 to 24 kHz in the time and frequency domain is illustrated realistically for the first time. The frequency multiplex principle is, of course, easiest to follow in the frequency domain, though it is interesting to take a closer look at the corresponding signals in the time domain. The carrier frequency can be easily recognised in the central and lower group of three. It is different and characterises the curve of the signal in the time domain although here we have a double sideband operation without a carrier. Above you see the structure of the circuit using DASYLab. Try to construct the system yourself, assembling it piece by piece and examining the curve of signals in the time and frequency domain.

A hint - in order to get back from the frequency to the time domain in the IFFT module you must select the setting "FOURIER synthesis". Finally, you should put together a signal composed of sinusoidal oscillations in the time domain.

Chapter 8: Classical modulation procedures

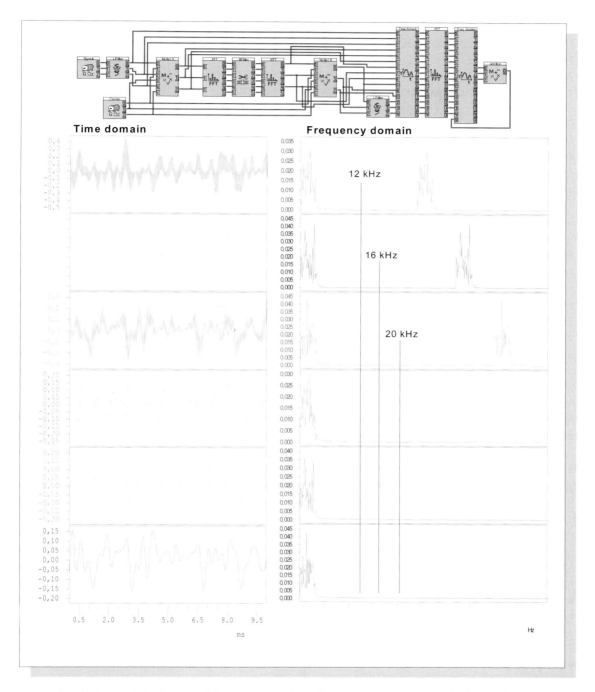

Illustration 157: **Simulation of the re-conversion of a pre-group in three telephone channels**

This realistic simulation using DASYLab shows the signal processing in the lower series of the block diagram. Apart from confirmation of the representation in Illustration 154 you also see (bottom) roughly the curve of the source signal in the time domain of the two upper groups of three.

This carrier frequency technique works on the following principle:

- 12 telephone channels with 300 - 3400 Hz bandwidth are assembled to form a base group, and five primary base groups to form a secondary base group with 60 channels. This is illustrated clearly in Illustration 155.

- According to this principle several secondary groups are combined to form a tertiary group and several tertiary groups to form a quaternary group, until altogether 10,800 channels are combined to form a bundle. This is also shown in Illustration 155.

- It used to be the case that for technical reasons relating to filters first pre-groups with three channels were formed, then four of these were combined to form a primary basic group.

In order to analyse and explain exactly the frequency multiplex principle for modulation and demodulation in single sideband modulation the pre-group formation shown in Illustration 154 will be first formed and then "demodulated" into three channels or converted by means of a DASY*Lab* simulation (Illustrations 156 and 157).

> Note:
> In these Illustrations high precision bandpass filters in the form of purely digital (computer-based) filters are used. The signal is transformed via an FT into the frequency domain where all the unwanted frequency data are set at zero. After that a re-transformation IFT into the time domain is carried out.

Mixing

As can be seen in the formation of pre-groups or basic primary groups the highest demands are made on filter technology in the case of single sideband modulation. This was always a problem in the history of radio technology, that is to filter out a station accurately from the frequency band of the closely packed stations.

In actual fact *adjustable* filters would be necessary. They cannot, however, - also for theoretical reasons - be realised with a constant bandwidth and quality. This is only successful in certain frequency ranges with filters (bandpass filters) given a constant conducting state region.

In traditional radio and television technology - everything will be different with digital radio and TV technology - a trick is used which leads indirectly to an adjustable filter. This works in principle as follows:

- By multiplication (of the whole frequency band by all transmitters) by an adjustable oscillator frequency (carrier) the frequency band (difference formation) can be converted to a lower range at will (intermediate frequency range)

- At the same time a second complete frequency band " at the very top" arises by sum formation which, however, is disregarded.

- In this intermediate frequency band (IF domain) there is a relatively high quality bandpass filter installed. This is usually a quartz crystal or ceramic filter. With a certain fine adjustment of the whole intermediate frequency band the desired transmitter lies precisely in the conducting state region of the IF filter and is thus selected. It is then demodulated and amplified.

- This controllable frequency conversion to an intermediate frequency range is called "mixing". It is, of course, also multiplication by an (adjustable) carrier, e.g. nothing but the conversion process of a modulated signal into a different frequency range (IF domain) and has, therefore, been given a name in radio and television technology.

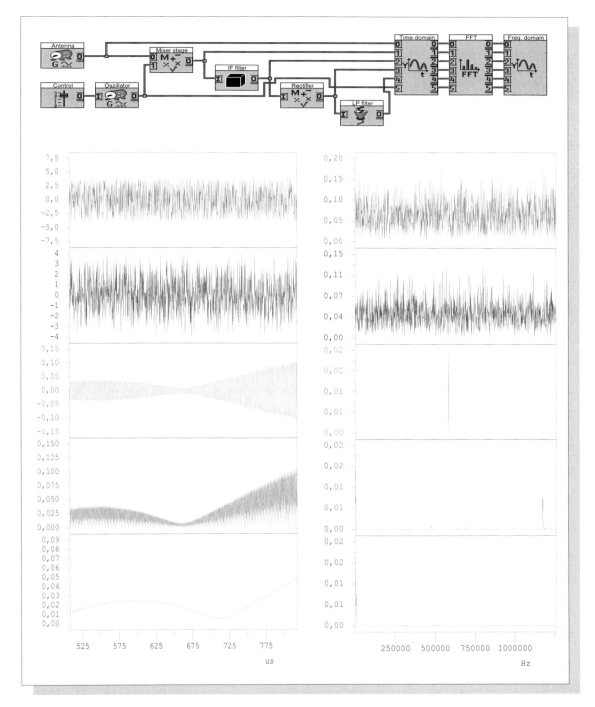

Illustration 158: **Simulation of an AM tuner for the medium wave range**

For the sake of simplicity a noise signal is used as an aerial signal, which contains all the frequencies, that is in the medium wave range from 300 to 3000 kHz (according to CCIR). Using a hand regulator the oscillator frequency can be selected in such a way that the "desired station" lies exactly in the conducting state region of the IF filter (blackbox). This IF filter is the ingenious bandpass filter from Illustration 156. The IF filter lies at 465 kHz and a wider filter was selected for the purposes of clearer illustration. The lowpass filter also exceeds the actual LF width of an AM receiver.

Surprisingly, the high frequency range can also be simulated by means of DASYLab. The axes are completely correctly scaled. In this domain real signals can be inputted and outputted by means of very expensive A/D and D/A cards. .Real time operations - as with a normal radio receiver - can at the moment hardly be realised.

Illustration 158 shows the simulation of a complete AM tuner as it is used in the medium wave range. For the sake of simplicity a noise signal is used here as an aerial signal which contains all the frequencies.

Frequency modulation FM

Apart from amplitude, frequency and phase also offer the possibility of having meaning superimposed on them. This means changing frequency or phase "in the rhythm" of the source signal.

Frequency modulation FM and phase modulation PM hardly differ at first sight as with every (continuous) phase displacement the "instantaneous frequency" is changed simultaneously. If the zero crossing of the sinusoidal carrier is displaced, the period length T* of the instantaneous frequency changes.

In the case of the FM things are easy at first. In Chapter 7 "Linear and Non-linear Processes" the VCO (Voltage Controller Oscillator) was already dealt with in connection with telemetrics. This VCO is simply an FM modulator.

As almost always in microelectronics there are analog and digital VCOs and various representatives within these two categories. Here is a brief survey:

Analog VCOs for the range 0 to 20 MH

These are ICs with internal voltage - controlled source of current. An (external) capacitor is charged and subsequently discharged linearly via an adjustable constant current generator whereby a near-periodic triangular voltage arises. The higher the current is set, the more rapidly the capacitor charges and discharges. The frequency of this triangular voltage is proportional to the current of the constant current generator. This triangular voltage is distorted by a transistor or a diode circuit in such a non-linear way that a sine-like voltage arises. Given the appropriate function generators this can be seen in the output signal: the sine has a small bend in the upper and lower areas. In the final analysis the linear curve of the triangular voltage is rounded off by a non-linear distortion.

Analog VCOs for the range 300 kHz to 200 MHz

Capacity diodes have the property of changing their capacity slightly depending on the voltage applied. This property is exploited in high frequency analog VCOs. A capacity diode lying parallel to an LC resonant circuit is activated by an alternating current, whereby the diode-capacity changes in the rhythm of the source signal. As a result the resonance frequency/ natural frequency of the resonant circuit changes minimally. To put it clearly: we have an FM signal.

Digital VCOs and FM modulators with DASYLab

DASY*Lab* has two modules with which FM signals can be generated directly. These signals can be outputted by means of the D/A converter of a multi-function card. With traditional cards it is only possible to generate FM signals in the LF domain in a real sense. We use these modules only to simulate, to establish the essential characteristics of FM signals.

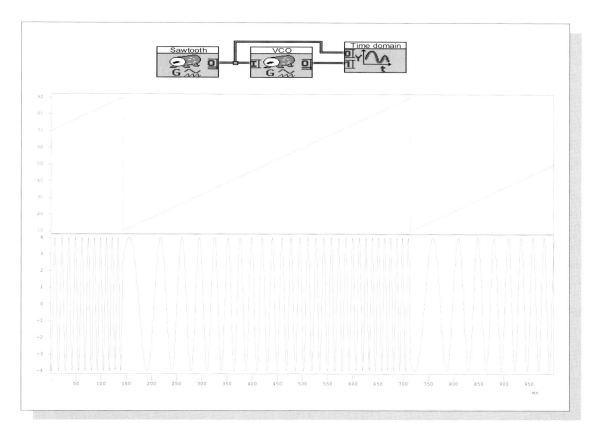

Illustration 159: **_DASYLab VCO as a frequency modulator_**

Using the example of a periodic sawtooth signal as a source signal the function of the generator module as a VCO or FM generator is shown here. In the experiment corresponding to this diagram the "instantaneous frequency" can be measured by means of the cursor. The result is: the level of the instantaneous frequency corresponds exactly to the instantaneous value of the sawtooth. According to this the instantaneous frequency (left) is 100 Hz at t = 0 ms and at the highest value of the sawtooth 250 Hz etc.

As this example shows it is very easy to set the desired frequency range. The source signal values -here from 0 to 250 - do not physically represent a voltage in V; rather these are digital signals, that is strings of numbers which only when they are transmitted to the D/A module - output via a multi-function card - have a realistic range of values.

The *generator module* is capable of generating an FM signal, an AM signal or a mixed signal of these two. A further possibility of generating an FM signal can be carried out by means of the module "formula interpreter", but requires mathematical skills (mathematical model of an FM signal).

Programmable function generators: digital VCO for the range 0 to 50 MHz:
Using a signal function chosen via the menu, a string of numbers - i.e. a digital signal - which corresponds to an FM signal is generated in a digital function generator. A subsequent D/A converter produces an analog FM signal from this. In the final analysis a program generates an FM string of figures via the signal processor.

The highest FM frequency is primarily determined by the maximum clock pulse frequency of the D/A converter. It should be noted that at least 20 "support positions" (figures) per period are necessary to generate an FM signal with a sine-like carrier. In the case of a maximum carrier frequency of 50 MHz the clock pulse frequency of the D/A converter would have to be 1 GHz! This value can only be achieved at present with 8 bit D/A converters. The FM function described is only one of many in the case of digital function generators. Controlled by a program they can generate any signal.

In Illustration 159 using the example of a periodic sawtooth signal the simple relationship in the DASY*Lab* generator module between the instantaneous value of the source signal and the instantaneous frequency of the FM signal is shown. From Illustration 160 onwards only sinusoidal source signals are used in order to obtain results which can be clearly interpreted. Finally, in Illustration 165 a band limited, "random" source signal is selected to show the curve of a realistic FM spectrum.

In the case of a sinusoidal source signal an FM spectrum arises which is perfectly symmetrical to a mid-frequency. This mid-frequency is set precisely using DASY*Lab* by a DC voltage (offset) superimposed on the source signal (see below).

It is not possible to see any rule for the curve of the amplitude of the spectrum. We see here in the FM an example of a typical non-linear process in the case of which predictions of the frequency spectrum are possible only to a very limited extent. In the case of FM, however, we are familiar with the maths (Bessel functions) by means of which it is possible to calculate or estimate the spectrum of a sinusoidal source voltage - and thus for all band limited signals (FOURIER Principle).

Real experiments with FM signals in the VHF range and the evaluation of Illustrations 159 to 165 nevertheless give a few interesting pointers:

- Unlike low frequency VCO signals (as in Illustration 159), real FM signals with a high frequency mid-frequency in the time domain all look almost alike at first glance, that is (apparently) all sinusoidal. See in this connection also Illustration 160.

   Note: In the case of an FM radio signal in the VHF range around 100 MHz the frequency only changes in the order of 0.1%, i.e one part per thousand.

- In Illustration 159 it is noticeable that the FM or VCO signal changes its instantaneous frequency in the rhythm of the instantaneous value of the source signal. What generally holds is that the higher the instantaneous value of the source signal the higher the instantaneous frequency of the FM signal. In the case of DASY*Lab* the instantaneous value of the control signal with which the FM generator is activated corresponds exactly in numerical terms to the "instantaneous frequency" of the FM or VCO signal.

- In order to generate a typical FM signal the actual source signal has an offset superimposed (e.g.1000) the value of which then corresponds exactly to the mid-frequency (1000 Hz). See in this connection the block diagrams in Illustrations 161 to 164.

- The information on the frequency of the sinusoidal source signal results from the intervals between the symmetrical lines of the FM spectrum. The FM is a non-linear process. As described in chapter 7 the integer multiples of the frequency typically appear. Here too!

Illustration 160: **_FM-signals with different frequency swings_**

In the top series the FM signal has a hardly discernible frequency swing and in the bottom series it has a very large frequency swing. The frequency swing is equivalent to the largest deviation of the instantaneous frequency of the FM signal from the mid-frequency (here 500 Hz). The double frequency swing does not, however, give the exact bandwidth of the FM signal because the instantaneous frequency is an imprecise concept. The greater the frequency swing, the greater the bandwidth of the FM signal.

The source signal is sinusoidal in all cases here and is exactly equivalent to the source signals in Illustration 161.

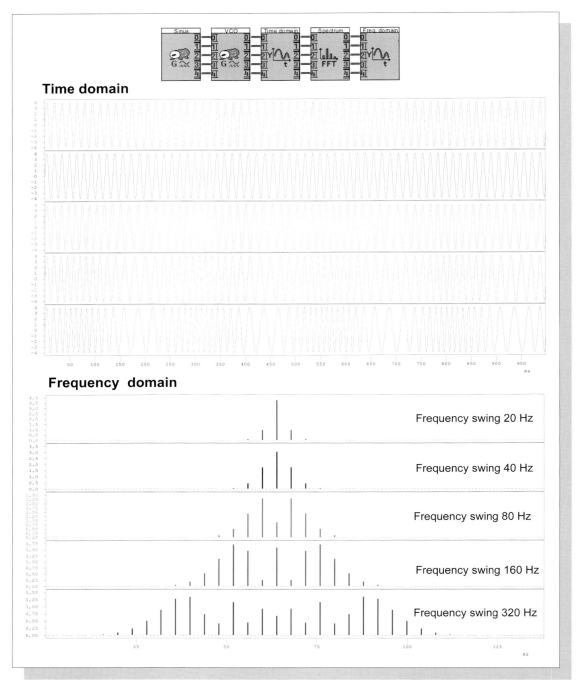

Illustration 161: ***Influence of the frequency swing on the FM spectrum***

In the DASYLab module "Generator" (option frequency modulation) there is a simple relationship between the time domains of the source signal and the FM signal. In the block diagram (top) you see the module "Offset" which adds a constant - here 1000 - to the source signal. Thus the mid-frequency is set at 1000 Hz. The instantaneous frequency of the FM signal fluctuates around this mid-frequency and at a maximum level around the amplitude of the sinusoidal source signal.

This maximum change is called the frequency swing. In the case of an amplitude of 20 V (top series) the frequency swing is 20 Hz, at the bottom, however, it is 320 Hz with an amplitude of 320 V. The bandwidth of the FM signals is always greater than twice the frequency swing.

The form of the envelope of the FM spectrum is subject to complicated laws (Bessel functions).

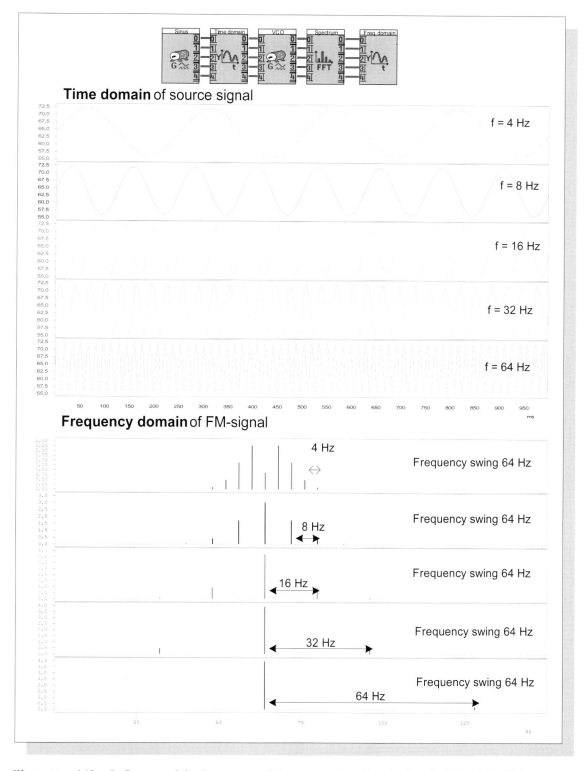

Illustration 162: **Influence of the frequency of the source signal on the bandwidth of the FM spectrum**

Apart from the frequency swing Δf_C the frequency of the source signal f_S also has an influence on the curve and width of the FM spectrum. This is demonstrated here for various frequencies of the sinusoidal source signal with a constant frequency swing $\Delta f_C = 80$ Hz.

All that remains now is to examine whether the choice of mid-frequency also influences the curve and bandwidth of the FM spectrum (Illustration 163).

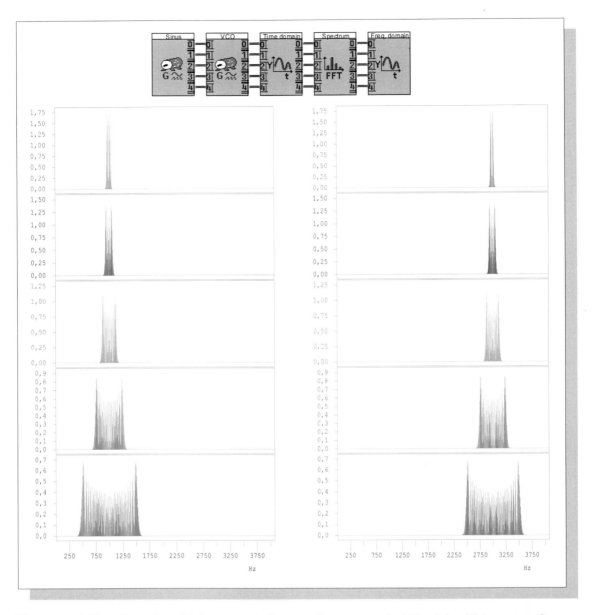

Illustration 163: **Does the mid-frequency influences the curve and width of the FM spectrum?**

We see here the same experimental arrangement and on the left hand side the same FM spectrum with the same mid-frequency of 1000Hz as in Illustration 160. On the right hand side the mid-frequency was set at 3000 Hz and the offset was set at 3000. All the other quantities such as the frequency swing and signal frequency remained unchanged. Result: there is no influence at all of the mid-frequency on the curve and width of the FM spectrum to be recorded. Thus the curve and width of the FM spectrum depend entirely on the frequency swing Δf_C and the source signal frequency f_S. It should now be possible to find a simple formula which describes the bandwidth of the FM signal by means of these two quantities. This is shown in Illustration 164.

- The better a change in the time domain in the instantaneous frequency can be seen, the wider the total spectrum of the FM signal (see Illustrations 161 and 162).

- A rule for the curve of the amplitude of the FM spectrum is not easy to see.

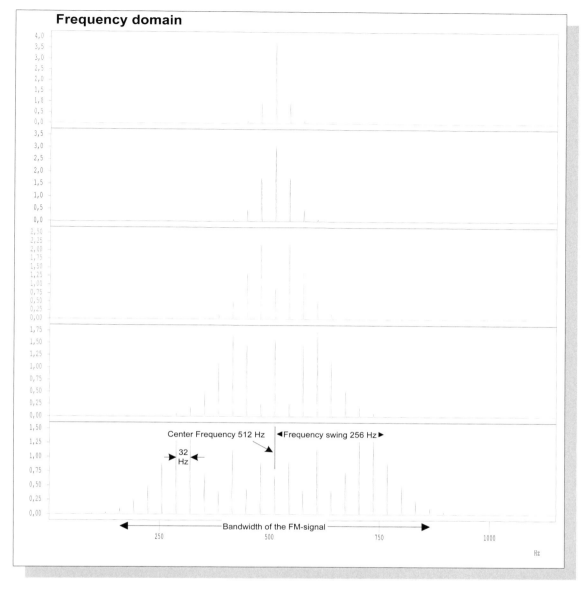

Illustration 164: ***Formula for estimating the bandwidth of an FM signal***

Here are once again the spectra of Illustrations 161 and 162. The interval between the lines corresponds to the frequency of the sinusoidal source signal (here 32 Hz). In the bottommost representation the quantities for the determination of the "essential FM bandwidth. A frequency swing of 320 - see Illustration 160 bottom left - results in one half of a spectrum of (320 + 2∗32)Hz. The total bandwidth results accordingly at 2(320 + 2∗32) Hz. For all other cases the following holds: $B_{FM} = 2(\Delta f_C + 2f_S)$.

The investigation of the influence of the frequency swing Δf_C, Signal f_S and mid-frequency f_C on the bandwidth B of the FM signal in the Illustrations 161 and 162 gives an unambiguous result. Only the frequency swing Δf_C and signal frequency f_S determine the bandwidth B_{FM} of the FM signal.

Illustration 164 leads finally to the formula for the bandwidth of an FM signal.

$$B_{FM} = 2(\Delta f_C + 2f_S)$$

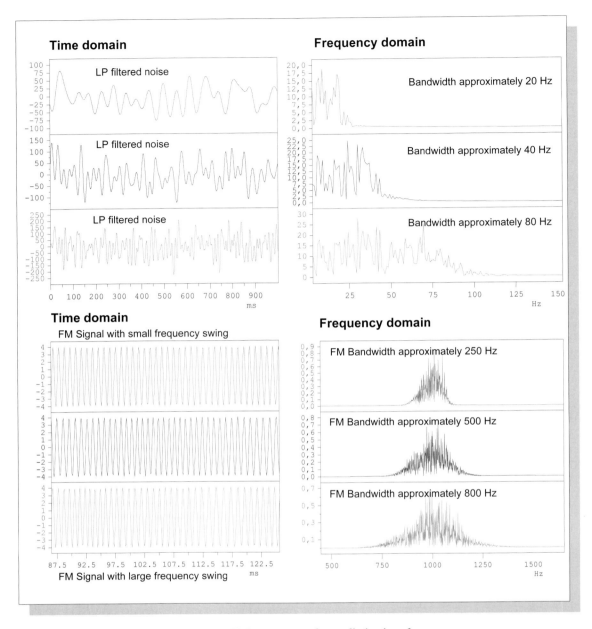

Illustration 165: ***FM spectrum of a realistic signal***

Up to now we have largely used a sinusoidal source signal in order to obtain straightforward usable results. A realistic signal has a stochastic component, that is, its future curve is not exactly predictable. As a result the spectrum has a certain irregularity reminiscent of a noise signal. In the time domain (left) this is not to be seen.

Here filtered noise signals of various bandwidths - see above - were frequency modulated. It can be seen clearly that the bandwidth of the FM signal depends on the bandwidth of the source signal. Although the same noise signal was lowpass filtered the real frequency swing depends on the bandwidth of the source signal. You see the reason for this top left in the time domain: the maximum instantaneous values increase with the bandwidth of source signal because the energy also increases with the bandwidth. The greater these values, the greater the frequency swing.

Unlike the AM and single sideband modulation, the FM bandwidth is at least twenty times the bandwidth of the source signal. The FM behaves extravagantly with the frequency band. If in spite of this the FM is used, for example in the VHF range, this modulation process must offer considerable advantages compared with the AM and single sideband modulation.

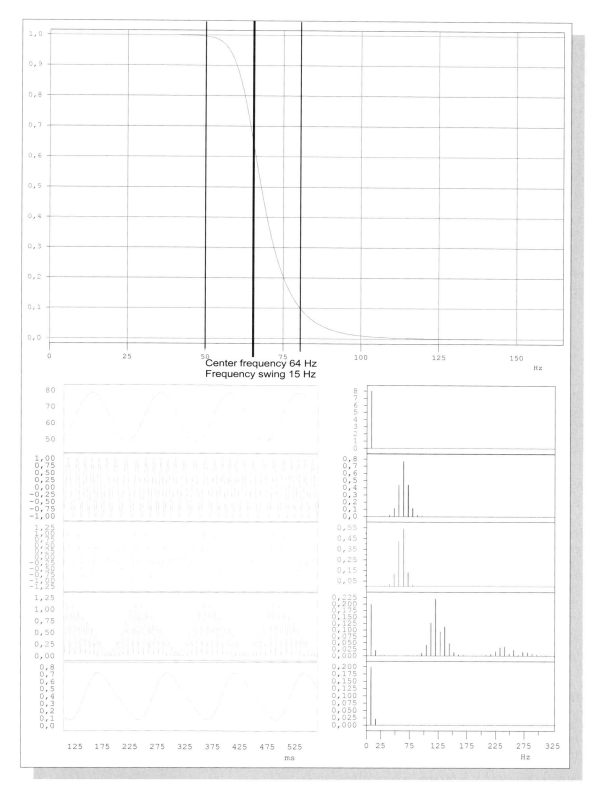

Illustration 166: **Demodulation of an FM signal in a filter edge**

Here the transition of a lowpass filter from the conducting state region to the blocking state region is used as a sensitive frequency voltage converter (f/U converter) for the demodulation of the FM signal. Because of the non-linear filter characteristic the retrieved signal is heavily distorted (the frequency swing is too great).

Demodulation of FM-signals

The information of the source signal is to a certain extent "frequency coded" in the case of FM. How can we get at this information in the receiver?

In the VHF range - that is in the range around 100 MHz - the frequency swing is less than 0.1 % (1 per thousand). On the screen of a rapid oscilloscope the signal looks like a pure sinusoidal carrier signal. It is the spectrum that betrays the FM signal. A *highly sensitive* frequency voltage converter (f/U converter, see example Telemetrics Illustration 121) is necessary, which registers the smallest change in frequency as a "voltage deviation".

Physics seems to be relevant here. Filters are the first hint. Filters have an edge at the boundary between the conducting state region and the blocking state region. Let us assume that this edge is rather steep. This would mean that if the bandwidth of the FM signal lay completely in this boundary domain a small change in frequency would decide whether the signal would be allowed to pass - thus have a larger amplitude at the output - or be largely blocked - i.e have a very small amplitude at the output of the filter.

> *The filter edge is thus a domain which reacts very sensitively to frequency changes*

Illustration 166 presents these facts in graphic form and is more expressive than a thousand words. In principle every filter edge represents a non-linear characteristic. Thus, unwanted non-linear distortions frequently occur at filter edges which reproduce the original signal after demodulation in a distorted way. Since the beginning of VHF radio in the fifties there was for decades nothing better available. For a long time development consisted in improving bit by bit a fundamentally unsuitable process. People attempted to linearise the filter edge by means of tricks in the technical design of circuits.

A new approach to this problem was needed. We enter a highly topical field of non-linear signal processing with this approach: the feedback of a part of the output signal on to the input of the system. In other words: *feedback systems or feedback control technology*.

Closed loop systems are an invention of nature. We find them everywhere, but rarely recognise them as such. As we shall deal several times in a later chapter with the feedback principle, we only mention here what is necessary in the context of the demodulation of FM signals.

The phase locked loop PLL

One of the main problems of signal transmission technology is the *synchronisation (in time) of transmitter and receiver*. If for example the picture on a defective television set begins to rotate so that a person's legs dangle helplessly from above while their head appears on the ground this is due to a synchronisation failure. If you turn the right knob the picture clicks into place again - the transmitter and receiver are synchronised again.

The television receiver has the task of "fishing out" the right clock pulse so that the reproduction of the picture is in sync with the transmitter. Nothing can do this better than the PLL - *the phase-locked loop*.

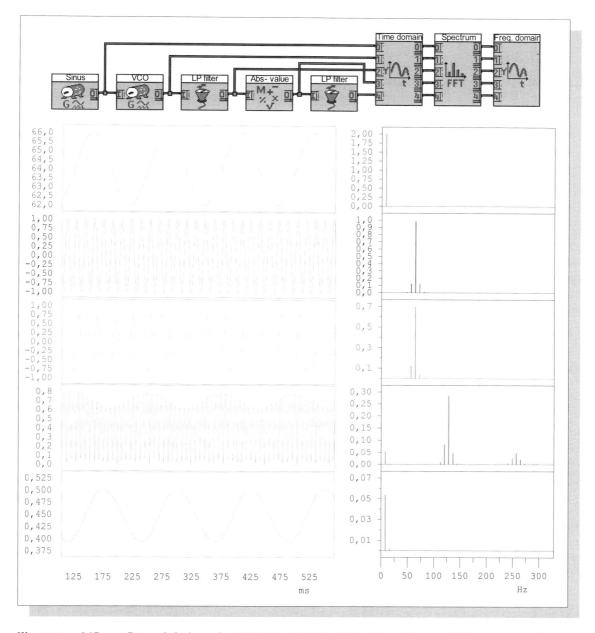

Illustration 167: **Demodulation of an FM signal at the filter edge with a small frequency swing.**

If this is compared carefully with Illustration 148 the smaller frequency swing can be recognised in the FM signal (top series), the FM signal (mirrored) at the filter edge and in the rectified signal. As a result the FM signal is mirrored only in the more linear part of the filter edge. The retrieved source signal has a much smaller amplitude (ca 0.05 V) but is only slightly distorted in a non-linear way. There is only a very small 2nd harmonic to be seen in the spectrum bottom right.

The PLL has been assembled from components with which we are familiar: multiplier, lowpass filter and VCO (voltage-controlled oscillator)). It would, however, probably be too difficult for you to describe the function of PLLs from Illustration 168 from scratch. There are two reasons for this. First, the multiplier is used here for a "special purpose", second, there is feedback. It is generally difficult to predict the behaviour of locked-loop systems because all locked loop systems are non-linear.

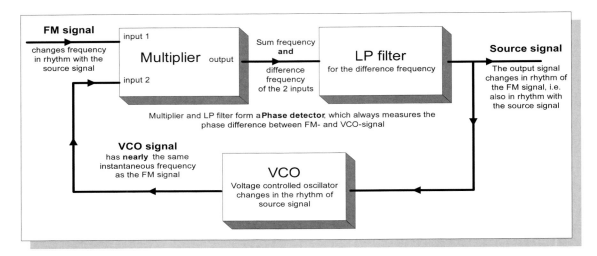

Illustration 168:   ***Simplified block diagram of an analog PLL***

Let us assume that the VCO signal had instantaneously (nearly) the same frequency as the received FM signal. The sum and difference frequency $f_{FM} + f_{VCO}$ and $f_{FM} - f_{VCO}$ appears at the output as the result of the multiplication. The lowpass filter filters out the sum frequency.

What frequency does the difference signal have as both frequencies f_{FM} and f_{VCO} are almost equally great? At the output of the lowpass filter a "fluctuating direct current" appears which ought to have exactly the level necessary to readjust the VCO to exactly the instantaneous frequency of the FM signal: the control signal is therefore the LF signal.

If you haven't quite understood - a diagram speaks louder than words (see Illustrations 169 -171).

In Illustration 168 the multiplier and the lowpass filter are intended to form a "phase detector". A phase detector compares two signals (of equal frequency) to see whether they are in phase i.e. completely in synchrony. This can for example be achieved by a comparison of the zero crossings of both signals. If the zero crossings are continuously displaced towards each other the two signals only have roughly the same frequency. (see Illustration 169 in this connection).

Let us begin in a straightforward way and leave out the feedback. In Illustration 169 the VCO is to be set at a frequency which corresponds approximately to the mid-frequency of the FM signal. The FM signal varies its frequency in the rhythm of the upper (sinusoidal) LF. In the diagram the zero crossings are compared. The sum and difference frequency appears at the output of the multiplier. The lowpass filter filters the sum frequency out and the "residual voltage" which is intended to control the VCO corresponds more or less to the LF signal (see Illustration 169). This voltage is greatest where the phase difference of zero crossings is greatest if the delay caused by the lowpass filter is included.

We have cheated a little. This is shown by Illustration 170, where the control voltage or the retrieved LF signal look very different from the original. The frequency domain shows what actually happens. The multiplication of both signals in the time domain results in a convolution in the frequency domain (sum and difference band). After the lowpass filtering not the original LF signal is left over, but an FM signal which lies on the "mid-frequency" $f_M = 0$.

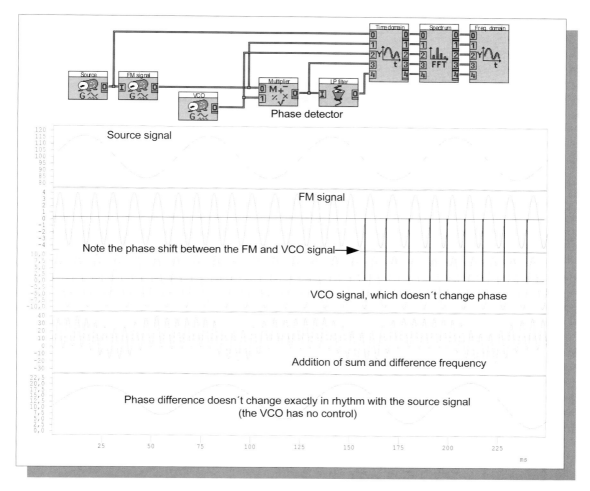

Illustration 169: **Basic mode of functioning of the phase detector**

In the second and third series the phase displacement between the FM signal and the VCO signal set at a constant frequency as the difference between the zero crossings is to be seen. This phase difference contains the difference frequency which appears together with the sum frequency at the output of the multiplier. The lowpass filters the sum frequency away.

The output voltage at the lowpass filter is the control voltage for the VCO and changes in the rhythm of the LF signal (top) because the instantaneous frequency of the FM signal changes in this rhythm.

This control voltage does not correspond to the original FM signal because the VCO is set at a fixed frequency and is not constantly being adjusted. If this control system is perfectly adjusted, a control voltage appears at the output of the lowpass filter which corresponds to the original LF signal.

The PLL attempts continually to synchronise the transmitter (LF signal) and the receiver (VCO signal). As the FM signal changes in the rhythm of the LF signal (source signal) the control voltage for the VCO must change in exactly this rhythm too.

The sensitiveness of this locked loop system is reflected in the extremely sensitive adjustment of the control voltage. This is shown in Illustration 171. The output voltage of the lowpass is amplified to the right level and provided with the correct offset. A tiny change in these quantities is sufficient to make the feedback system instable.

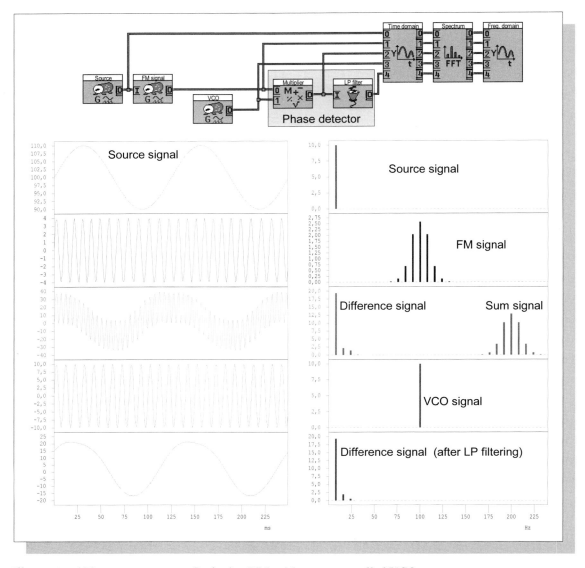

Illustration 170: **_Defective PLL without a controlled VCO_**

What happens if in the case of the PLL the feedback system is interrupted and the VCO is driven by a constant "carrier frequency"? This analysis is very simple. We are dealing with a simple multiplication in the time domain. In the frequency domain on the other hand the sum and difference signal is formed. If the sum signal is filtered out by the lowpass signal this does not leave the original source signal; rather a sideband of the "mid-frequency" f = 0 Hz appears in the frequency domain.

It is only the feedback system that brings back the source signal as it forces the VCO to synchronise itself with the instantaneous frequency of the FM signal. As this FM instantaneous frequency changes in the rhythm of the source signal this must also apply to the control voltage (the difference voltage at the output of the lowpass filter).

The mid-frequency is set with the offset and the frequency swing of the VCO with the amplifier. The module "delay" is prescribed in DASY*Lab* for feedback systems. The PLL only clicks into place after a certain time. A change in the second decimal place in the constant C in the two modules is sufficient to stop the PLL clicking into place. We see how sensitively the feedback system reacts to the smallest changes in the VCO signal.

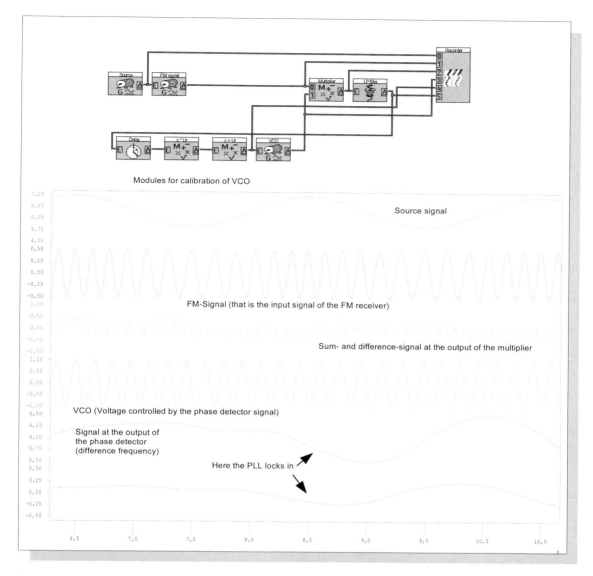

Illustration 171: **The PLL as a demodulator for FM signals**

In the top block diagram you will find several more components than in the default block diagram of Illustration 168. At the top left the FM signal which is to be demodulated is generated. The phase detector - consisting of the multiplier and the lowpass filter - shows the phase difference between the FM and the VCO signal. This output signal is processed for the adjustment of the VCO (setting the mid-frequency of the VCO by offset ($C + U_R$) and the frequency swing ($C * U_R$). The feedback system operates with a certain delay. You will see this by comparing the FM and the VCO signal. Feedback is only possible with DASYLab by using the module "delay".

If the FM signal were temporarily to show a higher frequency swing the PLL might also get out of step. The transmitter frequency spacing and the frequency swing are therefore fixed in the VHF range. The mid-frequency distance is 300 kHz, the maximum frequency swing is 75 kHz, the highest LF frequency is roughly 15 kHz ("HiFi"). Our bandwidth formula thus results in a transmitter bandwidth of roughly

$$B_{VHF} = 2 \, (\, 75 \text{ kHz} + 2*15 \text{ kHz} \,) = 210 \text{ kHz}$$

> *The importance of the PLL goes far beyond that of an FM demodulator. It is generally used where it is important to synchronise a transmitter and receiver as it can extrapolate the clock pulse from the received signal.*

Phase modulation

There is a clear relationship between the frequency and the phase of a sinusoidal signal which will be explained briefly once again here. Imagine two sinusoidal signals, one of 10 Hz, the other of 100 Hz.

Sine$_{10Hz}$ has a period length of

$$T_{10Hz} = 1/f = 1/10 = 0.1 \text{ s,}$$

Sine$_{100Hz}$ has a period length of

$$T_{100Hz} = 1/f = 1/100 = 0.01 \text{ s.}$$

Now imagine the rotating pointer in Illustration 24. This pointer rotates ten times faster at 100 Hz than at 10 Hz. This means also that the full circle (the phase) of 2π (360°) is completed ten times faster at 100 Hz than at 10 Hz. Thus the following holds.

> *The faster the phase changes (instantaneously), the higher the (instantaneous) frequency of a sinusoidal signal.*

In the case of *differentiation* as signal processing we have always required this kind of formulation:

> *The faster the input signal changes (instantaneously) the greater is the differentiated signal (instantaneously).*

On the other hand from Illustration 24 and page 39 we are familiar with the connection between angular velocity, phase and frequency:

$$\omega = \varphi/t = 2\pi/T = 2\pi f$$

This relationship is only absolutely correct if the frequency does not change and the pointer rotates with a constant angular velocity. In the case of the FM and PM this does not apply. Thus, precisely speaking, the following holds

$$\omega = d\varphi/dt$$

which simply - see above - says that the faster instantaneously the phase changes the higher the instantaneous frequency of a sinusoidal signal. Or in mathematical terms: the differentiation of the angle in time results in the angular velocity.

For the frequency therefore $2\pi f = d\varphi/dt$ and

$$f = 1/2\pi * (d\varphi/dt) = 0.16 * (d\varphi/dt) \text{ applies.}$$

The technical realisation of the phase modulation is not quite straightforward. But the above insights provide a simple method even if only a module for frequency modulation is available:

Chapter 8: Classical modulation procedures

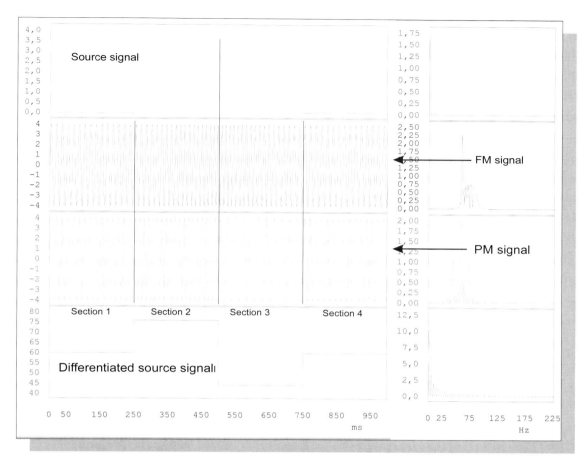

Illustration 172: **Comparison of PM and FM**

A signal is chosen as a source signal which has four segments: in segment 1 nothing changes; in segment 2 it rises linearly; in segment 3 it falls again linearly in the same way and in segment 4 finally it does not change (top series). Each time segment is 250 ms in length.

In the second series you see the corresponding FM signal. Thus in segment 2 the frequency increases linearly etc.

In the this series you see the PM signal. For this - see text for reasoning - the source signal was differentiated (bottom series) and connected with the module "FM generator". In the case of the PM signal the instantaneous frequency jumps altogether three times at the boundaries of the segments by a value, which depends on the gradient of the source signal.

The PM thus reacts much more critically to changes in the source signal, which actually follows from the insight: the faster the phase changes instantaneously, the higher the instantaneous frequency of a sinusoidal signal. This is also shown by the frequency domain on the right. The spectrum of the FM signal is wider than that of the FM signal. It is as if a source signal had been frequency-modulated with a wider frequency band - the differentiated source signal bottom - without at the same time increasing the processing speed of information.

As phase modulation is involved the phase angle φ(t) must change in the rhythm of the source signal.

If φ(t) is now differentiated, subsequently multiplied by 0.16 and then directed to the module "frequency modulation" we obtain a phase modulated signal $u_{PM}(t)$.

This is shown in Illustration 172. The PM was not successful compared with the FM in analog transmission technology. One of the reasons is the technically more difficult modulation. It has only achieved its real importance with the advent of modern digital modulation processes.

Immunity to interference of modulation processes

The safe transmission of information is the criterion of optimisation in information and transmission technology. Even in transmission systems (cables, radio relay, mobile telephony, satellite telephony) hampered by noise or other interference the safe transmission of information must be guaranteed under certain defined conditions - for example in a moving car. The graphic concept of „immunity to interference") derives from Lange's definition (F-H Lange: Störfestigkeit in der Nachrichten- und Messtechnik, VEB-Verlag, Berlin 1983).

> *Note:*
> A general distinction is made between additive and multiplicative interference. Additive interference can be "treated" with relative ease when it affects linear systems. No additional interference frequencies arise in the components of the system.
>
> The situation is different in the case of a non-linear conversion (e.g. modulation). Here - for example, as a result of multiplication - mixed products, i.e. multiplicative sources of interference in the most varied of frequencies occur. Above all in frequency multiplex systems (see, for example the V-10800 system in Illustration 155) defective non-linear components may lead to extremely complex interference which is difficult to find and eliminate by measuring.

The difference in resistance to interference can be explained using the example of AM radio in the medium wave range and FM radio in the VHF range.

There appears to be a perfect solution - the strength of the signal should be as great as possible compared with the strength of the interference. The relationship between the signal power P_{Signal} and the power of the interference P_{If} turns out to be the basic requirement of resistance to interference. This appears to make possible interference-free reception with any type of modulation.

But things are not so simple. Thus, the transmission performance of a mobile phone cannot simply be increased. In addition, in in the context of the double sideband AM with a carrier the wasted transmitter energy was highlighted. Even if there is no signal for transmission practically the whole energy of the transmitter is required - only a small fraction of the transmitter energy is reserved for the actual information (see Illustration 144 bottom right).

As early as 1936 Armstrong, the inventor of radio reception with mixer stage (see Illustration 158), increased immunity to interference of the FM compared with the AM, however at the price of a much greater bandwidth. This disadvantage at first prevented the FM being used in the MW range (535 - 1645 kHz) as within this relatively narrow frequency band roughly 1.1 MHz wide the number of radio channels was in any case relatively small. With a spacing of 9 kHz between transmitters this resulted in 119 different radio channels in Europe.

However, this difficulty disappeared with the technical exploitation of the very high frequency range after 1945. In the VHF range a transmitter mid-frequency spacing of 300 kHz was prescribed sufficient for FM. In addition to the greater resistance to interference of the FM in the very high frequency range there are fewer external sources of interference. Thanks to the limitation of the range by the horizon - as a result of the straight-lined, quasi optical propagation of the electromagnetic waves in the VHF range distant transmitters only produce interference under extreme conditions.

Theoretical investigations pursued the correct strategy in their demand for as great as possible an increase in the usable phase shift of the PM to enhance resistance to interference. It now had to be determined how much additional bandwidth in the PM was necessary and what the effects on the FM would be.

The difference between the two modulation types FM and PM is to be seen from Illustration 172. In the case of the FM the LF amplitude and the instantaneous value of the source signal and the instantaneous loudness is proportional to the phase shift. But the phase shift is independent of the instantaneous LF frequency. In the case of the FM this does not apply. With the same loudness, in the case of the FM the resistance to interference proportional to the phase shift deteriorates the higher the instantaneous low frequency of the source signal. This means that in the case of the FM the resistance to interference is in a reciprocal relationship to the modulating low frequency. It is greatest with the lowest modulating frequencies.

This defect can be removed by increasing the high modulation frequencies in the amplitude by a frequency response correction using a highpass filter. This is known as preemphasis. At the receiver this measure is reversed using a lowpass filter. This is called deemphasis.

The resistance to interference of the FM is limited to narrow band signals, in particular from external transmitters. But pulse-like industrial interference is also successfully suppressed. However, as soon as the usable field strength approaches the strength of the interference field the quality of transmission drops rapidly. In the case of VHF radio, interference of the transmission signal is noticeable in built-up urban areas. Several signals reflected from the walls of buildings overlap at the motor-car aerial in such an unfavourable way that the usable field strength may tend to zero. At the traffic lights it is often only necessary to move the vehicle a couple of feet to obtain a much better reception.

In the case of the FM and the PM the entire information lies in the zero crossings of the carrier signal. As long as this is preserved in the case of interference complete removal of the interference can be carried out. As soon as interference changes the zero crossings distribution of the modulated signal noticeably the usable flow of information is destroyed. See Illustration 173 in this connection.

Note:
Modern microelectronics creates much better solutions by means of digital signal processing. With the new digital radio technology (Digital Audio Broadcasting DAB) practically interference-free reception of HiFi quality will be possible in addition to many new technical possibilities thanks to the new modulation processes.

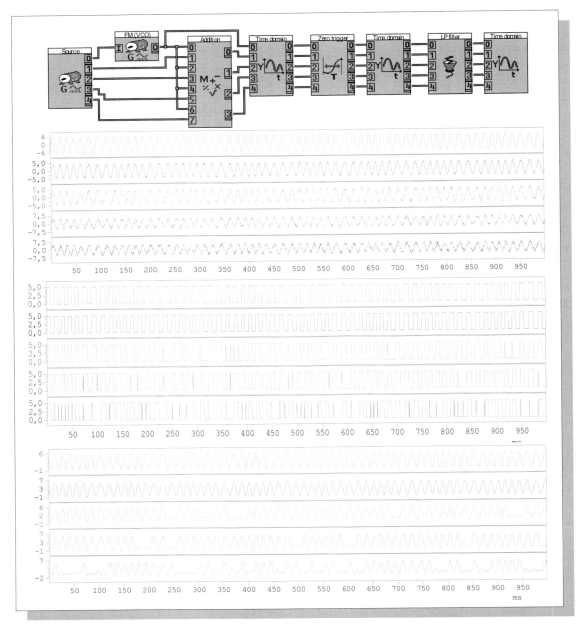

Illustration 173: ***PM and FM: the information lies in the zero crossings***

This thesis immediately suggests an experiment. It is to be demonstrated that the entire information lies in the zero crossings. Also: if the interference (here noise) increases more and more information is lost.

As the block diagram and the top signal block (time segment 1) show a pure PM or FM signal is first generated (which signal it really is could only be decided when the source signal is available). On four additional channels an increasing amount of noise is added to this signal.

The module "zero positions" now triggers on the zero positions of these five signals on to the rising and falling edge in turn. The entire middle signal block only shows the information contained in the zero positions. The distortions of the information produced by interference can be clearly seen. These zero point signals are now simply lowpass filtered. In the third signal block (time segment 3) the result can be seen. In the top series the complete undistorted signal was completely retrieved (time displaced). Thus, the thesis is proved. In the case of the next four signals the disturbances - like the noise - increase from the top to the bottom.
With more intelligent methods it would be possible largely to remove these disturbances too.

Practical information theory

To summarise it can be said that the FM and the PM are a good example of the tenet of information theory that increasing the bandwidth of the transmitter signal increases the resistance to interference.

Precisely the translation of theoretical results into practice by means of digital signal processing has in the last few years paved the way for mobile telephony, satellite telephony, digital audio broadcasting (DAB) and digital television technology (digital video broadcasting DVB).

The above mentioned basic tenet of information theory results from very complex mathematical calculations based on statistics and the calculus of probability. In the final analysis, the result is clear from the outset if we consider the Symmetry Principle **SP** (Chapter 5) and the equivalence of the time and frequency domain.

The resistance to interference of a transmission can be increased by increasing the period of transmission, for example the same information could be transmitted several times in succession or repeated. As the interference is basically stochastic the source signal could be regenerated successfully by forming a mean value as the mean value of (white) noise is zero.

> *Increasing the length of transmission by repeating the information several times in the time domain (time multiplex) with the same transmission bandwidth increases the immunity to interference.*

On account of the Symmetry Principle the reverse must also be true. We simply need to replace the word time by the word frequency and the word transmission length by transmission bandwidth and vice versa.

> *Increasing the transmission bandwidth by repeating the information several times over in the frequency domain (frequency multiplex) with the same transmission length increases immunity to interference.*

The explanatory model selected with time and frequency multiplex, i.e. time and frequency staggering of channels is unimportant. The immunity to interference is generally increased by time and frequency "stretching"

- because the "conservation tendency" of the information-bearing signal between two neighbouring time or frequency segments is increased (see Noise and Information, page 58 and Illustration 40) and

- the interference which has a stochastic effect in the time domain is also stochastic in the frequency domain.

Exercises on Chapter 8

Exercise 1 — Amplitude modulation AM

(a) Develop the circuit corresponding to Illustration 141 by means of the module "Cut out".

(b) Investigate the frequency spectrum according to Illustration 141 experimentally using DASY*Lab*. Set the values so that the instantaneous frequency begins at t = 0s and the other ends at t = 1 s. Therefore, select f = 2,4,8 Hz and all the instantaneous frequencies as a power of two. The block length and sampling rate should be set as in the past at the standard setting of 1024

(c) Create the circuit corresponding to Illustrations 142 and 143. As the source signal take lowpass filtered noise.

(d) Develop an AM generator with which by means of a manual regulator the carrier frequency $f_{carrier}$, the signal amplitude $\hat{U}_{LF}$ and the offset can be set.

(e) Develop an AM generator by means of the module "formula interpreter" which implements the AM formula given on page 222 and with which the degree of modulation m, f_{LF} and $f_{carrier}$ can be set. Select $\hat{U}_{LF} = 1$ V

Exercise 2 — Demodulation of an AM signal

(a) Develop the demodulation circuit corresponding to Illustration 144. Rectification should be shown by the „absolute value" process (module "arithmetics"), the RC circuit by lowpass filtering. See also Illustration 145

(b) Vary the offset and examine the spectra of the modulated signals according to Illustration 145

(c) Why is the double sideband with a carrier a disadvantageous process.

Exercise 3 — Single sideband modulation EM

(a) Design the circuit for the EM according to Illustration 150

(b) Set the parameters according to Illustration 150. Select as always the sampling rate and blocklength at 1024. The same carrier is used here for modulation and reconversion or demodulation.

(c) Why is the demodulated source signal time staggered?

(d) Design the circuit for the upper and lower sideband according to Illustrations 150 and 151.

Exercise 4 — Frequency multiplex processes

(a) Design the circuit for the formation of pre-groups and reconversion according to Illustrations 156 and 157. The bandwidth of the source signal - filtered noise - corresponds precisely to the bandwidth of a telephone signal of 300 -3400 Hz.

(b) Take your final test in the use of DASY*Lab* and design a suitable circuit for the primary basic group formation according to Illustration 138

(c) What are the program limitations of DASY*Lab* for simulation with reference to the creation of secondary, tertiary and quaternary basic groups?

Exercise 5   **Mixing**

(a) Design and simulate an AM tuner according to Illustration 158. You will have to set the sampling rate at roughly 2 MHz and the blocklength at at least 1024

(b) Afterwards create a realistic AM signal in the medium wave range from 525 to 1645 kHz which you superimpose on noise. Try to retrieve and demodulate the corresponding AM signal. Tip: make sure there is a favourable relationship between the signal and the level of noise.

Exercise 6   **Frequency modulation FM**

(a) Familiarise yourself with the function of the FM generator according to Illustration 159

(b) How can you set the FM mid-frequency and frequency swing in the module "source signal" (without using a manual regulator)? In these experiments select a sinusoidal source signal.

(c) Vary the frequency swing and the mid-frequency according to Illustration 159 using a manual regulator. Extend the circuit.

(d) Examine the composition and the influence in the frequency domain of the frequency swing, the frequency of the source signal and the FM mid-frequency on the width of the FM spectrum.

Exercise 7   **Demodulation of FM signals**

(a) Try to demodulate an FM signal on the edge of a lowpass filter

(b) Try to demodulate an FM signal with the same cutoff frequency as (a) on the edge of a highpass filter. How do the results differ?

(c) Repeat (a) and (b) with a realistic source signal (filtered noise)

(d) Design a phase detector according to Illustration 169

(e) Try to replicate the experiment carried out in Illustration 170

(f) On you own design a PLL as in Illustration 154 and try to set the control voltage for the VCO in such a way that the PLL works.

Exercise 8   **Difference between PM and FM**

(a) Develop the circuit corresponding to Illustration 172 and carry out the comparison of PM and FM yourself.

(b) Give reasons for the differences in the immunity to interference of PM and FM

(c) Why are *preemphasis* and *deemphasis* used in the case of FM

Exercise 9   **FM and PM: zero crossings as carriers of information**

Examine the thesis experimentally according to which the entire information of the PM and FM signals is contained in their zero crossings. Refer to Illustration 173.

Exercise 10   **Resistance to interference**

Give reasons why widening of the frequency band while retaining the length of transmission can increase resistance to interference as a general rule

Chapter 9

Digitalisation

Haha, you will probably think, now we are going to have an „Introduction to digital technology". Even at the first level of training in the vocational field of electrical engineering this subject is an indispensable element in the program of training.

It is such a appealing subject because it can be approached with no previous knowledge. There is a great deal of specialist teaching materials, experimental construction kits and simulation programmes for the PC available.

Digital technology does not always mean the same thing

There is not much point in following these well-trodden paths. There is, however, a much more important reason to transcend this kind of digital technology. It is very important for technical control and monitoring problems of all kinds but basically it only does one thing – it switches lamps or relays *on or off* or occasionally activates a stepper motor.

Let us take a set of traffic lights as an example – certainly an important application. It is confined – like all the examples dealt with in digital technology as a school subject up to now – to switching on and switching off processes, i.e. two conditions – *on or off*!

> *Whether using standard digital components (e.g. TTL series), freely programmable digital circuits such as GALs (Generic Array Logic) or FPGAs (Field Programmable Gate Arrays) or even microcontroller circuits – traditional digital technology in schools has confined itself up to now to switching on and switching off processes.*

In Chapter 1 we explained in detail why our approach has completely different objectives:

> *In this book and in present-day practice the central thing is the computer-based processing of real signals by means of "virtual systems" (i.e. programs) which as far as possible should like DASYLab be generated by graphic programming in the form of block diagrams.*

Digital processing of analog signals

Real signals in measuring, automatic control engineering, audio and video technology are first of all always analog signals. The trend is away from processing analog signals by analog systems. As has already been mentioned several times analog technology is now only of importance where it cannot be avoided.

> *Analog circuit technology is more and more to be found where it cannot be avoided – at the analog information source and drain and at the transition to the physical medium of the transmission path.*

The best example of this trend is the *internet*, the worldwide network of innumerable computer networks. Where do we still find analog technology here? The sound card is usually the source of information and drain for real analog signals. It contains a minimum of analog technology. Otherwise we only find analog technology at the transition to and from the transmission medium Cu cable, optical wave guide, terrestrial relay radio and satellite telephony.

Any kind of (analog) communication can be effected via the internet. Video webphoning, i.e. world wide video telephony at local rates, is the latest step in this development.

> *In schools we are entering new territory with the digital processing of real analog signals. At the present time there are no ministerial guidelines or curricula which only as much as mention the most important subject of modern information technology, i.e. DSP or digital signal processing.*

At present digital technology ends in schools with the A/D or D/A converter, that it the point at which analog signals are converted into digital signals and vice versa. This should be changed as quickly as possible.

DSP has made possible completely new, fantastic signal-processing systems. An excellent example comes from the field of medical diagnostics in computer tomography, in particular NMR-tomography (Nuclear Magnetic Resonance). The living human being is recorded segment by segment by means of measuring technology and is presented pictorially. By linking the data from each "segment" using the computer it is possible given sufficient computer capacity to navigate through the body of the living person, to represent bones and tissue separately and recognise and define certain tumours precisely.

It would be absolutely impossible to achieve this amazing feat using analog circuit technology or analog computing technology. Another example is *DASYLab* that we are using. Highly complex signal processing systems can be programmed graphically and a large number of details can be "visualised", i.e. represented pictorially.

> *Digital processing of analog signals (DSP) makes new applications of information technology (IT) possible, which could not be realised using analog technology. Using a computer, pure signal processing can be carried out virtually at will with a degree of precision which borders on the limits of nature as defined by physics.*

DSP also leads to the standardisation of signalling processes. DAB and DVB (Digital Audio Broadcasting and Digital Video Broadcasting) use signalling processes which are also used in other modern technologies, for example, mobile telephony. Modern information technology will in the final analysis become simpler and less complicated.

> *Not only will standard chips characterise hardware, standard processes will characterise the software of signal processing systems.*

The gateway to the digital world: the A/D converter

If DSP opens up such impressive perspectives, we still have to enter this territory. The central issue is to make equivalent digital signals – strings of numbers – from analog signals. The term A/D conversion has come to be used to designate this process.

There are many different processes and variants for this conversion process. At this point the main thing is to understand the principle of the A/D converter.

> *A/D (and D/A) converters are always hardware components on which a series of analog and digital signalling processes take place.*
>
> *In contrast to practically all other signalling processes in DSP systems A/D (and D/A) conversion cannot be carried out using a computer by means of a virtual system – i.e. a program.*
>
> *At the input of the A/D converter lies the (real) analog signal and at the output the digital signal appears as a string of numbers.*
>
> *The numbers are outputted in the dual number system and are further processed.*
>
> *The conversion process takes place in three steps:* **sampling**, **quantization** *and* **coding**.

The principle of an A/D converter is now to be simulated using DASY*Lab*. The process selected works according to the „counter coding principle" (see Illustration 174).

In the top series a small segment from an LF signal (for example, a speech signal) is to be seen, below it the time pulse with which " measurement samples" can be taken. These and two other signals are produced by the component and the module "function generator". The two upper signals are connected with the module "Sample & Hold". The features of this process can be seen in the third row – the measurement at the point of time of a needle pulse (see time pulse in the second series) is calculated and stored, that is, "held" until the next measurement is sampled.

Each of these measurement values must be converted into a discrete number in the time between two measurements. In the fourth series a periodic sawtooth curve is to be seen which runs in synchrony with the time pulse (sampling frequency). In the subsequent "comparator" the step-like Sample & Hold signal is compared with this periodic sawtooth. At the output of the "comparator" there is a signal of level 1 („high") as long as this sawtooth voltage is smaller than the instantaneous Sample & Hold value. If the sawtooth voltage exceeds the instantaneous Sample & Hold value, the output signal of the output signal of the comparator jumps to 0 ("low").

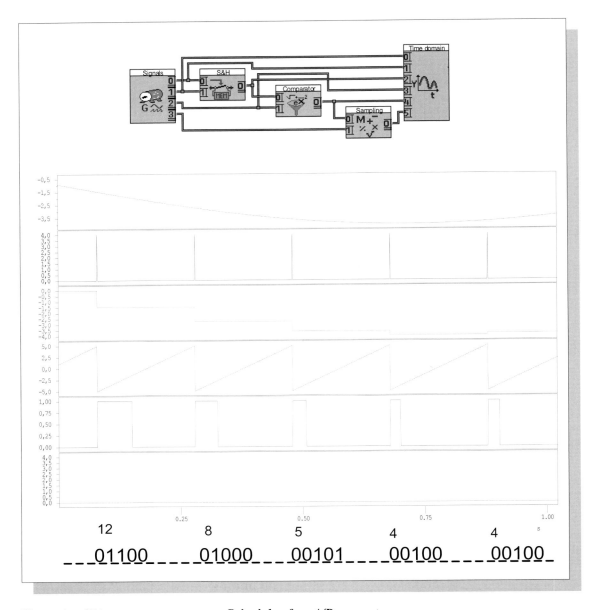

Illustration 174: **Principle of an A/D converter**

Here the processes sampling and quantization are shown in detail. The coding as a string of dual numbers is to be found at the bottom. In the serial form of digital output signal represented the clock pulse frequency is 5 times the sampling frequency because every measurement is coded with 5 bit.

In the top series you see a small segment from the analog input signal. You will find further details in the text. Note that this a pure simulation.

The information on the size of the instantaneous measurement lies in the pulse length at the output of the comparator. It can be clearly seen that the pulse length increases (linearly) in accordance with the increasing height of the "step curve". This form of information storage is called *pulse duration modulation* (PDM) or *pulse length modulation* (PLM). Please note that this signal is a value-discrete but time-continuous signal. The information is present in analog form.

The PDM signal represents a rectangular time window whose pulse length depends on the instantaneous measured value. This signal opens and closes a gate. At the input of the gate there is a periodic needle pulse sequence with 32 times the frequency of the above sampling process. Depending on the pulse length more or fewer pulses pass through the gate and, to be precise, between (minimum) 0 and (maximum) 32. The number of pulses is always discrete, i.e. 16 or 17 but never 16.23 ... . Now the information on the instantaneous measurement is no longer present in analog continuous form but in discrete form. This is the process of *quantization*, as explained in Illustrations 138 and 139.

In the present case a maximum of 32 different measurements can be taken and registered. The pulse groups in the lower series – not shown here- are connected with a binary counter which gives the number of impulses as a dual number, for example 13 pulses are equivalent to the dual number 01101. This is the *coding* process.

Note:

- In practice the "gate" described here is merely an AND circuit which is set at "high" at the output when (instantaneously) both inputs are at "high".

- This simulation is a 5 bit A/D converter ($2^5 = 32$) for 32 different numbers can be coded using equivalent 5 bit combinations.

- The process of number coding described here is very simple but it is hardly ever used in practice. In the case of a 16 bit A/D converter as used in audio technology the pulse frequency would have to be $2^{16} = 65536$ times as high as the sampling frequency.

- There are two possibilities in the case of the A/D of outputting strings of dual numbers:

    - *Parallel output*: here there is an output for every bit. In the case of a 5 bit combination there would be 5 outputs. Parallel output is usually internal in chips or systems, for instance between the A/D converter and the signal processor.

    - *Serial output*: using a shift register the bit combination is connected with a single channel. As a result the pulse frequency in the above example is increased to five times the sampling frequency. In transmission technology serial outputting is used almost exclusively.

- The digital signal is present in serial form on an audio CD, that is the entire music consists of a sequence of 0 and 1 ("low" and "high") which seems more or less random.

Principle of a D/A converter

At the inputs of a D/A converter there is a dual number which is converted internally into a discrete analog value. The output signal of a D/A converter is thus, to be precise, a step-like signal. In the case of a 5 bit D/A converter 32 different "steps" would be possible.

The "jump height" at the edges of the steps is always an integer multiple of the minimum value defined by the quantization process (see in this connection Illustrations 138 and 139).

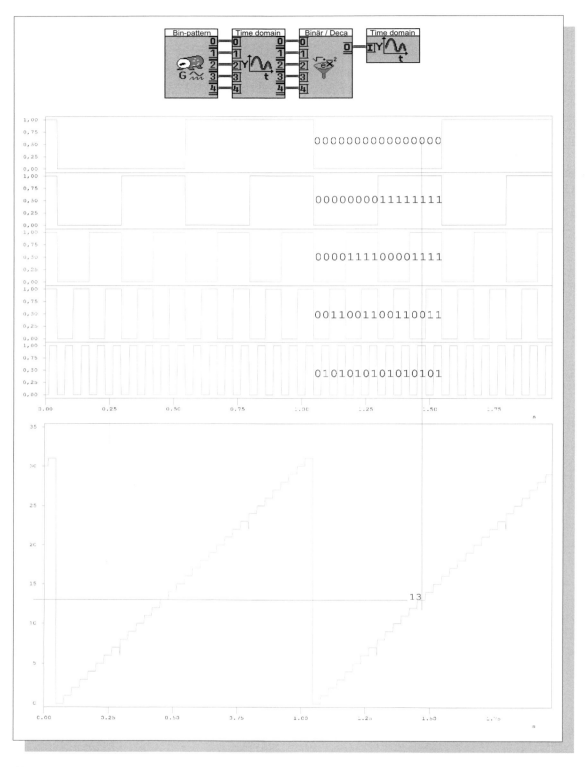

Illustration 175: **Principle of a D/A converter (DASYLab simulation)**

Five retangular signals form a binary pattern generator in this simulation which increments from 00000 to 11111 and from 0 to 31. From the top to the bottom the frequency is doubled in each case. This produces a linear rising step curve or a periodic sawtooth. Exactly as in practice there are several interference pulse peaks to be seen. The D/A converter is designed purely according to a formula in which every bit is multiplied 1, 2, 4, 8 or 16 depending on its weight.

> ## Reminder: Positional notation of numerical systems
>
> The Babylonians had a number system with 60 (!) different numerals. We need it still today in the measurement of time: 1 h = 60 min and 1 min = 60 s
>
> Our decimal system uses 10 different number from 0 to 9. For example the number 4096 is as agreed an abbreviation of
>
> $$4 * 1000 + 0 * 100 + 9 * 10 + 6 * 1 = 4 * 10^3 + 0 * 10^2 + 9 * 10^1 + 6 * 10^0$$
>
> Thus the number 10 represents the **basis** of our numerical system. The **position** of the number indicates its "value", hence "positional notation of numerical systems"!
>
> The decimal system has a set of values which range from 0 to 9. This basic set is repeated over and over again to create large numbers.
>
> In digital signal processing the Binary Number System is used. There are various reasons for this:
>
> - Electronic components and circuits with two switching conditions ("on" and "off") can be implemented most simply.
>
> - Signals with two states are the most noise-resistant, because it is only necessary to distinguish between "on" and "off"
>
> - The mathematical basis is very straightforward. The basic table for binary numbers is very simple:
>
> $$0 * 0 = 0; \quad 1 * 0 = 0; \quad 1 * 1 = 1$$
>
> How much school years were needed in order to learn our multiplication tables?
>
> The base of this number system is 2. The binary number 101101 written out in full is equivalent
>
> $$1 * 2^5 + 0 * 2^4 + 1 * 2^3 + 1 * 2^2 + 0 * 2^1 + 1 * 2^0 =$$
> $$1 * 32 + 0 * 16 + 1 * 8 + 1 * 4 + 0 * 2 + 1 * 1 = 45$$
>
> Hence : $101101_2 = 45_{10}$
>
> In words: "101101 to the base of 2 corresponds to 45 to the basis 10"
>
> A little test: what would the number 45_{10} be in the "three number system", i.e. on the base of 3?

Illustration 176:   ***Significance number systems***

The only thing required in this book is in principle the four basic arithmetical operations. However, you should also be proficient in them in the binary number system.

In Illustration 175 a periodic sequence of dual numbers beginning at 0 and ending at 31 is connected with a – purely computer-based – D/A converter. Accordingly, this results in a periodic sawtooth and step-like signal curve at the output. For the instantaneous value 01101_2 at the input the value 13_{10} results at the output. See in this connection Illustration 159 (significance number systems)

The D/A converter simulated here is a module in which mathematical formulae can be inputted. The formula inputted in this example is

$$IN(0) * 16 + IN(1) * 8 + IN(2) * 4 + IN(3) * 2 + IN(4) * 1$$

or alternatively

$$IN(0)\,2^4 + IN(1) * 2^3 + IN(2) * 2^2 + IN(3) * 2^1 + IN(4) * 2^0$$

Thus the bit with highest weight lies at the input IN(0) and the lowest at IN(4). Please note that this is a pure simulation intended to demonstrate the underlying principle.

Analog pulse modulation processes

The pulse duration modulation in the case of the A/D converter shown in Illustration 177 is of great importance in microelectronic measuring and automatic control technology. Alongside these analog pulse modulation processes – as a result of the continuously modifiable pulse length the information is present in analog form – there are other important analog pulse modulation processes within the framework of measuring and automatic control technology, which are as follows

- Pulse *amplitude* modulation PAM
- Pulse *duration* modulation PDM
- Pulse *frequency* modulation PFM
- Pulse *phase* modulation PPM

These pulse modulation processes are of practically no importance in transmission technology. They serve usually as intermediate processes in the conversion of analog "measurement values" into digital signals.

The PAM simply describes the sampling of an analog signal as shown in Illustration 174. The Sample & Hold process is simply a variant of the PAM.

Characteristic of the other three analog pulse modulation processes is the conversion of analog measurement values into analog time length. As very small and precise units of time are available in microelectronics thanks to quartz technology the comparision of times can be carried out extremely accurately using the instruments of microelectronics.

In Illustration 177 the different analog pulse modulation processes are shown. The A/D conversion according to Illustration 174 was modified and added to:

- The PPM signal was obtained by triggering on the negative edge of the PDM signal. The pulse width of the PPM signal was set in the menu of the trigger module.

- The PFM signal was obtained by a frequency modulator. To this end the offset and amplitude of the LF signal - output 5 of the function generator module – were changed accordingly.

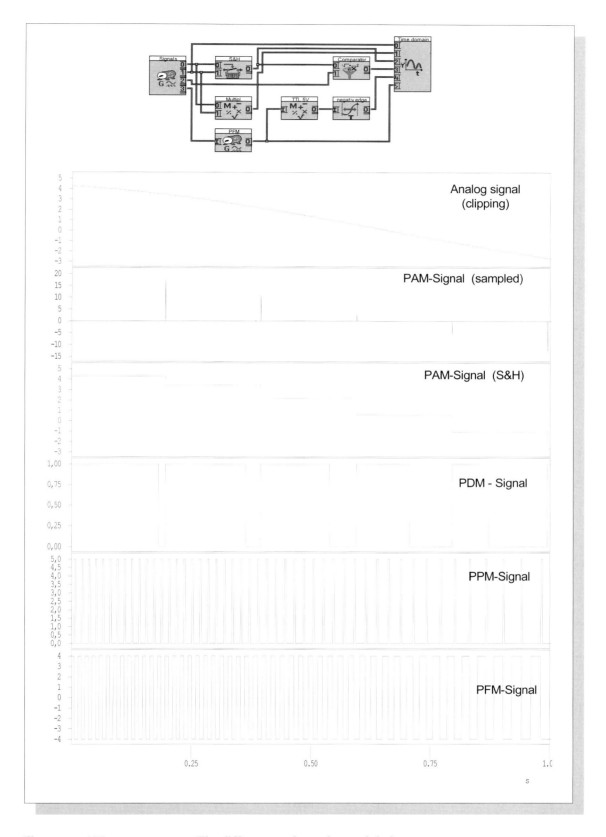

Illustration 177: **The different analog pulse modulation processes**

In order to obtain the signals shown the circuit for the simulation of the A/D converter was modified and added to.

DASYLab and digital signal processing

We have worked with DASY*Lab* from the first chapter onwards, that is in the final analysis we have carried out computer-based signal processing. In the chapters dealing with basic principles this was not specially mentioned. For educational reasons the impression was given that we were dealing with analog signals. Almost all the Illustrations accordingly appear to show analog signals.

Now, however, the moment of truth has come. Computer-based signal processing is always digital signal processing (DSP). Here is a brief reminder of what this means.

Illustration 178 once again shows the situation by means of an example. A digitalised audio signal is read from a disk and shown on a screen. Optically it appears to be a continuous analog signal. If however, using the lens function a tiny segment is zoomed out, individual measuring points can be see which are joined up by straight lines. These straight lines are produced by the program to make it easier to see the course (of this originally analog signal). They are not present for signal processing.

These measurement points can be brought out more clearly, if so desired. Thus, each point can be represented additionally by a small cross or triangle. The representation of the points as vertical bars is very graphic.The height of a bar is equivalent to a measurement. We should emphasise once again:

> *A digital signal consists pictorially in the time domain of a discrete equidistant sequence of measurements which reproduce the curve of the (original) continuous analog signal more or less accurately.*

In a less pictorially graphic way Illustration 179 represents this digital signal much more accurately. For this purpose the module "list" is used. Here it can be seen clearly

- at what time intervals measurements (samples) of the analog signal were taken and

- with what degree of accuracy the measurements were registered (quantization – see Illustration 139)

As a result of quantization the measurements are value discrete, i.e. only a finite number of different "graduated" measurements can occur (see also Illustration 138).

> *Digital signals are time and value discrete in contrast to the time and value continuous analog signals. This results in particular requirements which must be taken into account in the processing of digital signals. Without the necessary background knowledge mistakes are unavoidable.*

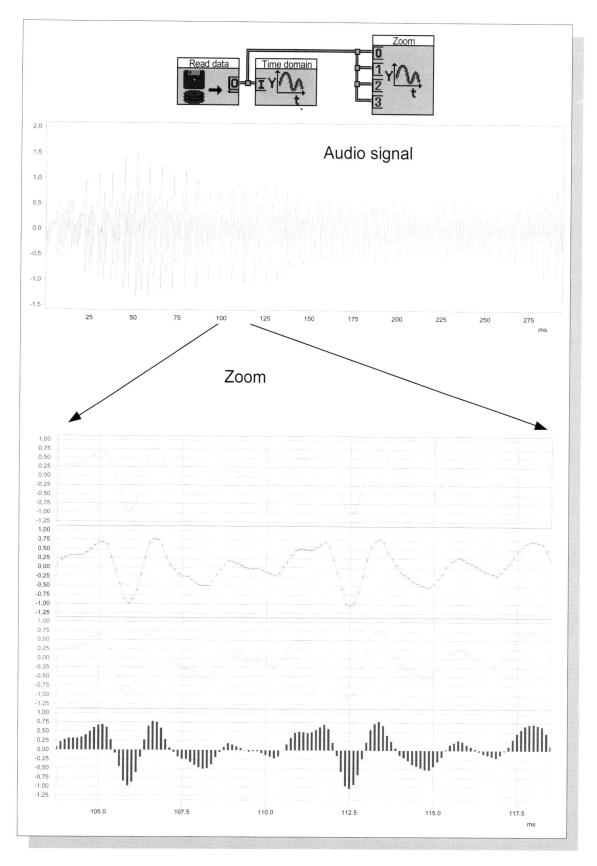

Illustration 178: **_Possibilities of presenting a digitalised audio-signal using DASYLab_**

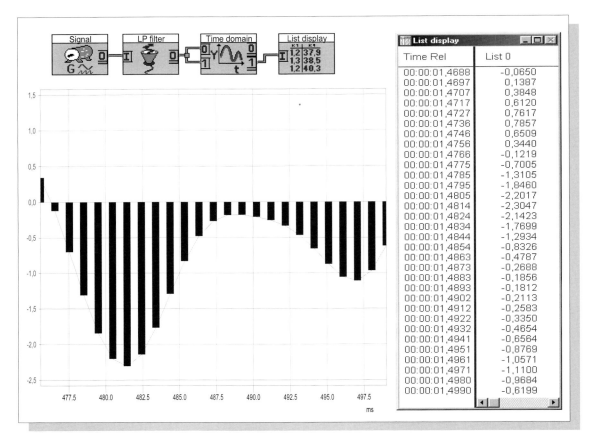

Illustration 179: **A digital signal as a series of numbers**

Important though a visual representation for a (physical) understanding of signal processing may be, digital signal processing DSP in the time and frequency domain is simply the processing by computer of series of numbers. Fortunately, the computer presents all the results of its computations in a perfect graphic form. And as man thinks in pictures the processes can be understood and analyzed without mathematics.

The precision of the representation of series of numbers can be seen here clearly. The time information is given to four decimal places, the fifth figure being rounded up. A sample of the analog audio signal was taken roughly every μs. The measurements themselves are accurate only to four decimal places because a 12 bit A/D converter was used for the recording which only allows $2^{12} = 4096$ different measurements. One should not believe everything that the display shows to five decimal places.

Digital signals in the time and frequency domain

A digital signal consists only of "samples" of the analog signal which are taken at regular, usually very brief intervals.

The most important question is – how does the computer know about the signal curve between the measurement points given that, theoretically at least, an analog signal even between any two small time segments has an infinite number of values?

The following comparisons provide first indications of a more precise answer:

- Any photo or any printed picture consists of a finite number of points. The particle size of the film determines the resolution of the photo and in the final analysis its information content

Chapter 9: Digitalisation

- A television picture and any film conveys the impression of a continuous "analog" change in the sequence of movement although only 50 individual (half) pictures are transmitted per second in the case of television.

It is a good idea – as in Chapter 2 – to begin with periodic digital signals. Basic features of digital signals and sources of error in the computer-based processing of digital signals will now be explored step by step using suitable experiments.

The period length of digital signals

How can the processor or the computer know whether the signal is in fact periodic or non-periodic as it only stores the data or measurements of a given "block length"? Can it sense what the signal was originally like and what it would have been like later? Of course not. For this reason we will now examine experimentally without a great deal of prior conjecturing how the computer or DASY*Lab* copes with this problem.

Note:

- The block length n does not indicate a time but only the number of the intermediate stored measurements

- Only the inclusion of the sampling frequency f_S results in something like the "signal length" Δt. If T_S is the period of time between two sampling values, $f_S = 1/T_S$ applies. Hence

    "Signal length" $\Delta t = n * T_S = n/f_S$

    For Illustration 180 for instance "signal length" $\Delta t = n * T_S = n/f_S = 32/32 = 1$ s

- The block length and sampling rate are set with DASY*Lab* by the menu item A/D

- The signal length Δt will always be 1 s if the sampling rate and the block length are set at the same level in the menu item A/D

- It is striking that the block length n in the menu item A/D is always to the power of 2 e.g. $n = 2^4, 2^5, \ldots 2^{10}, \ldots 2^{13}$ or $n = 16, 32, \ldots 1024 \ldots 8192$. There is an important reason for this. Only than the frequency spectrum can be calculated very quickly via the FFT algorithm (FFT: Fast FOURIER transformation).

In Illustration 180 both the block length n and the sampling frequency f_S are set at 32. A (periodic) sawtooth of 1 Hz was selected as a signal as we are very familiar with its spectrum from Chapter 2. At the top the signal is to be seen in the time domain and at the bottom in the frequency domain. The length of the signal Δt is the same as the period length.

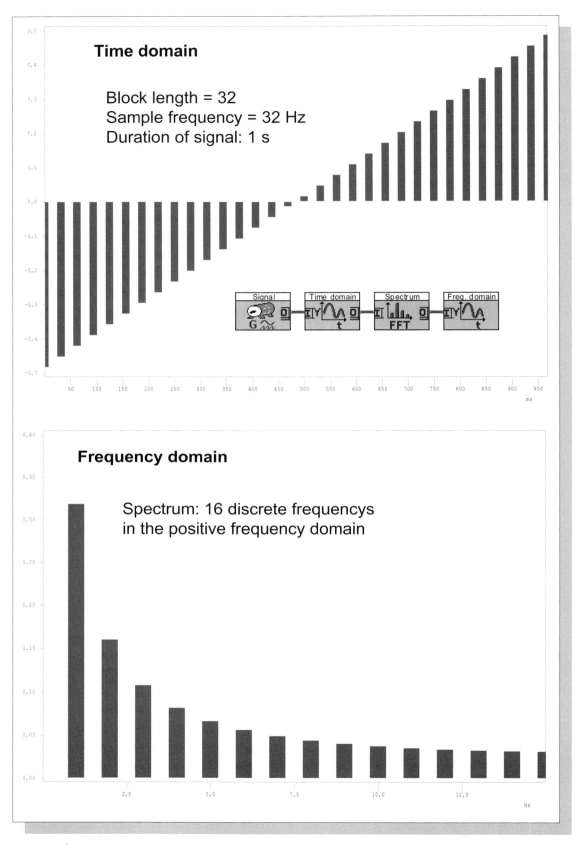

Illustration 180: **Digital signal in the time and frequency domain (sawtooth 1Hz)**

A short block length (n = 32) was purposely selected here. The individual values can be represented as columns and the time discrete nature of digital signals is thus clearer. In addition, all the changes or possible defects compared with analog signal processing can be recognised more easily. The yardstick for this investigation is provided by the three fundamental phenomena to which we have so far been introduced: FOURIER Principle **FP**, Uncertainty Principle **UP** and the Symmetry Principle **SP**.

In the time domain two different times can be seen:

- The signal length Δt (= 1 s),
- The time interval between two measurements T_S (= 1/32 s)

As a result of the Uncertainty Principle **UP** these two times must influence the frequency spectrum:

- The signal length of $\Delta t = 1$ s results on account of the Uncertainty Principle in a frequency uncertainty of (at least) 1 Hz. This explains why the lines or frequencies of the amplitude spectrum in Illustration 180 have a spacing of 1 Hz. The computer does not know that the signal is really periodic.

  Note:

  DASY*Lab* illustrates this frequency uncertainty very graphically by the width of the column. In this case the line spectrum is more than a "column spectrum".

- The time interval between two measurements - here 1/32 s – gives the shortest time span in which the signal can change. In the amplitude spectrum the highest frequency displayed is 16 Hz. A sine of 16 Hz changes twice per period. In the first period half it is positive, in the second negative. As Illustration 181 very clearly shows a sine of 16 Hz is capable of changing 32 times per second from the positive to the negative region. This is an initial explanation as to why only the first 16 frequencies of the sawtooth are given.

It is important that you should check all these figures in the relevant Illustrations (or by means of the interactive experiments on the CD). This is why the diagrams are large and contain verifiable figures.

The most important questions – see above – have so far only been touched on. The results of systematic experiments will be shown on the next few pages which will provide clarfication as to what the computer or the program of digital signal processing (DSP) actually perceives of the real analog signal. The illustrations together with the caption texts are really self-explanatory. But you must examine them closely. As additional texts summaries and preliminary findings follow. The first important conclusion in this context is mentioned again below.

In the time domain the (time and value discrete) digital signal differs from the (time and value continuous) real analog signal in that it only contains "samples" of the real signal taken at regular intervals.

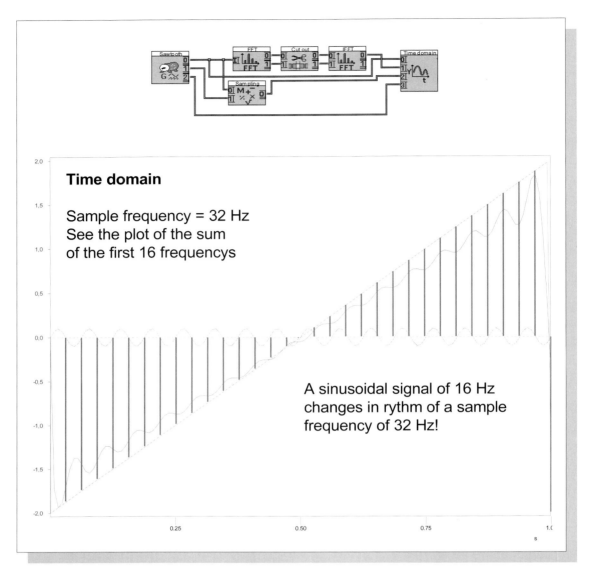

Illustration 181: **Block length, sampling frequency and bandwidth of the spectrum displayed**

Using a rather ingenious circuit it will be indicated why in Illustration 180 with a sampling frequency of 32 Hz the highest frequency of the spectrum displayed is 16 Hz. For this purpose the sawtooth is connected with an almost ideal lowpass filter of 16 Hz in the circuit diagram at the top. The highest frequency 16 Hz – as in the spectrum – is at the output of the lowpass. The sum of the first 16 frequencies has the input signal superimposed (sawtooth 1 Hz) and the 32 sampling values of this sawtooth. It can be clearly seen that the sawtooth signal of 16 Hz is able to model the shortest time change in the sampled signal of 1/32 s. To put it another way, because a sine changes twice per period, a sine of 16 Hz changes its polarity 32 times per second.

This immediately suggests the following questions:

- Can the original real signal be reconstituted retrospectively from the bits and pieces of the digital signal?

- Is it contained exactly in the digital signal or is the information only partly present?

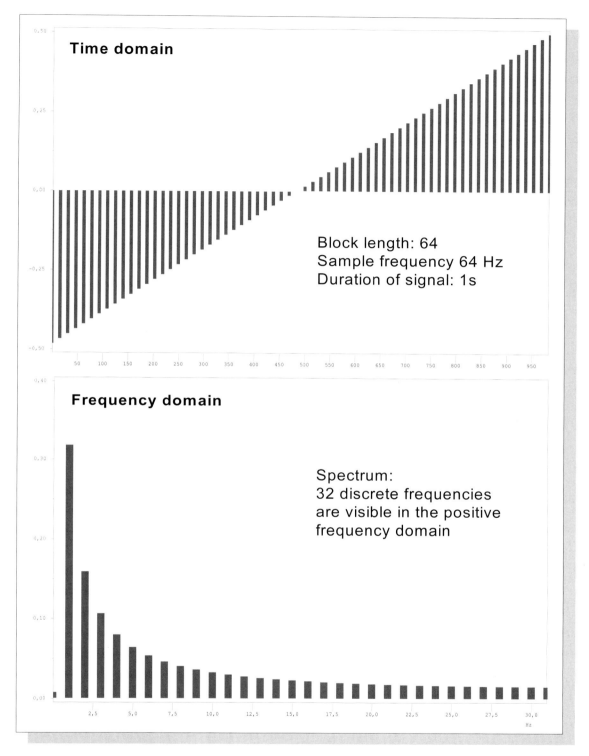

Illustration 182: ***Block length and number of frequencies of the spectrum***

As an extension to Illustration 180, the sampling rate is doubled (n = 64) in Illustration 182, and in Illustration 183 the sawtooth frequency is doubled to 2 Hz with the sampling rate at n = 64. The following preliminary conclusions can be drawn from the three Illustrations.

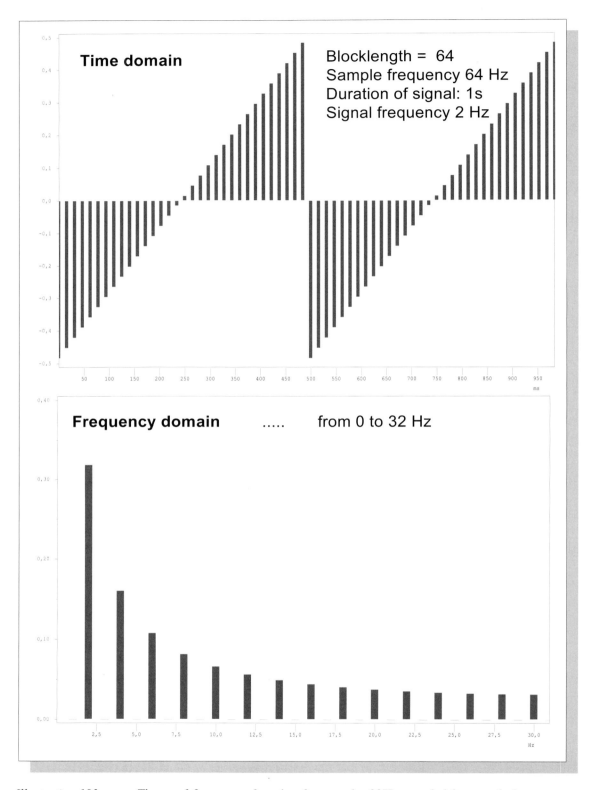

Illustration 183: ***Time and frequency domain of sawtooth of 2Hz sampled for exactly 1s***

As in Illustration 182 the spectrum in this illustration comprises 32 frequencies at n = 64. On account of the sawtooth frequency of 2 Hz the spectrum only contains the integer multiples of 2, i.e. 2,4,6,8, ……Hz. The columns are only half as wide as in Illustration 180 because the resolution has doubled.

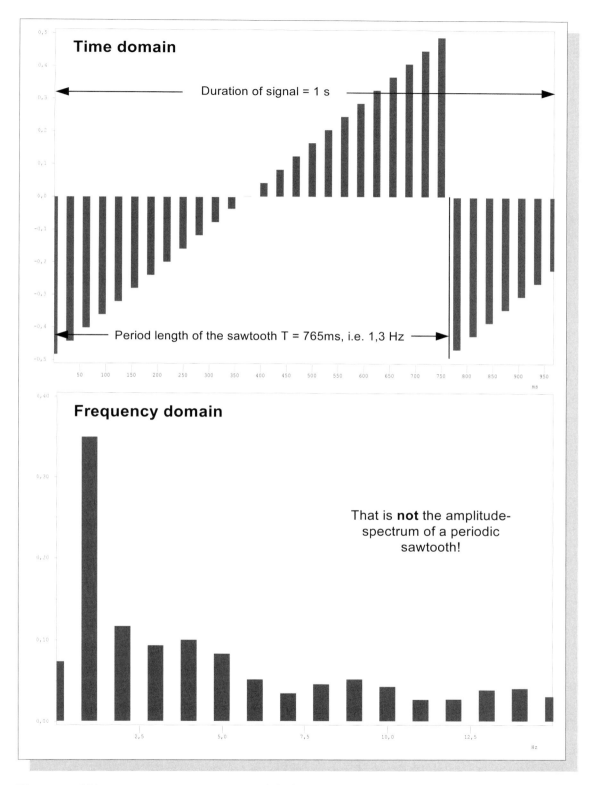

Illustration 184: ***Digital non-periodicity***

Here too the signal length is 1 s and the sampling frequency 32 Hz with a block length n = 32. However, the frequency of the sawtooth is 1.3 Hz here and as a result does not fit into the time grid of the signal segment. The amplitude spectrum has a completely irregular curve and is not identical to that of a periodic sawtooth. On the other hand, it is a discrete line spectrum and must therefore belong to a periodic signal.

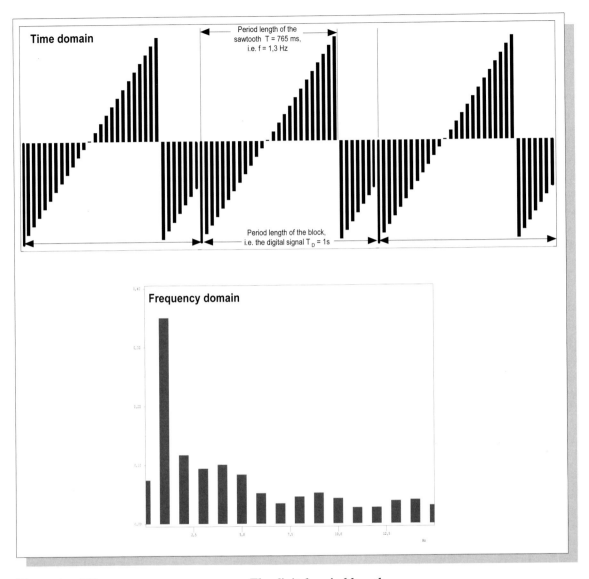

Illustration 185: **The digital period length**

The above illustration reveals how the "chaos" in the frequency domain in Illustration 184 arises. In digital signal processing the signal segment to be analyzed is always regarded as periodic. The length of the signal segment is always equal to the period length.

Precisely for this reason the spectrum of digital signals is always a line spectrum and if you look closely the spacing of the frequencies is exactly the inverse value of the period length T_D. Thus $\Delta f = 1/T_D$ applies. In Illustrations 180 to 184 the signal length was always 1 s and thus the spacing of the frequencies or the "frequency columns" was exactly 1 Hz.

The number of the visible (positive) frequencies in the spectrum is always exactly half the sampling rate n. Remember that other n/2 information is present – that of the negative frequencies or the phase position of the frequencies (phase spectrum).

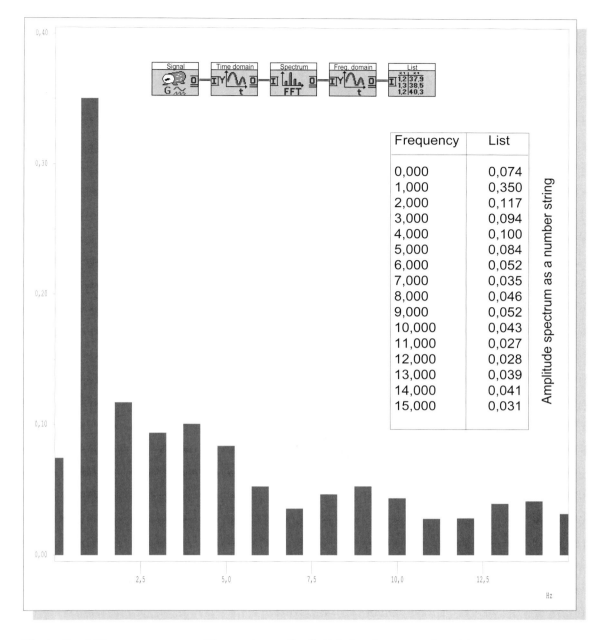

Illustration 186: **The spectrum of a digital signal as a number string**

Here the spectrum from Illustration 185 is shown once again together with the corresponding number string, which the module "list" shows on request.

Strictly speaking the frequency spectrum is described by two number strings: The list of the frequencies and the list of the amplitudes of the existing sine-wave oscillations.

The period length of digital signals is always equal to the length of the whole signal segment which is being analyzed and processed. This results from the block length divided by the sampling frequency (sampling rate):

$$\text{Signal length } \Delta t = n * T_S = n/f_S$$

> *This also explains why digital signals have discrete line spectra. The spacing of the frequencies is*
>
> $$\Delta f = 1/T_S$$

The actual reasons are much more straightforward:

> *The digital signals to be processed by the processor must generally be regarded as periodic in the time domain because the data of the frequency domain – a string of numbers – consists of a limited number of discrete numbers. As a result of this characteristic the spectrum of digital signals must necessarily be seen as a line spectrum and thus the spacing of the lines of this spectrum depends directly on the block length and sampling frequency.*

The following considerations are important for the next few experiments:

- If the block length and the sampling frequency are equal the signal length is 1s

- At $n = 32 = f_S$ the spectrum displayed ranges from 0 to 16 Hz,
  At $n = 64 = f_S$ the spectrum displayed ranges from 0 to 32 Hz
  At $n = 256 = f_S$ the spectrum displayed ranges from 0 to 128 Hz usw.

The periodic spectrum of digital signals

An important phenomenon is now to be gone into in more detail which was explained in Illustration 85 by means of the Symmetry Principle **SP**.

Not only in the time domain must every digitalised signal be regarded as periodic - the period length T_D is simply the length of the temporarily stored signal – the signal is also periodic in the frequency domain. Here are once again the reasons for this:

> *Real periodic signals always have a line spectrum. The spacing between the lines is constant.*
>
> *As a result of the Symmetry Principle **SP** the reverse must also be true: lines (at equal intervals) in the time domain necessarily imply periodicity in the frequency domain. As all digital signals consist of such "lines" as a result of sampling they must have periodic spectra. These periodic spectra consist in their turn of lines or discrete values (string of numbers) which again explains the periodicity in the time domain.*
>
> *Thus: lines (at equal intervals) in the one domain result in periodicity in the other. If both domains consist of lines (at equal intervals) from the point of view of the computer both domains must be regarded as periodic.*

Chapter 9: Digitalisation

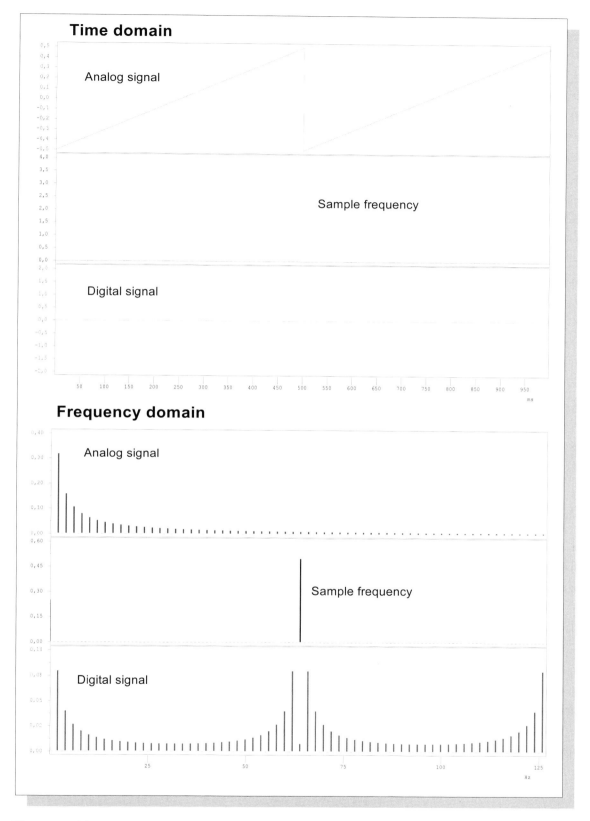

Illustration 187: ***Visualisation of the periodic spectra of digital signals***

Using a trick – see text – the area above the spectrum of Illustration 183 is shown here.

At first sight this seems a strange business but it is simply the necessary consequence of a single property of digital signals – they are *discrete in both domains*.

The Sampling Principle

We are not yet at the end of the tunnel as a new problem arises as a result of the periodicity of digital signals in the frequency domain. Where were these periodic spectrums to be seen in the previous illustrations of this chapter? Using a systematic ingenious experiment we should try to see how this problem can be solved. That would bring us to the end of the tunnel.

Illustration 187 in the top series shows an analog periodic sawtooth of 2 Hz and below this the sampling signal (a periodic δ–pulse sequence) and at the bottom the digital signal, at the top in the time domain and at the bottom in the frequency domain.

If you look closely you will see that there is a frequency domain of 0 to 128 Hz, in contrast to Illustration 183, in which it extended only from 0 to 32 Hz.

Nevertheless the digital signals from Illustrations 183 and 187 agree in the time domain (apart from the level of the measurements). On the other hand, the spectra of the digital signals look completely different. The spectrum from Illustration 183 appears as the first quarter of the spectrum from Illustration 187 from 0 to 32 Hz.

Now to the trick we have used. In the case of the experiment in Illustration 187 a block length of n = 256 and a sampling frequency of 256 Hz were selected at the top in the menu item A/D. As already explained a frequency domain of 0 to 128 Hz results. In the simulation circuit shown there a block length of n = 32 was set "artificially" by the periodic δ-pulse sequence of 32 Hz. Up to now only the frequency range from 0 to 16 Hz was shown with this value.

As a result of this trick we can now see what there is above (and indirectly as a result of the Symmetry Principle also below) the frequency band of Illustration 183, i.e. above 32 Hz. The spectrum from Illustration 183 is repeated constantly, sometimes in the "inverse" position (lower sideband), sometimes in the regular position (upper sideband) convoluted or mirrored from the frequencies of the periodic δ–pulse sequence. In this connection see also Illustration 136. Altogether the spectrum from Illustration 183 is contained in Illustration 187 (bottom) four times (4∗ 32 Hz = 128 Hz).

There is a problem however. As the spectrum of the analog signal shows, the sawtooth has an extremely wide spectrum. As we already know from Chapter 2, this bandwidth tends towards infinity. For this reason the frequency bands mutually overlap whereby the influence of the immediately adjacent bands is the greatest.

This means that the spectra in Illustrations 180 – 187 certainly contain errors. This also implies:

> *If a signal is falsified in the frequency domain, the same also occurs in the time domain as both domains are inseparably linked.*

Chapter 9: Digitalisation

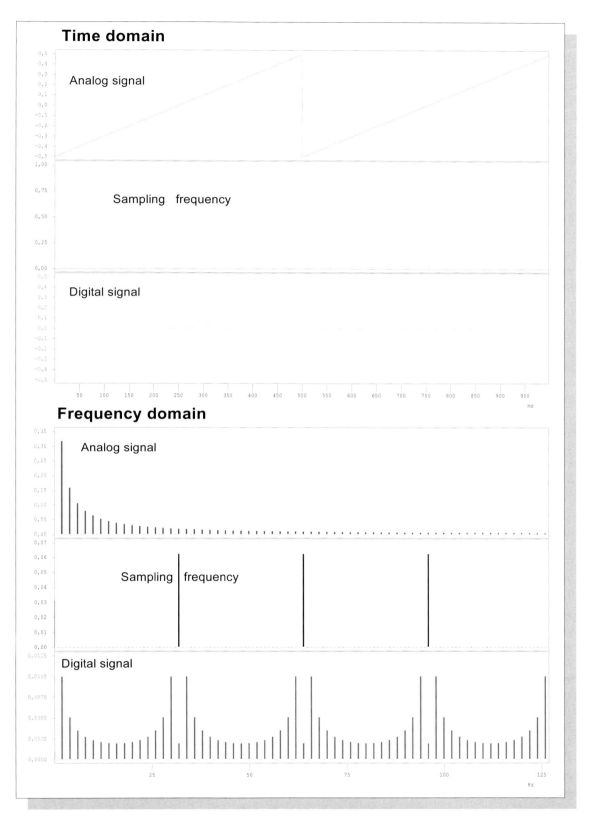

Illustration 188: ***Overlapping of the frequency bands of periodic spectra***

Compared with Illustration 187 the "artificial" sampling rate was halved. You will find the interpretation in the text.

This can be clearly seen in Illustration 188 if you compare the amplitude spectrum of the analog signal with the first 16 frequencies of the digital signal. Below the lines or amplitudes are higher in relation to the first frequency, above all in the middle between two frequency bands. As Illustration 188 shows the error becomes clearer the smaller the sampling frequency f_S (here the frequency of the periodic δ–pulse sequence). A comparison of Illustrations 187 and 188 shows the following: if the sampling frequency is doubled, the interval between the frequency bands doubles. But there is still overlapping but to a lesser degree.

A good way of curing such errors is to increase the sampling frequency and the block length at the same time (this means that the signal length does not change). Now a really interesting aspect emerges. How far apart would these frequency bands lie if the sampling frequency and the block length gradually tended towards infinity. Correct – they would lie "infinitely far" apart. But then we would simply have an *analog* signal with a continuous curve in the case of which all the sampling values would lie very close together. And the spectrum would look exactly the same as in the two illustrations in the top series.

> *From a theoretical point of view analog signals represent the limiting case of a digital signal with which the sampling frequency and the block length tend towards infinity.*

What possibility could there be to avoid the distortion by the overlapping of the frequency bands in digital signal processing? Illustration 189 shows the solution. Here the (frequency bandwidth limited) Si function is used as an analog signal. The bandwidth of this signal shown in the illustration is 10Hz (in the positive region) and the sampling frequency is 32 Hz. Between the frequency bands there is now a respectable gap and there is hardly any overlapping. Every frequency band contains the complete undistorted information on the original analog signal of 1 s length. Thus the next preliminary result is:

> *In order that an analog signal can be processed digitally without distortion its spectrum must be bandwidth limited.*

> *In other words: before an analog signal is digitalised it must be bandwidth limited by means of an analog filter. This takes place by means of a so-called "antialiasing filter", as a rule an analog lowpass circuit.*

A practical example of such an antialiasing filter is a microphone. The cutoff frequency of cheap microphones is practically below 20kHz. As a result the electrical signal generated or converted by it is directly bandwidth limited. Thus for experiments a microphone is a very suitable source of signals. Special antialiasing filters are extremely expensive.

> *The antialiasing filter (it would be better to write it anti-aliasing) prevents the overlapping of adjacent frequency bands of digital signals.*

Chapter 9: Digitalisation

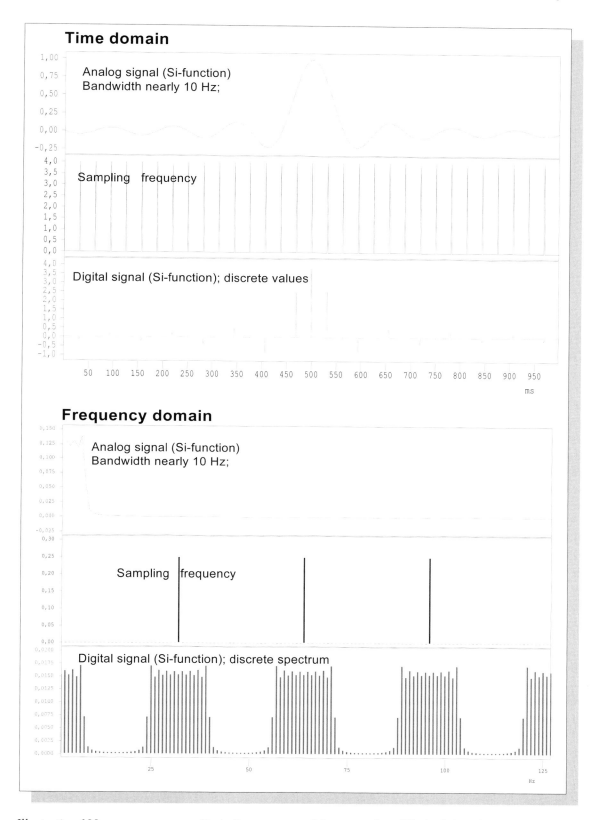

Illustration 189: **_Periodic spectrum of frequency band limited signals_**

At the top you see an analog signal (Si-function) with a frequency band limited to roughly 10 Hz. With a sampling rate of 32 Hz there is hardly any overlapping of adjacent frequency bands.

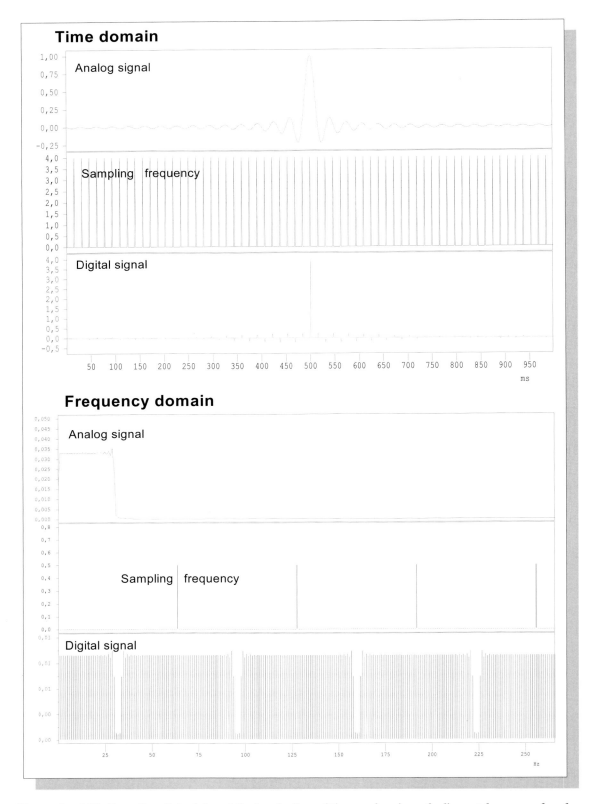

*Illustration 190: **Sampling Principle: at the borderline of the overlapping of adjacent frequency bands***

Here the bandwidth of the analog signal is roughly 30 Hz, the sampling frequency is 64 Hz. It can be clearly seen that the limiting case of "non-overlapping" of periodic spectra is practically given here.

Illustration 190 lets the cat out of the bag. This clearly shows what relationship there must be between the sampling frequency f_S and the highest frequency of the analog signal and the cutoff frequency of the antialiasing filter, so that the frequency bands of the digital signal just avoid overlapping. This relationship is fundamental for the whole of digital signal processing DSP and thus represents the fourth principle of this book.

> *Sampling Principle (called „sampling theorem" in the literature): the sampling frequency f_S must be at least twice as big as the highest frequency f_{max} occurring in in the analog signal. Thus the following holds:*
>
> $$f_S \geq 2 * f_{max}$$

The reasons for this can be clearly seen in Illustrations 189 and 190. As the spectrum of the analog signal is convoluted in the inverse and regular position (lower and upper sideband) at each frequency line of the sampling signal the frequency lines of the sampling signal must be at least twice as wide apart as the spectrum of the analog system is wide.

In Illustration 190 the highest frequency of the analog signal is roughly 30 Hz. The sampling frequency is 64 Hz. The adjacent frequency bands either do not overlap or overlap slightly. Only as long as they do not overlap can the analog signal be retrieved from the digital signal.

In order to check sampling principle once again the same value should be selected for the sampling frequency f_S as for the highest frequency f_{max} occurring in the analog signal. All the sidebands – consisting of „inverse" lower and „regular" upper sideband – are as a result of overlapping only half as wide as they should be.

Retrieval of the analog signal

Illustration 191 shows how the original analog signal can be retrieved at the output of the D/A converter from the digital signal – as long as the sampling principle SP was adhered to – in the case of the non-frequency limited sawtooth signal. This takes place in a very simple way. All the frequency bands apart from the lowest frequency band must be filtered out by a lowpass filter (corresponding perfectly to the antialiasing filter). Then only the spectrum of the analog signal is left and thus the analog signal in the time domain.

> *The retrieval of the analog signal from the correponding digital signal takes place by means of the lowpass filtering because in this way the spectrum of the analog signal and at the same time the signal in the time domain can be retrieved.*
>
> *To be precise, D/A converters produce a step-like curve (see Illustration 175) which largely reproduces the curve of the analog signal. The lowpass filter deals with the fine adjustments.*

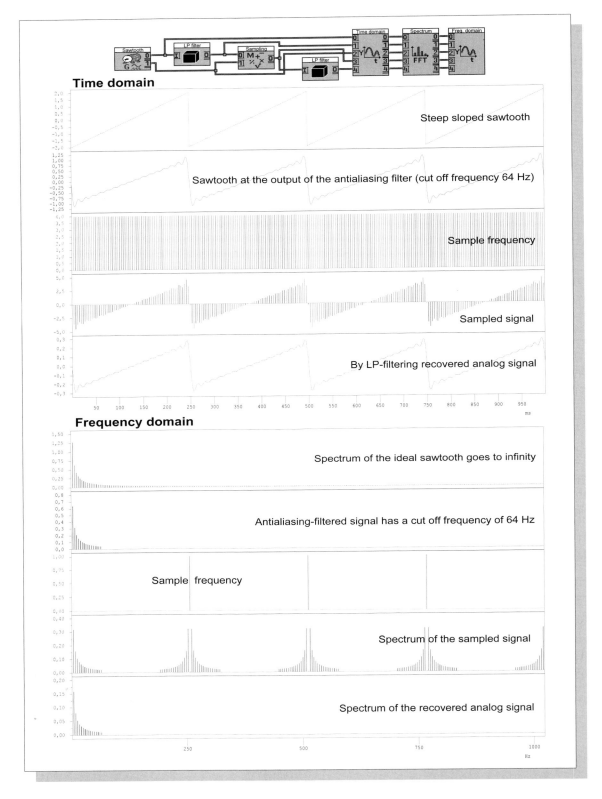

Illustration 191:   *Principle of digital signal processing with analog antialiasing filters.*

In this example everything is as it should be. The non-limited frequency bandwidth sawtooth was limited by means of an antialiasing filter to 64 Hz. The (virtual) sampling rate or the sampling frequency f_S is at 256 Hz. In these relation both values should lie in practice. Instead of the analog antialiasing filters digital filters were created using DASYLab.

Non-synchronicity

There are other traps, some of them rather hidden, which may lead to erroneous results. They occur for example by the incorrect selection of the parameters block length n and sampling frequency f_S. Illustration 192 shows a case of this kind. Initially, everything seems to be OK. The analog signal is limited to a band of roughly 30 Hz, the sampling frequency and the blocklength are set at 512, that is the signal lasts exactly 1 s. Let this stored signal now be sampled by multiplication by an equivalent pulse sequence with a (smaller) sampling frequency f_S of 96 Hz, which according to the sampling principle ought to be sufficient.

But the spectrum of the periodic δ-pulse sequence or the (virtual) sampling frequency f_S points to inconsistencies. Actually it ought to contain only the frequency 96 Hz and its integer multiples. However, it contains – albeit with much smaller amplitudes – the frequencies 32 Hz, 64 Hz etc, i.e. all the integer multiples of 32 Hz. At each of these frequencies the spectrum of the analog signal is convoluted or mirrored, whereby the amplitude of these frequencies indicates how strrong the (unwanted) sidebands occur to the left and right.

As a result the sampling principle is practically invalidated, because sidebands overlap here nevertheless. In Illustration 192 the overlapping of these small sidebands and the large ones can be clearly seen. At all events, virtually inexplicable erroneous measurements occur because these are non-linear effects.

> *Note:*
> The question as to where the "intermediate frequencies" 32 Hz, 64 Hz etc in Illustration 192 come from must first be clarified. This is a consequence of the non-synchronicity between the block length n (here n = 512) and the sampling frequency f_S of 96 Hz. A pulse frequency of 96 Hz does not "fit" in the grid of 512 dictated by the block length. It is non-periodic within the grid as 512/96 = 5 and 32 left over. This non-synchronicity or non-periodicity gives rise to combination frequencies with the frequencies of 96 Hz, 192 Hz etc which are actually to be accepted. Sum and difference frequencies result with the form 0 ± 32 Hz, 96 ± 32 Hz etc and also with the form (n * 96 ± m * 32 Hz) with n, m = 0,1,2,3,....

How could one prevent a remainder from being left over in the division? What values should be selected for the sampling frequency f_S? The block length should always be an integer multiple of the sampling frequency f_S. As the block length is pre-determined and always is a power of 2, i.e. 32, ..., 512, 1024 etc powers of two are also only possible for the sampling frequency.

> Example:
> In the case of a block length of 1024 only the following sampling rates should be selected: ...32 ($=2^5$), 64, 128, 256, 512, 1024 ($=2^{10}$)

> *The sampling rate/sampling frequency f_S and the block length n should always be in an integer relationship to each other.*
>
> *The most straightforward case is when both quantities are selected of equal size.*

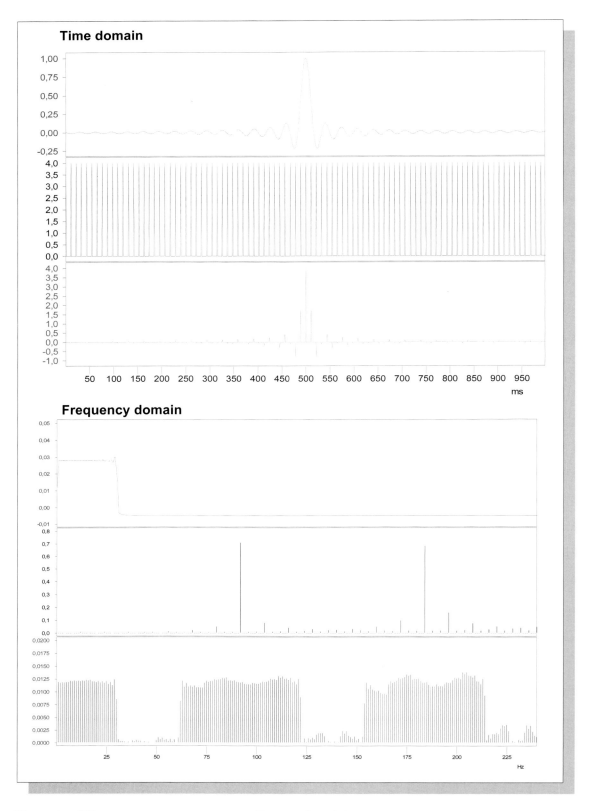

Illustration 192: **Non-synchronicity**

Non-synchronicity arises when the block length is not an integer multiple of the sampling frequency.

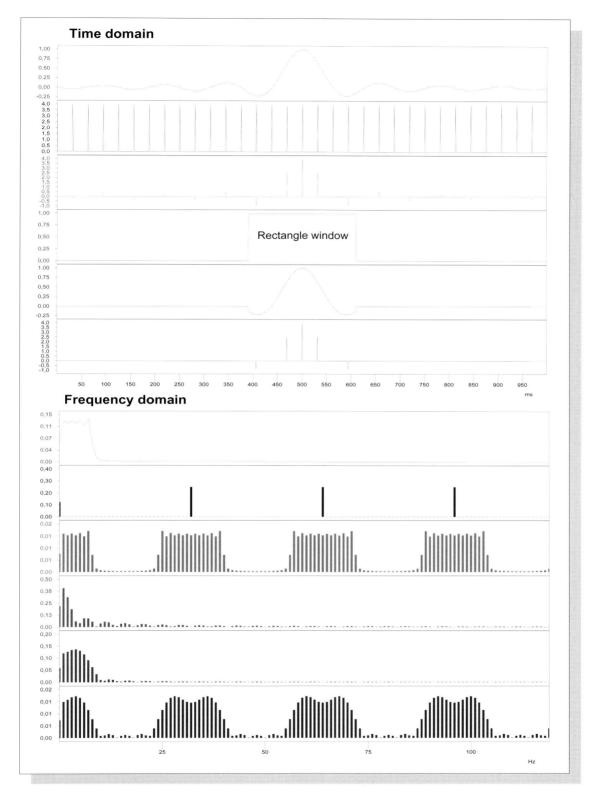

Illustration 193: ***Signal distortion as a result of signal windowing***

In the case of longer signals the signal must be processed in segments. This gives rise to errors as shown in Illustration 51. Using the example of the rectangular window we show once again how the spectrum can be changed or distorted. Typical of rectangular windows is the "ripple effect" of the spectrum (bottom) compared with the correct spectrum in the third series (bottom).

Signal distortion as a result of signal windowing

In conclusion we should like to return to the digital processing of long-lasting analog signals – for instance an audio signal.

> *The digital processing of long-lasting real analog signals always implies – as in the process of hearing – the signal processing of equally long overlapping signal segments.*

Here is a brief summary based on the descriptions in Chapter 3 (see Illustrations 51-54) and in Chapter 4 (see Illustrations 61 - 65):

- Long-lasting signals must be analysed and processed in blocks. This block length must always be presentable as a power of two (e.g. $1024 = 2^{10}$) because the FFT is only optimised for these block lengths.

- The block length and sampling rate – in accordance with the Sampling Principle – should as far as possible be synchronised with each other so as not to conflict with period length T_D of the digital signal. This always applies if the signal processing includes the frequency domain. Remember that only integer multiples of the basic frequency $f_D = 1/T_D$ can be shown in the spectrum. Select the sampling rate f_S if possible as a power of 2.

- The windows must strongly overlap otherwise information is lost which was contained in the separate time segments. The information is contained in the overall signal.

- The problem can be solved by recourse to our fundamental principles. As you know information-bearing signals can according to the FOURIER Principle be understood as consisting of nothing but sinusoidal signals of a certain bandwidth. If the transmission of these sinusoidal signals is guaranteed this also holds for the information they are transporting.

- For long-lasting signals it is only possible to use a window function in which the windowed signal begins and ends gently. Rectangular windows produce signal steps which have nothing to do with the original curve (see Illustration 193). GAUSSian windows are an appropriate example.

- The overlapping required can be visualised and assessed via a frequency-time landscape. Thus it can be seen from Illustration 66 that a shorter overlapping of the windows would not give more information on the frequency-time landscape. In Illustration 65 there is by contrast a need for more information. Apparently the overlapping of the windows is still too far apart.

- The overlapping required can be more precisely estimated by means of the Uncertainty Principle:

    - A window with the length Δt necessarily produces a frequency uncertainty $\Delta f \geq 1/\Delta t$. The selection of the "window length" determines the frequency resolution – independently of the bandwidth B of the signal!

    - The defining of the bandwidth B also determines the highest frequency f_{max} which is to be recorded in terms of information. The fastest time change, which the signal can contain, is also determined via this. As a result of **UP** the fastest change in time in the range $\Delta t \geq 1/B$ as B ranges from 0 to f_{max}. The overlapping of the windows must take place at the interval τ.

Example:
Let a window have the length $\Delta t = 100$ ms. Then the frequency uncertainty Δf is at least 10 Hz. If the bandwidth of the signal is 20 kHz the overlappings should be effected at time intervals of $\tau = 1/20000$ s = 50µs.

- Overlapping at shorter intervals would not give more information because the signal length (of the windows) as a result of UP simply cannot produce more. Longer intervals would (given the assumed bandwidth of the signal) mean a loss of information.

> *Only the bandwidth B of the signal determines the interval τ with which the windows should overlap. The desired frequency resolution Δf is determined only by the signal length Δt of the windows.*

Check list

In view of the many possibilities for errors in the digital processing of analog signals you may now be somewhat confused. For this reason here is a kind of check list. You should go through this in the context of practical applications.

- First try to determine the bandwidth B of the present analog signal. This can be done in several ways:

    - The physics of signal generation allow certain conclusions with regard to the bandwidth. This presupposes a lot of experience and understanding of physics.

    - You may know the bandwidth B as a result of the source of the signal – e.g. a microphone.

    - The fastest instantaneous time change of the signal can be estimated from the highest frequency f_{max}

    - Set the highest possible sampling rate of your multifunction board in the DASY*Lab* menu – to ensure that you fulfill the Sampling Principle – and represent the frequency range by means of an FFT

    - If all these things do not give absolute certainty connect an (analog) antialiasing filter between the signal source and the multifunction board, i.e. practically an analog lowpass filter.

    - Its cutoff frequency should be set as low as possible but as high as necessary (to avoid loss of information).

- You may find empirically established figures in the literature (specialist articles) on your particular problem.

- The transfer function of the antialiasing filter always influences or changes your signal. The higher the quality – i.e the "more rectangular" the filter, the slighter the influence.

• Select the sampling rate as high as possible so that the sampling principle is overfulfilled.

- In this way the sidebands are far apart in the spectrum as in Illustration 191

- In this case you do not need an antialiasing filter of the highest quality with almost rectangular edges (very expensive) but can possibly use a lower quality filter – for example an RC circuit. The edge of the filter lies outside the bandwidth of the (analog) signal if the cutoff frequency is selected accordingly.

• If long-lasting signals are present which have to be processed in segments, the following rules apply:

- You set the frequency resolution ("uncertainty") with the "window length" between the signal source and the multifunction boardt. $\Delta f \geq 1/\Delta t$ applies.

- The bandwidth B of the analog signal or the antialiasing filter defines the time interval τ with which the windows must overlap. $\tau = 1/B$

Note:

With a sound card with two analog inputs and two analog outputs the sampling frequency cannot be selected at will. The highest sampling frequency of the sound card is 44100 Hz, followed by 22050, 11025, 8000, 4000, 2000 Hz …etc. If an incorrect sampling frequency is selected an automatic correction informs you.

As the frequency range only extends to 20 kHz even in the case of high-quality microphones the sampling frequency is guaranteed up to 44100 Hz.

A microphone and a sound card give you a high quality and inexpensive system for the reception, processing and reproduction of real analog signals in the audio-sphere.

Exercises on Chapter 9

Exercise 1

(d) Why can there not be a software-based – i.e. computer-based conversion of real analog signals into digital signals?

(e) What does the file of a digital signal of this kind contain and how can the information be presented pictorially?

Exercise 2

(a) Transform the circuit simulation of an A/D converter in Illustration 174 by means of DASY*Lab* in such a way that the accuracy is 4 bit or 6 bit instead of 5 bit ($2^5 = 32$)

(b) Instead of the AND process use a "Relay".

(c) How could the string of numbers in the binary system be represented by means of DASY*Lab*? What would have to be added to the circuit in Illustration 174.

Exercise 3

(a) Explain the principle of a D/A converter in Illustration 175, i.e. the term and the formula in the mathematics module (binary/deca).

(b) Explain the conversion of numbers in the decimal system into the numbers of a different system with the base 2, 3 and 4 based on Illustration 176

Exercise 4

(a) Where are analog pulse modulation processes used in practice?

(b) Why are these processes so important in measuring technology?

(c) Compare the pulse modulation signals in Illustration 177 with reference to their bandwidth.

Exercise 5

(a) Explain the general difference between an analog and digital signal in the time domain

(b) What graphic possibilities does DASY*Lab* provide for digital signals (see Illustrations 178 and 179)?

(c) Digital signals can only be analyzed in segments. What problems result from this?

(d) Explain the influence of the two quantities block length n and sampling rate f_S on the representation of a digital signal in the time domain. How should these quantities be selected so that the digital signal on the screen is as similar as possible to the analog input signal?

(e) Explain the term "digital period length T_D"

(f) When is the spectrum of a periodic analog signal reproduced correctly via digital signal processing by means of FFT?

Exercise 6

(a) Why is it not possible to process signals with steps – e.g. sawtooth signals – correctly using digital processing?

(b) Why analog signals must be bandwidth limited before they are processed digitally?

(c) What effect does the bandwidth of an analog signal have on the required sampling frequency?

(d) Formulate the sampling principle for band limited analog signals which are to be processed digitally?

Exercise 7

(a) Explain why a digital signal necessarily has a periodic spectrum

(b) Why must every digital signal – stored over a block length n – be understood as periodic in the time domain over the block length n?

(c) Analog signals can be interpreted as the limiting case of a digital signal. What does this limiting case look like?

(d) By means of what trick using DASY*Lab* was the frequency domain in Illustrations 187–193 expanded to such an extent compared with Illustrations 180 – 186 that the periodicity of the spectra can be recognized.

Exercise 8

(a) What sources of error can arise if the block length and the sampling frequency are not in an integer relationship to each other?

(b) Cut out from a longer speech signal a short segment, for example, of 0.07s in length and limit the bandwidth of this segment on account of the sampling principle. Analyze the frequency domain of this signal with different sampling rates and block lengths. Discuss the differences in the spectral representations. Which of these clearly show errors? What is behind this?

Exercise 9

You wish to analyze spectrally a long-lasting (bandwidth limited) audio signal without any loss of information via a Frequency-time landscape.

(a) How do you define the frequency resolution of this frequency-time landscape?

(b) How is the bandwidth of the signal guaranteed in this representation

(c) Explain the selection of the window function you prefer.

Chapter 10

Digital filters

Filters are extremely important in signal processing. Filters were already dealt with in Chapter 7 – as an example of linear processes. Although the focus was on analog filters digital filters were also dealt with. The advantages of digital as opposed to analog signal processing can be shown using the example of *digital* filters.

Filters – whether analog or digital – are considered from a theoretical point of view to be very complicated. Practitioners tend to use books of tables to select the circuit and dimensions of the components used (e.g. resistances and capacitors including permitted tolerances for the analog filter they require). In the case of digital filters this is similar at first sight: depending on the type of filter the appropriate "filter coefficients" are required.

The aim of this chapter is to clear up these difficulties especially for digital filters, to explain clearly the way they work and the computer processing. As for any other process of digital signal processing the following also holds for digital filters: the signal – in the form of a string of numbers – is processed by computer. You should be in a position to design and use digital filters of the highest quality with the help of DASY*Lab*.

Hardware versus software

Analog filters differ completely from digital filters both in the approach and in their form although they are both intended to do the same thing: to filter out a certain frequency range and effectively suppress all the others.

One difference is immediately noticeable:

> *An analog filter is a circuit – usually with operational amplifiers, and discrete components such as resistances, capacitors ("hardware").*

> *By contrast a digital filter is of a virtual nature, a program ("software") which calculates from the string of numbers corresponding to the input signal another string of numbers corresponding to the filtered signal.*

Note: special filter types such as surface wave filters, quartz-crystal and ceramic filters and filters with mechanical resonators (as used in carrier frequency technology) are not to be dealt with here.

How analog filters work

The function of analog filters is based on the frequency-dependent behaviour of the components capacitor C and inductivity L (coil). From a mathematical point of view the relationship between the voltage u and and the current flowing through i in these two components is described by differentiation or integration. The physical behaviour can be described more simply and in a way that is easier to understand as follows:

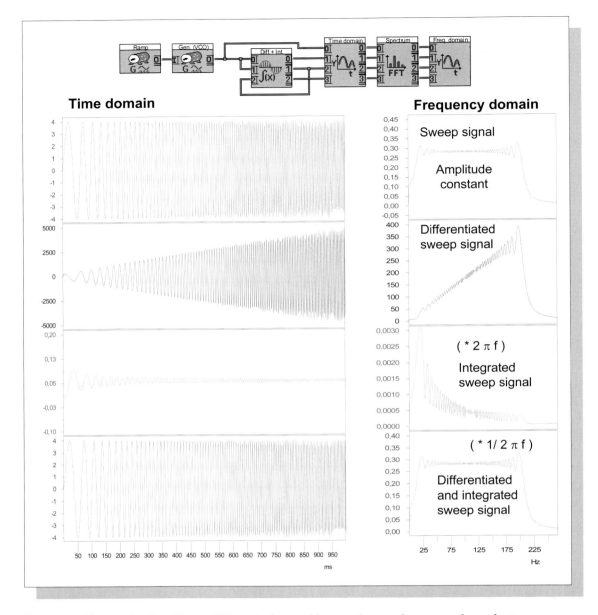

*Illustration 194: **Analog filters: differentiation and integration as frequency-dependent processes***

In the differentiation of a sweep signal the amplitude increases linearly (proportionally) to the frequency and in the case of integration it is inversely proportional to the frequency. This can be seen in the Illustration both in the time and frequency domain. Note the effect of the Uncertainty Principle in the frequency domain (see in this connection Illustration 97).

*In the case of an analog resonance circuit consisting of a coil and a capacitor or an analog bandpass the voltage **u** and current **i** are linked by these two processes differentiation and integration. Only in this way can the resonance circuit effect be achieved.*

- The faster the current **i** in a coil changes the greater the instantaneous induced voltage **u** (*law of induction*)

- The faster the voltage **u** changes at a capacitor the greater the current **i** which the condenser discharges or charges (*law of capacity*).

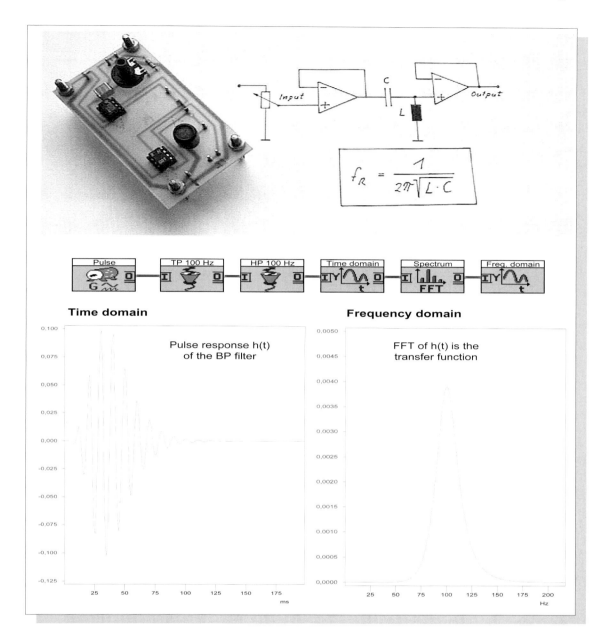

Illustration 195:   ***Analog bandpass (resonance filter) and its representation using DASYLab***

Above you see an analog bandpass circuit (resonance filter) in which an L-C-series resonant circuit together with two operational amplifiers form the filter. The operational amplifiers are connected as "impedance converters". Here the input resistance tends towards infinity (as a result the coil doesn't "notice" that its signal is being "drawn off") and the output impedance tends towards zero (this means that no additional resistance is switched into the series. In the case of the resonance frequency f_R the resistance (or the "impedance") is smallest, the current is consequently greatest and the voltage drop at the coil and the capacitor reaches its maximum value, which may be much greater than the input voltage. This is selected at so small a value via the input voltage divider that the operation amplifier is not overdrived.

These are all problems which do not exist with DASYLab or digital signal processing. Below you see a DASYLab circuit equivalent to the analog resonance filter. The resonance filter is represented by a series circuit of lowpass and highpass with the cutoff frequency 100 Hz. The pulse response h(t) is a "short sinusoidal signal" of 100 Hz which as a result of the Uncertainty Principle goes hand in hand with the "spectral uncertainty" around 100 Hz of the filter curve.

As already dealt with in detail in Chapter 7 (see Illustration 120) in the case of sinusoidal input signals a sine appears at the output of a differentiator the amplitude of which is proportional to the frequency. In the case of integration – as the reverse of differentiation – the amplitude is inversely proportional to the frequency.

A simple resonant circuit is particularly frequency-selective, i.e sensitive to a change in frequency. This is a serial or parallel circuit consisting of a coil and capacitor or inductivity L and capacity C. In Illustration 195 there is a reasonably good analog resonance filter (bandpass) in conjunction with two operational amplifiers.

Thus we come to the real problem: there are no good analog filters. They may be good enough in one respect or represent a compromise between edge steepness, ripple effect in the conducting state region and (non-) linear phase curve in the conducting state region. The reasons for this were explained fully in Chapter 1 („Clarification of objectives", under Analog components). Among other things the following holds:

- Real resistances, capacitors and particularly coils show mixed behaviour. Thus a coil consists in purely physical terms of a serial circuit of inductivity L and the ohmic resistance R of the coil wire. In the case of very high frequencies a capacitor C between the parallel wire coils becomes apparent.

- Analog components can only be produced with limited accuracy and in addition they are temperature-dependent etc.

To summarize, analog filters do not do what ought to be theoretically possible because "dirty effects" make this impossible. Thus the resonance behaviour of the bandpass in Illustration 195 depends largely on the quality of the coil. The smaller the resistance of the coil the sharper the frequency selection.

Three types of analog filter predominate whose particular advantages and disadvantages were described in Chapter 7 and particularly in Illustrations 131. In the future they will only be used where they cannot be avoided:

- Analog filters will always be used to limit the analog signal in frequency before it is processed digitally. Note: this frequency limiting can also be effected by a "natural" lowpass, for example, by means of a microphone. The human voice is clearly limited in frequency.

- Signals of high and the highest frequencies can at present not be digitally filtered on account of the limited speed of A/D converters.

FFT filters

We have often used the first type of purely digital – i.e computer – filter with DASY*Lab* (Illustrations: 25, 26, 27, 109, 110, 151, 152, 156, 157, 158, 181, 191).

This principle which is easy to understand - but which from a computer point of view is complicated - is shown once again in Illustration 196. The signal or a block of data is transformed into the frequency domain by means of an FFT transformation.

Chapter 10   Digital filters

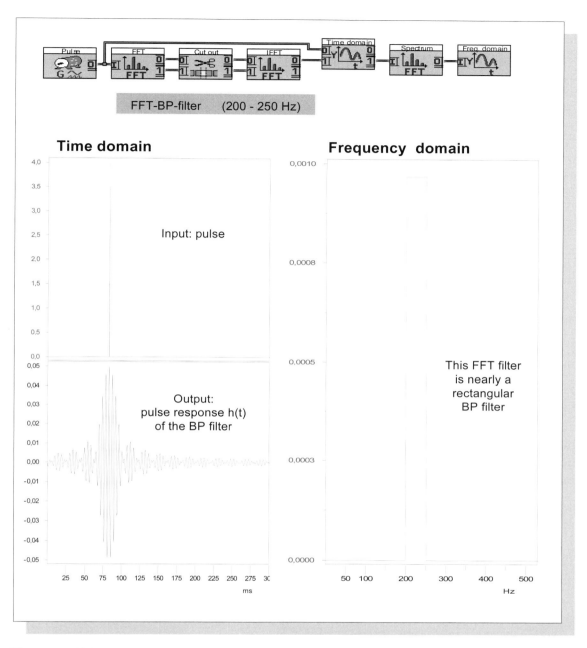

Illustration 196:   ***Bandpass as an FFT filter***

You can see here in the example of a bandpass filter how high-quality FFT filters can be. The criterion for a very good – i.e. virtually rectangular filter – can be seen in the time domain – the pulse response looks roughly like an Si-function – multiplied by the mid-frequency of the filter in the case of the bandpass.

The edge steepness is limited only by the Uncertainty Principle, i.e by the length of the input signal (string of numbers, or data blocks) in the time domain. It was 1 s here (the illustration above shows only a time segment of 250 ms).

The complex FFT of a real signal should be selected by means of the menu. The FFT module has one input, but two outputs. Both outputs provide the real and the imaginary part of the spectrum (see Chapter 5 under "Inverse FOURIER transformation IFT and GAUSSian plane").

In a physical sense this means that two pieces of information are essential for any frequency and any sinusoidal signal: i.e. amplitude and phase.

Subsequently the frequencies are "cut out" using the cutout module (i.e. are set at zero) which are not to be allowed to pass. The same values must be set on both channels (real and imaginary).

> Note:
> In the cutout module you will find the values 0 to 8192 preset under "data of sample". 8192 gives the largest possible data block length for FFT processing. How can the desired frequency range be set? Look carefully at Illustration 109, for example. If you select the sampling frequency equal to the block length (e.g. 1024) the number range – e.g. from 0 to 40 – gives the conducting state region of the filter in Hz. In this case the time length of the data block is exactly 1 second and this is the most straightforward state of affairs. Otherwise you have to be extremely careful. If you select a frequency above the area allowed by the Sampling Principle (in this example that would be above 512 Hz!) you may find the position of the filter at a completely different point in the frequency range. Try this out using the circuit in Illustration 196.

Afterwards you go back into the time domain via an IFFT (inverse FFT). Here "complex FFT of a complex signal" should be selected from the menu options. In addition, you must also indicate that you wish to go back into the time domain. Finally, the signal is to be made up of the frequencies which have been allowed to pass through (sinusoidal oscillations), that is a FOURIER *Synthesis* is to be carried out. The filtered signal is only present in its correct form at each upper output.

Advantages of FFT filters:

- How well an FFT filter of this kind works can be seen from Illustration 110. Individual frequencies are filtered out of a noise signal, the bandwidth is 1 Hz with a date block length of 1s and represents the absolute physical limit resulting from the Uncertainty Principle **UP**. The edge steepness of the filter thus depends on the length of the signal or the data block.

- A further advantage is the absolute *phase linearity*, i.e the form or the symmetry of the signal in the time domain is not changed. Compare in this context the properties of analog filters in Illustration 131 with those of the FFT filter in Illustration 196

Disadvantages of FFT filters

- The large amount of calculation required for the FFT and IFFT. FFT means "Fast FOURIER Transformation". The algorithm was published in 1965 and is much faster than the ordinary DFT (Digital FOURIER Transformation) by the use of the Symmetry Principle. FFT and – in addition – IFFT are nevertheless still compute-bound in comparison to other signalling processes.

Chapter 10 Digital filters

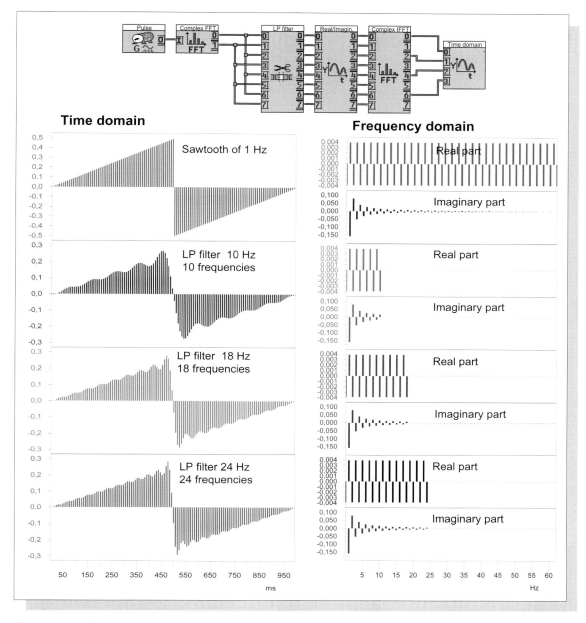

Illustration 197: **FFT-filtering of a sawtooth signal**

This shows clearly how simply FFT filters work in principle: the frequency range which is of interest is literally cut out using the "cutout" module. After re-transformation into the time domain by means of an IFFT the filtered signal appears.

- It is unsuitable for long lasting signals. As shown in Chapter 3 under the heading "Frequency measurements in the case of non-periodic signals" and in Chapter 4 "Language as an information carrier" these would have to be cut up into segments using an appropriate window, then arranged with overlapping and filtered. This would be too error-prone and compute-bound for nothing more than filtering.

An (apparent) inconsistency of the FFT filter should be explained. In Illustration 196 the filter appears as "non-causal" as a result of the representation. The output signal is present on the screen before the input signal arrives at the input!

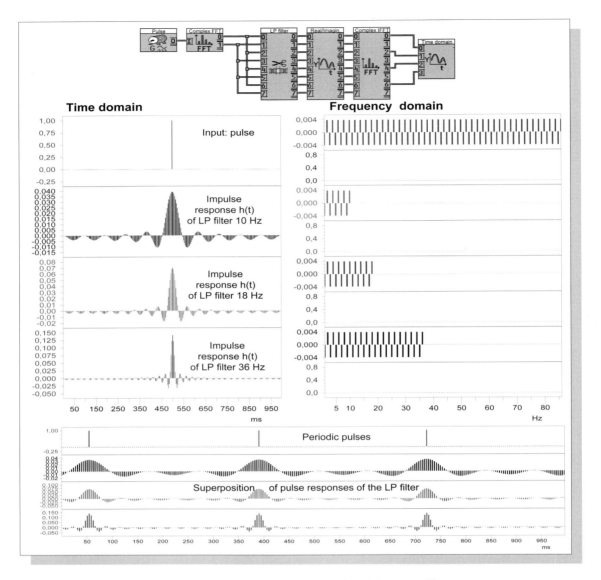

Illustration 198: **Pulse response h(t) of FFT lowpass filters**

It can again be clearly seen how in the case of "rectangular-like" filters a pulse response h(t) similar to an Si function results. The "ripple effect" of the Si-function corresponds to the highest frequency that the filter lets through.

As we are dealing here with digital signals which are always discrete in the time and frequency domain and are therefore also periodic (see Chapter 9), the "segment" from an Si function is in reality a segment from a periodic Si function. To be precise, as a pulse response you see a period length of the complete pulse response.

The reason for this is described in Chapter 9 under the heading "The periodic spectrum of digital signals". A digital signal in the time and frequency domain consists of lines and is time and frequency discrete. As a result of the Symmetry Principle the signal being processed is regarded and represented as periodic in the time and frequency domain. To be precise, what is filtered at the input of the measuring system is different from the analog signal segment, i.e. it is a periodic signal extending into the past. Under this aspect the filter is causal.

Note: In the Version 6.0 of DASY*Lab* there is a so-called FFT filter which (at present) does not correspond to the procedure shown here. It simply makes "cutting out" in the frequency domain easier.

Digital filtering in the time domain

Hopefully you have not lost the ability to feel amazement at the unexpected! You will now be introduced to digital filters which

- require little calculation,
- avoid the route via the frequency domain (FFT – IFFT),
- In principle and in practice have no fixed blocklength/signal length,
- for this reason can filter signals of any length directly,
- are completely phase-linear,
- can be designed with the desired edge-steepness (the Uncertainty Principle is the only physical limit) and
- make do with the three most elementary (linear) signalling processes: *addition, multiplication by a constant* and *delay*.

Perhaps you already sense how that might be possible. The input signal in the case of a lowpass characteristic simply has to be "deformabled" in such a way that it has a "ripple effect" which, for instance, in the case of a lowpass filter is equivalent to the highest (cutoff) frequency of this lowpass (see in this connection for example Illustrations 49 and 191). All the prerequisites for this have, of course, already been dealt with. They are summarized briefly in the following:

All digital signals represent a discrete sequence of (weighted) δ-pulses (see in this context Illustrations 178 and 179).

- The pulse response of a virtually ideal, i.e. rectangular lowpass filter must look roughly like the Si function (see, for example, Illustrations 48 and 49)

- The pulse response of a virtually ideal i.e. rectangular bandpass filter is always roughly an amplitude-modulated Si function (see Illustrations 109 and 196). The mid-frequency of this bandpass is equivalent to the carrier frequency from Chapter 8, under the heading "Amplitude modulation".

- A sampled Si-function can be seen as the pulse response of a digital lowpass (with periodic, virtually rectangular spectra) (see Illustrations 205 and 206). The Sampling Principle must be adhered to to avoid these spectra overlapping.

The following conclusion results:

> *As a digital signal consists purely of weighted δ-pulses a signalling process is required which generates a discrete but time-limited δ-pulse sequence which is as similar as possible to the Si function from each of these δ-pulses.*

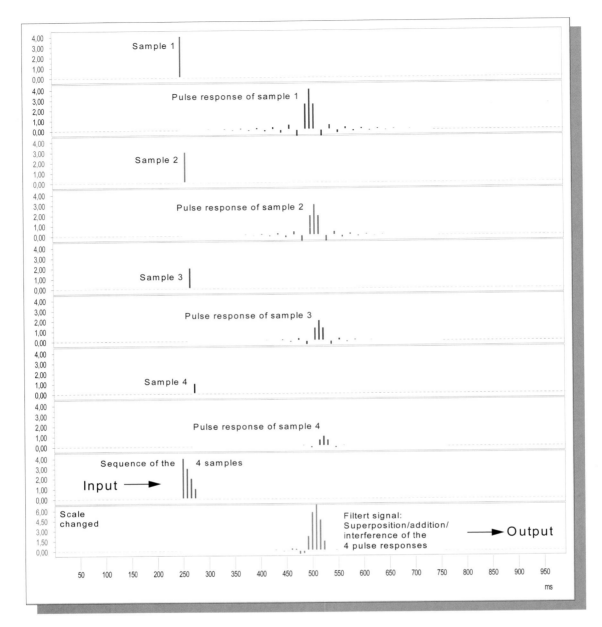

Illustration 199: **Digital filtering in the time domain by overlapping the pulse responses**

The digital signal or the string of numbers consists of discrete "measurements" which reproduce the instantaneous curve of the signal. Every measurement is a weighted δ-pulse the pulse response of which in the case of a rectangular-like filter curve has an Si shaped curve. As the overlapping (addition) of the discrete measurements corresponds to the instantaneous curve of the signal, the addition of their pulse responses must give the instantaneous curve of the filtered signal.

Illustration 199 shows this in a particularly straightforward way. Here there are four equidistant δ-pulses of different amplitudes and – below each one – the discrete sinusoidal pulse responses. The former are intended to represent three "measurements" of the present signal at the imput of a digital lowpass with their discrete pulse responses at the output. Below you see the sum as superposition of these three pulse responses. It gives the filtered instantaneous curve of the input signal. The cutoff frequency of the lowpass can be deduced from this "ripple effect".

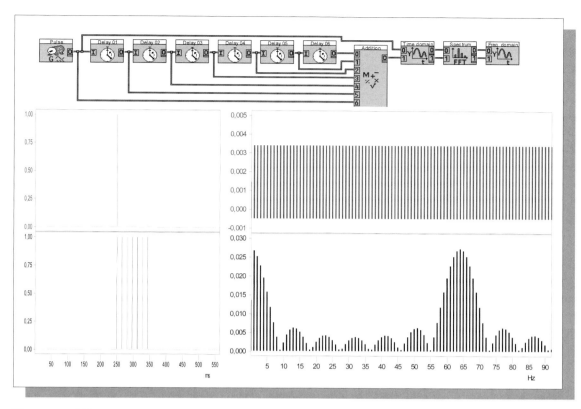

Illustration 200: **Generation of a longer lasting pulse response**

By means of an extremely simple circuit with the two elementary signal processes delay and addition, a digital signal of any length can be generated from a single δ-pulse. The last pulse has passed through all 6 delay processes etc. The addition of the seven time-displaced pulses result in the bottom signal.

In the case of a rectangular output signal in the time domain one is not surprised at the (periodic) Si-shaped curve in the frequency domain. But in actual fact we wanted things the other way round: a rectangular curve in the frequency domain (filtered). What must the pulse response look like for reasons of symmetry and how would we have to complete the circuit?

> *The addition (or sum) of the - time-displaced – si-shaped pulse responses of the particular type of filter result in the filtered output signal.*

By the way, you encountered a digital filter of a special type ("comb filter") in Illustration 116 and 81 and 82. There you see what role the delaying process plays in the case of digital filters. Instead of the constant amplitude spectrum of a single δ-pulse the spectrum in the case of two δ-pulses are cosine-shaped. You will find the explanation for this in the text of Illustration 81. Certain frequencies at regular intervals ("comb-like structures") are not allowed to pass.

Digital filters also obey the Uncertainty Principle: the smaller the bandwidth of the filter the longer the pulse lasts and vice-versa. In the case of a filter the pulse response is necessarily longer than the signal at the input. This must be made technically possible by the circuit. First a straightforward circuit should be developed which makes several δ-pulses of the same level from one δ-pulse i.e. makes a longer pulse response from a single pulse and therefore must function like a filter. This is shown by Illustration 200.

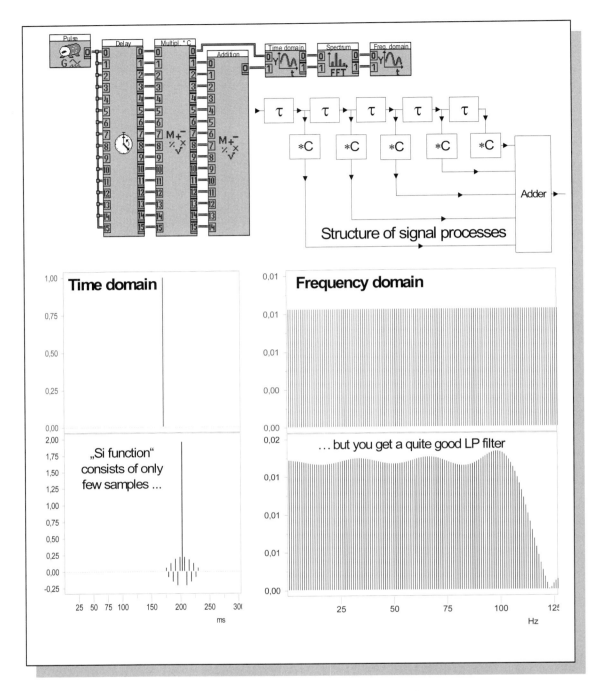

Illustration 201: *Making a pulse response into a Si-function*

At the top you see the circuit which makes something like a (discrete) Si-function from a δ-pulse, first as a DASYLab circuit and then as a principle block diagram next to it on the right. Below this you see the Si-function in the time domain and next to it the spectrum which already suggests quite reasonable lowpass properties.

In other words – the schematic circuit (top right) represents the technical circuit structure of a digital filter and is content with the three elementary (linear) processes – addition, multiplication by a constant and addition. What type of filter, what quality of filter and what conducting state region result, depends on the number of delays and multiplications (here n= 5) and from the filter coefficients which correspond to the value of the different constants C_n. There is another important influence: the sample frequency/rate plays the role of a „bandwidth scaling factor" (see Illustration 205).

You will be familiar with the course of the amplitude spectrum. The curve is based on the response of an Si-function as in numerous examples in Chapter 2 (e.g. Illustration 38), but is here periodic because it is discrete in the time domain.

The Symmetry Principle tells us that a pulse response which is almost like an Si function would have to result in a rectangular-like filter curve. But how can an Si shaped pulse response be generated by a modified circuit compared with Illustration 200. This is in principle shown by Illustration 201. By multiplication of the individual δ-pulses by *certain constants* ("filter coefficients") the pulse response is transformed as far as possible into an Si function like form.

This circuit has a very straightforward structure and is content with only three elementary (linear) signal processes – *addition, multiplication by a constant* and *delay*.

But how do we arrive at the right coefficients? A potential but complicated possibility would be – as shown in Illustrations 189 and 190 – to sample an Si-function with a periodic δ-pulse sequence and to have these values stored as a list. In this way the correct constant for each module could be inputted but in a very laborious way.

Please note in Illustrations 200 and 201 that at the output of the adder (summation) the weighted δ-pulse at the lowest input is the first to appear at the output of the adder and the uppermost which passes through all the delays appears last.

Convolution

5 or 15 weighted δ-pulses according to Illustrations 201 and 202 are not sufficient to produce an Si-shaped curve. 256 δ-pulses would be better. Then the circuit would be so complex that it would not fit on the screen. And setting 256 coefficients manually would be slave labour.

But this is not necessary as the schematic circuit according to Illustration 201 embodies an important signalling process – convolution – which is available with DASY*Lab*, also in the educational version, as a special module.

Convolution as a signalling process was already mentioned in Chapter 7 in the section "Multiplication of two signals as a non-linear process". See above all Illustration 136 and the relevant text. In this case, however, it was a question of a convolution in the frequency domain as a consequence of a multiplication in the time domain. This is a multiplication in the frequency domain ("rectangular filter") and – as a result of the Symmetry Principle – a convolution in the time domain.

The following result should be noted:

- A *multiplication in the time domain* produces a *convolution in the frequency domain*. Important example: The sampling of an analog signal with a periodic δ-pulse sequence as in Illustration 136. The spectrum of the analog signal is "convoluted" at each frequency of the δ-pulse sequence.

- For reasons of symmetry the following must hold:
  A *multiplication in the frequency domain* (as with the filter) produces a *convolution in the time domain*.

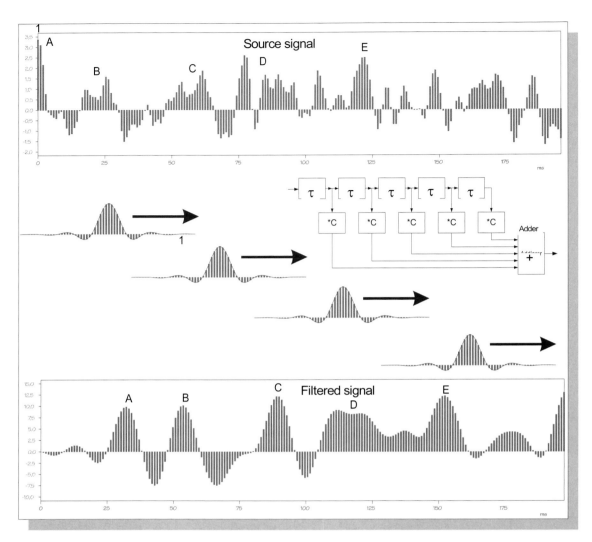

Illustration 202: **Illustrating the process of convolution in the time domain**

The block diagram shows the development of the convolution process but it is difficult to internalise this pictorially. Imagine the (discrete) approximated Si-function as a template which moves from the very left to the very right by the top signal, stopping briefly at each step from measurement to measurement. At each position of the template the convolution operation represented by the block circuit is carried out.

At first only the first two values of the signal and "template" overlap, designated in the representation by means of "1". Then in the next step 2. 3 ... up to a maximum of 64 values (length of the template is here $n = 64$). At every step a kind of "weighted average value calculation" is carried out, i.e. in the frequency domain a lowpass filtering process. A maximum of 64 different values are adduced to calculate the average value, i.e. 64 different measurements are simultaneously within the block circuit diagram.

Note the Si-shaped beginning of the filtered signal. It is a result of the pulse-like beginning of the unfiltered signal. The letters designate the comparable segments of both signals.

Creating a good filter means from a mathematical point of view the multiplication of the frequency spectrum by a rectangle-like function. This means however that in the time domain convolution must take place with an approximated (above all time-limited) Si function because rectangles and Si functions are inseparably linked by a FOURIER transformation (see Illustration 84).

While multiplication is a familiar mathematical operation this is not true of convolution. It is therefore important to illustrate it by suitable processes of visualisation. The basis for this is the combination of the three basic linear processes of delay, addition and multiplication by a constant. The signal flow in convolution produces a block diagram with a very simple structure which is again underlined in Illustration 202.

Note the delay between the input signal and the filtered output signal (see designation A, B....., E). It can be seen that all rapid changes in the imput signal are "swallowed up" by the filter or disappear via the weighted averaging.

Unlike the FFT filter the output signal is here strictly causal, i.e. something only appears at the output after a signal was connected with the input. All in all, there are the following advantages compared with the FFT filter:

- The filter process takes place in the time domain and as a result of the elementary processes involved does not require a great deal of calculation. As a result real time filtering in the audio field is now perfectly possible. The amount of calculation necessary increases "linearly" with the block length of the Si-function or with the precision or quality of the filter.

- It is possible to filter continuously – i.e. not in blocks. As a result all the problems which occurred with the "overlapping windowing" (see Chapter 4: "Frequency-time landscapes") of signal segments do not arise.

- The filter functions causally like an analog signal.

Note:
The type of filter described here is called FIR filter (finite impulse response) in the literature. This type of filter generates a pulse response of a finite length (e.g. in the case of a block length of n = 64 or n = 256). In addition so-called IIR filters (infinite impulse response) are also used. Here the same elementary processes are used but as a result of feedback effects fewer processes – i.e. less calculation – are required all in all for the design of the filter. However, the phase curve is no longer linear. IIR filters are not dealt with here.

Case study: Design and application of digital filters

The right instruments seem to be available in the convolution module to deploy effective digital filters. As filtering corresponds to multiplication in the frequency domain, the equivalent operation in the time domain represents convolution. Both processes are completely equivalent as far as their effect is concerned if the convolution function is the IFFT of the filter function.

The variety of design alternatives offered by DASY*Lab* make one want to design a convenient development tool for digital filters where the filter range can be set on the screen and the filter coefficients of the Si-shaped pulse response appear at the touch of a button.

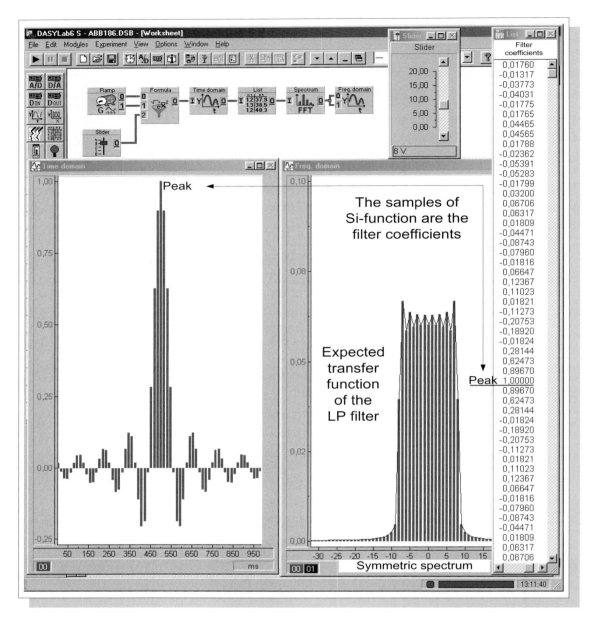

Illustration 203: A "filter development site"

The maximum number of filter coefficients is here n = 1024. The number desired must be set at the top in the menu under A/D. In this case n = 16 was selected so as to avoid the list becoming too long and to show that even with this small number the result (in the case of a relatively broadband TP filter) is quite acceptable. These 16 filter coefficients can be seen at the bottom of the diagram as the vague suggestion of an Si function. The FFT shows the characteristics of this lowpass in the frequency domain in the central section of the diagram.

Values from n = 64 are recommended to obtain a good lowpass and from 256 for a bandpass. The narrower the frequency range of the filter, the longer the Si shaped pulse response has to last according to the Uncertainty Principle, and the more coefficients are normally required.

The following points are especially important:
(a) Never select the bandwidth of a filter higher than half the final sampling frequency (see text), otherwise you will contravene the Sampling Principle.
(b) The further apart the bands of the total periodic spectrum are to be the higher the sampling frequency selected must be.

Solution:

- An Si-function generator is created by means of the formula module in which the Si function can be selected "at will" at the inputs of the module using a slider module.

- The possibility of multiplying the Si function by the mid-frequency of the bandpass should be provided for. The module "list" is connected with the output of the formula module. It indicates the filter coefficients set. These can be copied from the list on to the clipboard.

- As it is difficult to allocate the relevant frequency range to each Si-function at the output of the list an FFT is carried out with the subsequent indication of the frequency domain. It can now be clearly recognised in the frequency domain how a change set by the slider makes itself felt (see Illustration 203).

The string of numbers of the filter coefficients must be represented in a particular form as a "vector file". For this purpose an Editor is used which is to be found under Accessories under Programs in the Start menu:

(1) The list is placed in the Clipboard via the menu (by clicking the left mouse key on to the list option) Edit and List to Clipboard.

(2) Start the Editor – under Accessories in the Windows Programs overview.

(3) Activate the menu of the *Convolution* module and then *Help*. The design of the vector file is described here. Examine the example and note the structure of the vector file.

(4) Write the header of the file (see Illustration 204), copy the filter coefficients from the clipboard, delete everything apart from the "string of numbers" and add EOF (end of file) at the end.

(5) Store the vector file first as a *.txt-file in a filter folder. Then change the file ending from "txt" to "vec" in *Explorer*.

(6) Now load this *vec-file* into the *Convolution* module and the digital filter is complete.

In designing a filter the following facts which result from the Uncertainty Principle and the Sampling Principle are of fundamental importance:

- If you select for example a block length and sampling frequency of $n = 128$ for the pulse response of the planned filter ($n = 128$ filter coefficients) you can at first only select a maximum filter bandwidth of 64 Hz (more precisely, from -64 to $+64$) on account of the Sampling Principle. As the block length and the sampling frequency in the filter experimentation site should always be selected as equal, the pulse response then lasts exactly 1 second.

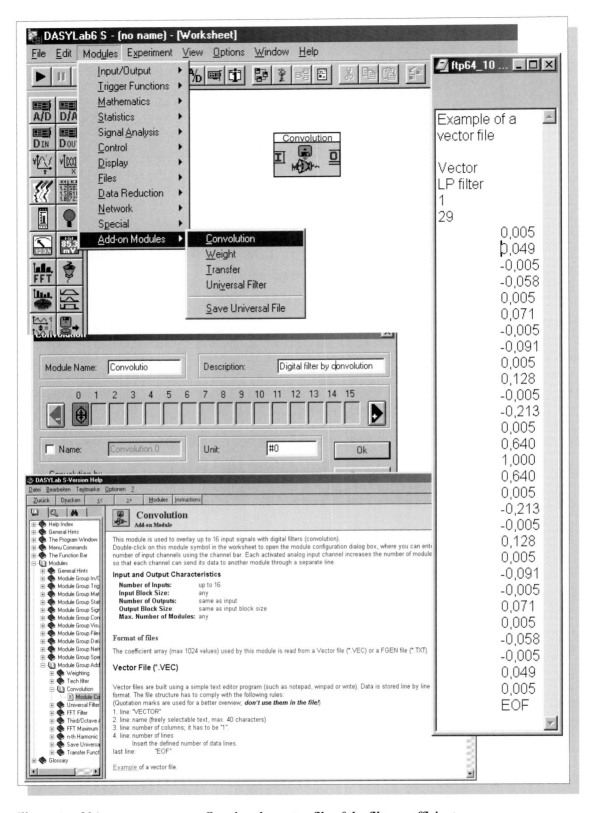

Illustration 204: ***Creating the vector file of the filter coefficients***

As with any DASYLab module the Help function provides a description of the procedure for creating the filters – "vector file". Familiarise yourself with the way the editor works and how the file *.txt can be renamed *.vec in Explorer.

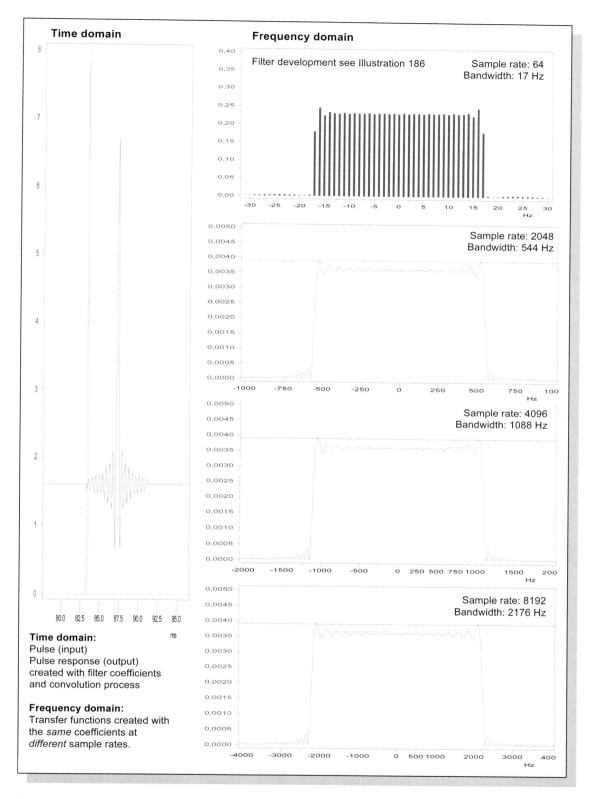

Illustration 205: **Bandwidth of a digital signal**

In the case of the "filter development site" the block length and sampling frequency are equivalent to the number of filter coefficients. By contrast, the sampling frequency of the convolution module is much higher in practice. The real bandwidth of the filter results from the relationship between both frequencies, here for example 8192/64 = 128 at the bottom (128 *17 (top) = 2176 Hz bandwidth).

- If the system in which the convolution module is contained works at a sampling (rate) frequency of for example 8192 the time for the pulse response used for the convolution is shortened by 64/8192 = 1/128 s. According to the Uncertainty Principle the frequency band widens to a maximum of 128*17 Hz = 2176 Hz. The filter bandwidth of digital filters therefore depends on the final sampling frequency of the system.

- This vector file is assigned to the convolution module via its menu.

The number of filter coefficients should be between 16 and 1024. The higher the number the greater the quality of the filter and/or the intervals between the bands of the periodic spectrum and the greater the amount of calculation. Initially you should in your design select the sampling rate and the block length as equal.

> *In the design of digital filters it must be known from the outset what the final level of the sampling frequency of the system will be. If the sampling rate of the system in which the digital filter is to be used is n times greater than the sampling rate than in the original setting of the filter coefficients, the Si-function represented by the filter coefficients will also be run off n times faster. Thus the bandwidth B of the filter is increased by the factor n.*

Avoiding ripple content in the conducting state region

The preceding representations clearly show ripple content in the conducting state region of the filter. This effect can be clearly seen, for example, in Illustration 49. The cause is not difficult to guess. Although it is attempted to reproduce the Si function as accurately as possible by means of the filter coefficients, this is only successful for a segment of the Si function which is limited in time. Theoretically the Si-function extends "infinitely" towards the left and right, into the past and future. By this segment the true Si function is cut out as if with a rectangular window (see also Illustration 51) and as a result steps do not arise. This specific Si function would have to begin gently at zero and end at the end of the segment.

There is no problem achieving this. For this purpose a time window is built in addition into the "filter development site" which makes this zero beginning and zero end possible. The success of this measure shows that it is appropriate. It is now possible to achieve a very smooth curve in the conducting state region. However, the filter edges now seem a little less steep.

All in all, this "filter development site" makes it possible to develop digital lowpasses and bandpasses of the quality desired (by means of a suitably large number of coefficients and a reasonable amount of calculation). These digital filters can be deployed directly by means of the *convolution* module in DASY*Lab*.

Precisely speaking, these filter coefficients and digital filters can be used with any computer system. It is simply necessary to transmit the schematic circuit of the digital signal or convolution module into a small program in the given programming language.

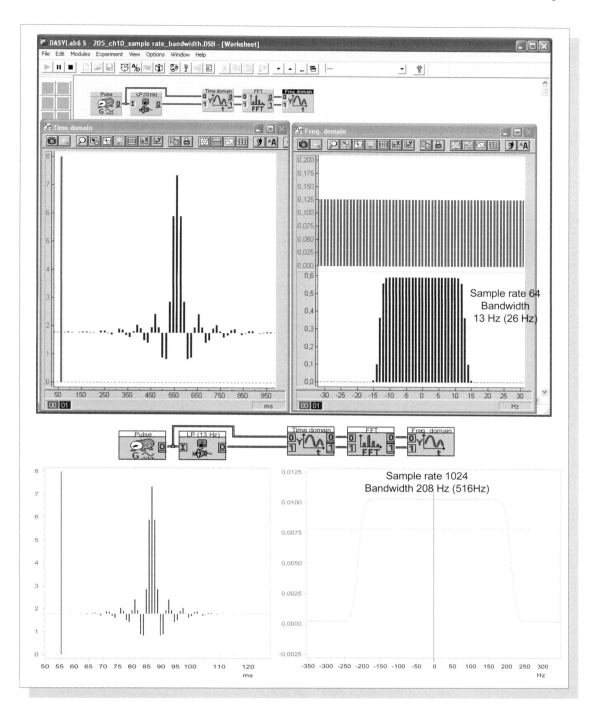

Illustration 206: **Digital lowpass with "smooth" conducting state region**

By adding a suitable window (e.g. a "Hamming" window) the Si-function begins and ends gently, i.e. there are no steps at the beginning or the end of the time segment of (here) 1s. But the curve of the Si function is changed as a result, as the whole signal segment is "weighted" by it. The effect is to make the curve in the conducting state region straighter and the edge steepness decreases somewhat.

While the bandwidth of the lowpass at the top is 13 Hz, in the circuit at the bottom it is 64 Hz although the same filter coefficients were used. The reason is the much higher sampling rate of 1024 compared with 64 top). The pulse response of the filter (Si function) is as a result runs 16 times faster, it lasts only 1/16 of the original time. According to the Uncertainty Principle the bandwidth must be 16 times as large: 16*13 = 208 Hz.

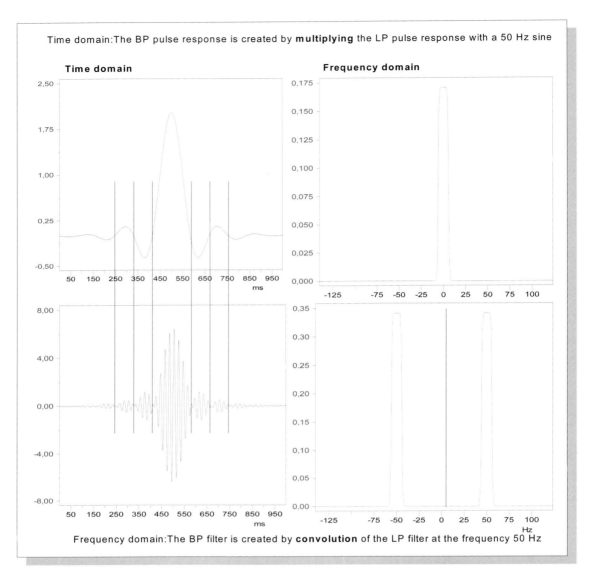

Illustration 207: ***Development of a digital bandpass filter***

This shows how the pulse response of a bandpass differs from that of a lowpass. In the final analysis it is a the pulse response of a (symmetrical) lowpass which is multiplied by the mid-frequency of the bandpass.

First create the pulse response for a lowpass which should have the same bandwidth as the bandpass. The bandwidth in the above case goes from roughly –12 Hz to + 12 Hz. Here we see how important the symmetrical representation of the frequency domain is in the "filter development site".

Then select on Channel 1 of the generator a sinusoidal signal with the mid-frequency of the bandpass (here 100 Hz) and the amplitude 1 instead of an offset of 1. Then you will see a state of affairs as shown above. Illustration 209 shows these common features in greater detail.

A prerequisite for the successful development of digital filters are the correct basic conditions:

- First establish how high the sampling frequency in the planned DSP system can be selected. The higher the better. You then have the opportunity to make the intervals between the periodic filter spectra of the digital signal as large as possible.

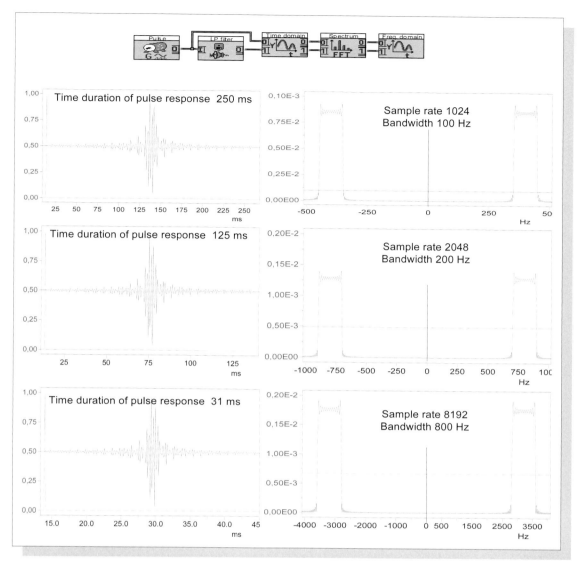

*Illustration 208: **Importance of the sampling rate for the filter curve***

The relations are once again illustrated using the example of a bandpass. In the above circuit the same coefficients file was used for a bandpass in the convolution module or the digital filter. The top pulse response and the top filter curve corresponds to the state of affairs in the "filter development system". In the case of a bandpass a blocklength of at least 256 should be selected.

If in the above circuit the sampling rate is increased to 8192 (compared with 256) the bandwidth and midfrequency increase by the factor 8. The "filter form" however does not change. Because it is determined exclusively by the filter coefficients.

- The number of the filter coefficients is a measure of the quality of the filter. With DASY*Lab* a maximum of 1024 filter coefficients are possible in the convolution module. If the number is greater so is also the amount of calculation necessary. This can lead to difficulties in real time processing.

- The sampling frequency must also be at least double the highest cutoff frequency of the bandpass. The Sampling Principle also applies here, of course. The highest signal frequency which passes through the bandpass and not the bandwidth is decisive.

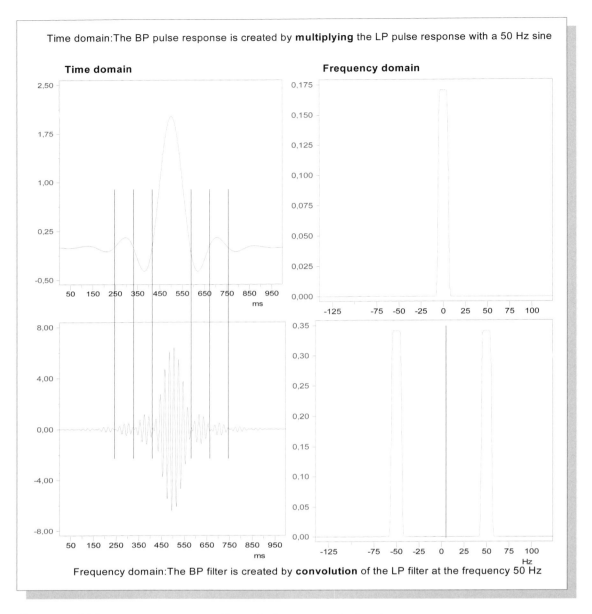

Illustration 209: **_From the lowpass to the bandpass_**

An equivalent bandpass – i.e. with the same filter form was realised from the above lowpass by the multiplication of the pulse response of the lowpass by the mid-frequency of the bandpass. This relationship can be seen in the pulse response of the bandpass. The envelope of this pulse response is the above Si-function.

Set the mid-frequency of the bandpass on Channel 1 of the generator. Select a sinusoidal voltage of equivalent mid-frequency instead of the constant c = 1 so that the bandpass does not amplify the the filtered signal.

Exercises on Chapter 10

Exercise 1

What difference between an analog and a digital filter is immediately striking?

Exercise 2

(a) Which components in an analog filter determine its frequency-dependent behaviour and how can this be described?

(b) Why are there no high-quality analog filters?

(c) Where is it certain that analog filters will be used in future?

Exercise 3

(a) Describe the structure and way of functioning of the FFT filter consisting of three components (modules)

(b) What must be taken into account in the case of the FFT filter to make sure that the desired conducting state region of the filter is set correctly.

(c) Why does the selectivity or edge steepness of the FFT filter extend to the physical limit imposed by the Uncertainty Principle?

(d) What are the advantages and disadvantages of the FFT filter?

Exercise 4

Experiment with the bandpass circuit as in Illustration 196 and establish how the envelope (Si function) and the mid-frequency changes with the pulse response. How can the filter properties of the bandpass be seen clearly from the pulse response in the time domain?

Exercise 5

At first glance the FFT filters appear not to be "causal" as the output signal – the pulse response – seems to appear before the input signal – the δ-pulse – appears at the output. Why can the computer not do this differently?

Exercise 6

(a) What idea leads to the digital signal which filters directly in the time domain and thus avoids transformation into the frequency domain and back again my means of the FFT?

(b) Explain the principle of digital filters on the basis of the overlapping of Si shaped pulse responses of all discrete measurements (see Illustration 199).

(c) A filter restricts the bandwidth and for this reason a correspondingly long lasting discrete pulse response also occurs in the case of digital filters. What signalling processes are required to produce a longer lasting discrete pulse response " artificially" from a δ-pulse?

Exercise 7

(a) Design the schematic circuit diagram of a digital filter (FIR).

(b) How can an Si shaped pulse response be trimmed by means of a schematic circuit.

(c) What role do the filter coefficients of a digital signal (FIR) play?

Exercise 8

(a) Why is convolution such an important signalling process? Why does it make filtering in the time domain possible?

(b) How can the digital filtering of a longer lasting signal be described graphically by convolution (see Illustration 202)?

Exercise 9

(a) Describe the concept of a "development site" for digital filters.

(b) How is a high-quality digital filter produced with DASY*Lab* using the convolution module?

(c) The filter coefficients say nothing about the real bandwidth of the digital signal. What information is decisive in this respect?

Exercise 10

(a) How does the ripple content come into being in the conducting state region of the digital filter and how can it be avoided?

(b) How can digital bandpasses be created by means of the "development site" for digital signals?

(c) Why cannot digital highpasses be easily created by means of DASY*Lab*?

Exercise 11

Describe what framework conditions apply for the development of digital filters.

Chapter 11

Digital transmission technology I: source encoding

Modern microelectronics offers applications in digital signal processing (DSP) which a few years ago would not have been thought possible. Mobile telephony is only one example, the global internet another. The general public is not so aware of the fascinating applications of medical technology and the radio and television technology of the future: *Digital Audio Broadcasting* DAB and *Digital Video Broadcasting* DVB.

A single new development of this kind can completely transform the market, ruin companies or catapult them forwards. Take the example of ADSL.

ADSL (asymmetric digital subscriber line) is a new development of this kind which makes a data highway out of the old copper twin wire of an underground cable. ASDL transports via a traditional telephone line up to 125 times that of an ISDN channel, i.e. up to 8 Mbit/s from the network to the subscriber and roughly 700 kBit/s from the subscriber to the network. The original ISDN channel on this line remains in existence and can be used additionally. This is sufficient for 4 digital television channels which can reach the subscriber simultaneously in real time.

The good old telephone line had been declared defunct. Investments by Deutsche Telekom on a scale of 250 bn euros would have been necessary to connect every household by glass fibre as a medium of transmission. ASDL will make these huge investments unnecessary in the foreseeable future and the competitors of Deutsche Telekom are in for a rough ride. It owns the "last mile" – the telephone line - to the subscriber.

Illustration 210 shows the huge appetite of multimedia applications for transmission capacity in which pictures, animations, speech, and music apart from text are used. Video applications have the biggest appetite. A so-called CCIR 601video signal requires 250 Mbit/s for the video data stream alone. The required bandwidth is more than even high speed networks can cope with. The ATM technology which is mainly used at the moment – the data from the transmitters are split up into small data packages of 53 Byte and inputted into a transmission channel in the sequence in which they arrive (asynchronic transfer mode) – usually provides only 155 Mbit/s.

DSP (digital signal processing) provides the solution. (Mathematical) signal theory has created procedures which make possible the effective compression of data. In addition, it offers solutions as to how data can be protected efficiently from interference such as noise. Both processes – compression and error protection encoding – can be applied to real signals using the computer. They have for instance made it possible to transmit needle-sharp video pictures from the furthest stars of our solar system over many millions of kilometers. The transmitter of the space probe had a power of only 6 W.

Memory capacity of different media
(at a resolution of 640 * 480 pixel)

Text : A symbol represents a 8 * 8 pixel pattern
(ASCII encoded 2 Bytes are needed)
Memory per screen page = 2 Bytes * 640 * 480 / (8*8) = 9,4 kBytes

Pixel picture: For example a picture is made of 256 colors, i.e. 1 Byte per pixel

Memory capacity per picture: 640 * 480 * 1 Byte = 300 kByte

Language: Language in phone quality is sampled with 8 kHz and quantized with 8 bit. A data stream of 64 kBit/s results. That shows a

Memory capacity of 8 kByte/s

Stereo-audio-signal: A data rate of 2 * 44100 * 16 bit /8 = 176,4 kByte is needed.

A memory capacity results from 172 kByte/s   (1kByte are 1024 byte)

Video sequence: A video film consists of 25 complete pictures per second. The luminance and chrominance (brightness and color information) of every pixel is encoded together in 24 bit or 3 Bytes. The luminance is sampled with 13,5 MHz and the chrominance with 6,75 MHz.

8 bit encoding shows: (13,5MHz + 6,75 MHz) * 8 bit = 216 Mbit/s

Data rate: 640 * 480 * 25 * 3 Bytes = 23,04 MByte

This results in a memory capacity per second:
23,04 MByte*/1,024 = 22,5 MByte i.e. 22,5 MByte/s

Source: http://www-is.informatik.uni-oldenburg.de/

Illustration 210: ***Transmission rates of important multimedia applications***

Speech, music and video turn out to require a great deal of storage space and bandwidth. How can real signals be compressed? How might it be possible to establish and remove the redundancy of real signals while preserving the important information?

The answer to this is complex and the following comments deal with strategies and some of the processes which are used in modern transmission technology.

Encoding and decoding of digital signals and data

The general expression for the numerical representation of symbols (e.g. letters or measurements) and for the systematic modification of digital signals or data is the term "*encoding*".

Illustration 211 shows an important example of encoding. In this "ASCII encoding" all the important signs (symbols) in written communication are allocated a number between 0 and 127. 128 (= 27) different signs can only be coded in this way. A 7- Bit code is sufficient for this purpose. Since 1963 this type of encoding has been a worldwide standard for computers.

> Note:
> More than 128 different signs still cannot be transmitted via the internet and all other computer networks. German umlauts (e.g. ä) and the sign "ß" are not included. If only one bit more had been taken originally (8 Bit = 1 Byte) the possible number of signs would have been twice as large. A 2 byte (16 bit) code (65,536 signs) would have produced a really universal code for the electronic networking of the whole world. But it is the good old 7 bit ASCII that makes its way through the networks.

The coded (digital) signal can be partly or completely transformed back into ins original form in the receiver. These two processes are often called encoding and decoding. The word code is used for the actual encoding process.

In digital transmission technology the term encoding is used above all in connection with the following processes:

- A/D and D/A converters (really A/D encoders)
- Data compression
- Error protection encoding
- Encoding of digital data

Compression

The compression of digital signals is the general expression for processes (algorithms) or programs which transform a simple data format into an optimally compact format. In the final analysis, the number of bits and bytes are to be reduced as far as possible. Decompression reverses this process.

In this sense the ASCII encoding – see Illustration 211 – is badly encoded. Each of the many symbols has the same length of 7 bits, no matter whether it occurs frequently or very, very rarely. Thus, in text the space separating two words is by far the most frequent. Small letters are more frequent than capitals. The letter "e" is certainly more frequent than "x".

ASCII - Code

0	null	32	space	64	@	95	`
1	start heading	33	!	65	A	97	a
2	start of text	34	"	66	B	98	b
3	end of texte	35	#	67	C	99	c
4	end of xmit	36	$	68	D	100	d
5	enquiry	37	%	69	E	101	e
6	acknowledge	38	&	70	F	102	f
7	bell, beep	39	'	71	G	103	g
8	backspace	40	(	72	H	104	h
9	horz. table	41	)	73	I	105	i
10	line feed	42	*	74	J	106	j
11	vert. tab, home	43	+	75	K	107	k
12	form feed,cls	44	,	76	L	108	l
13	carriage return	45	-	77	M	109	m
14	shift out	46	.	78	N	110	n
15	shift in	47	/	79	O	111	o
16	data line esc	48	0	80	P	112	p
17	device control 1	49	1	81	Q	113	q
18	device control 2	50	2	82	R	114	r
19	device control 3	51	3	83	S	115	t
20	device control 4	52	4	84	T	116	t
21	negative ack	53	5	85	U	117	u
22	synck idle	54	6	86	V	118	v
23	end xmit block	55	7	87	W	119	w
24	cancel	56	8	88	X	120	x
25	end of medium	57	9	89	Y	121	y
26	substitute	58	:	90	Z	122	z
27	escape	59	;	91	[	123	{
28	file separator	60	<	92	\	124		
29	group separator	61	=	93	]	125	}
30	record separator	62	>	94	^	126	—
31	unit separator	63	?	95	_	127	del

Illustration 211: **ASCII encoding**

This is a long established standard for the representation of special signs, letters, and numbers in a digital form (as a number or bit pattern). Every printable symbol has a number between 32 and 127 allocated to it, while the numbers from 0 to 31 are control numbers for an obsolete form of communication technology and are practically no longer needed.

Usually all the ASCII signs are stored with one byte (8 bits). The non-standardised values from 128 to 255 are often used for Greek letters, mathematical symbols and various geometrical patterns.

A good compression program for text should determine first the frequency with which symbols occur and categorize them according to the probability of their occurrence. In this sequence the most frequent symbols should have the shortest bit patterns (i.e. the shortest numbers) and the rarest should have the longest ones. This would result in a much smaller overall data files without any loss of information.

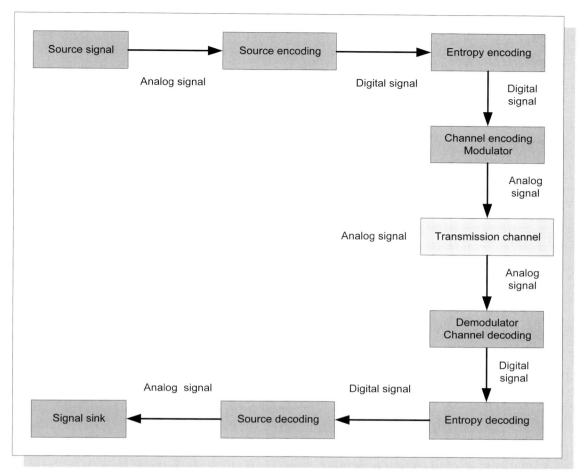

*Illustration 212: **The transmission path from the point of view of encoding***

The block diagram shows three different coders and decoders. The source encoder could be an A/D converter the output code of which is already "particularly efficient" (see delta and sigma-delta converters (encoders)). The entropy encoder (see Huffmann encoding) attempts to optimise this code in terms of compactness without loss of information. This compactness is reduced again somewhat by the channel encoder as a result of redundancy in the form of additional bits being added. The aim is to preserve error protection for the signal. The receiver should be able to recognise an erroneous sign and and correct it independently.

Low-loss and lossy compression

In view of the large number of compression processes which are used at the moment, it is not surprising that a universal and optimum process cannot exist. The most suitable process depends in each case on the signal or type of data.

Certain signals or data must be compressed without any loss, i.e. the signal which is recovered by the receiver must be absolutely identical with the original. Not a single bit must change. This applies to software and test data files.

On the other hand a certain loss of quality and therefore information can be tolerated in the case, for example, of audio and video signals. This is also called data reduction. Thus there is information contained in audio-signals which we cannot perceive. It is irrelevant. In this context the expression *irrelevance reduction* is used. We have got used to "poor-quality" television pictures. Lossy compression processes are used here.

The differentiation between these two kinds of compression is also important. Lossy processes compress much more efficiently than low-loss procedures. From a signalling and theoretical point of view more lossy compression is characterised by a larger noise element. Noise here embodies the "disinformation", i.e. the tolerated loss of information.

A/D conversion with its part processes sampling, quantization, and encoding can be seen as a lossy compression method (see Illustrations 138, 139 and 174). The selection of the sampling rate, the number of quantisation steps and the encoding method have a great influence on the quality and compactness of the digitalised signal.

In order to know for what signal or type of data a specific compression process is advantageous a few compression strategies will be explained using examples.

RLE encoding

RLE (run length encoding) is probably the simplest but sometimes the optimal compression process.

> *Principle:*
> more than three identical consecutive Bytes are coded by their number.
>
> *Example*:
> "A-Byte" A; AAAAAA is coded as MA6. M is a marker byte and designates an "abbreviation" of this kind. In this example there is a reduction of 50%. The marker byte must not be present in the source text as a sign, otherwise it would have a "double" meaning.
>
> *Application*:
> RLE is particularly suitable for data sets with long sequences of the same signs, for example, black and white drawings. It is therefore frequently used for fax formats in which large white areas are only occasionally interrupted by black letters.
>
> *Note*:
> Data files with frequently changing bytes are highly unsuitable for this process.

Huffman encoding

> *Principle:*
> it is based on the morse alphabet principle. The shortest codes are allocated to the symbols which occur most often (e.g. letters) and the longest to the most infrequent ones. The symbols of the source are coded and *not* the data to be transmitted. This is called entropy-encoding. It is loss-free.
>
> *Procedure:*
> It must first be determined what symbols occur (for instance, in the text). Then

with what frequency (or more precisely, what probability) they occur. The Huffmann algorithm produces a "code-tree" which results from the probability of the individual symbols. The code tree is used to establish the code words for the individual symbols.
The process is described in detail in Illustration 213.

Decoding:
in order that the receiver can recognise the original data from the byte sequence the Huffmann tree must also be transmitted. When "descending from above" you always land on one of the 7 different symbols selected in Illustration 213. Then you have to jump back up to the top and branch out to the left and right until the next signal is reached.

Note:
The longer for instance the text is, the less room is taken up by the transmission in addition of the Huffmann tree.

LZW encoding

This process is named after its originators Lempel, Ziv and (later) Welch. It is probably the most commonly used process for all-purpose compression. Thus it is used in the ZIP compression of files and of many graphic formats (e.g. GIF). Compression factors of 5:1 are usual.

Illustration 214 shows top left the content of the string of numbers before and after each step in the encoding process. In the first step the longest pattern found will only be a single letter which is to be found in a standard dictionary. In our example this is "L". In the same step the next letter to be examined is "Z" and is attached to the L.

The chain of signs which is produced in this way is definitely not in the dictionary and is entered for the first time under the Index (256). After this in the last step the "longest" chain of signs found – i.e. the "L" – is removed and is issued at the same time (see "Recognised pattern"). In this way "Z" becomes the first sign of the new string.

Here everything begins again from the beginning. "Z" is now the longest known pattern. "ZW" is stored under the Index (257) in the lexicon. "Z" is now removed, issued and a new round begins. "W" is now the longest chain of signs entered in the lexicon so far. "WL" is now included under the index number (258), the "W" is deleted from the entry and issued.

It is only now that something interesting happens with regard to the desired compression. The longest known pattern is now a sequence of signs which was previously entered in the lexicon for the first time ("LZ" with the index number 256). In the next step, the index of the pattern from the lexicon is issued and not just two individual signs.

As the dictionary contains 4096 (= 2^{12}) the entries get longer and longer in the case of a very long suitable entry string (this is the file which is to be compressed). More and more often longer sequences of signs for which short indexes are transmitted belong to the higher indexes.

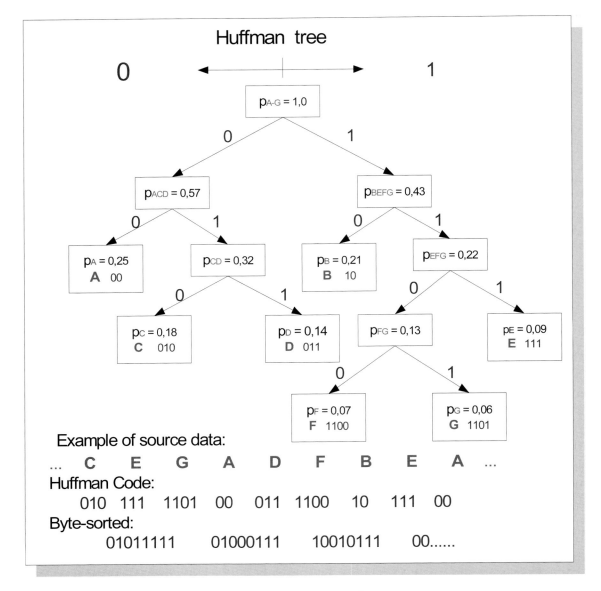

Illustration 213: ***Huffman encoding***

7 different symbols, here designated by the letters A to G, are to occur in a file. Let A occur most frequently with a probability of 25% or 0.25. B follows with 0.21, C with 0.18, D with 0.14, E with 0.09, F with 0.07 and G with 0.06. The idea is now to allocate the shortest code to A and the relatively longest to G. As A and B are almost equally probable it would make sense to give them the same code length.

The Huffmann algorithm generates the Huffmann tree. In the "leaves" of the Huffmann tree the probability p for the relevant symbol is entered. At the branching nodes you see the sum of the probabilities. At the top there is probability p = 1 or 100%, because one of the 7 symbols has to be selected!

If – moving from top to bottom – you branch off to the left, 0 results, and to the right a 1 in each case. By sorting the symbols according to probability of occurrence the algorithm knows in what sequence the branchings lead to the symbols A to G.

Sorted according to bytes the compressed signal arrives at the receiver. Now decoding takes place, beginning each time at the top. After 00 A is reached. The pointer jumps back to the top again, after 111 E is reached etc. The prerequisite is the transmission of the Huffmann tree in addition to the compressed document.

LZW encoding

The standard dictionary contains the 256 different Byte-pattern. Every symbol occurring here in the input corresponds to one specific pattern among these byte-patterns!

Standard dictionary
0
1
.
.
254
255

Input	Recognized pattern	New dictionary entry	
LZWLZ78LZ77LZCLZMWLZAP	L	LZ	(=256)
ZWLZ78LZ77LZCLZMWLZAP	Z	ZW	(=257)
WLZ78LZ77LZCLZMWLZAP	W	WL	(=258)
LZ78LZ77LZCLZMWLZAP	LZ	LZ7	(=259)
78LZ77LZCLZMWLZAP	7	78	(=260)
8LZ77LZCLZMWLZAP	8	8L	(=261)
LZ77LZCLZMWLZAP	LZ7	LZ77	(=262)
7LZCLZMWLZAP	7	7L	(=263)
LZCLZMWLZAP	LZ	LZC	(=264)
CLZMWLZAP	C	CL	(=265)
LZMWLZAP	LZ	LZM	(=266)
MWLZAP	M	MW	(=267)
WLZAP	WL	WLZ	(=268)
ZAP	Z	ZA	(=269)
AP	A	AP	(=270)
P	P		

Output: LZW(256)78(259)7(256)C(256)M(258)ZAP

LZW decoding

Input symbol	C	New dictionary entry		P
L				L
Z	Z	LZ	(=256)	Z
W	W	ZW	(=257)	W
(256)	L	WL	(=258)	LZ
7	7	LZ7	(=259)	7
8	8	78	(=260)	8
(259)	L	8L	(=261)	LZ7
7	7	LZ77	(=262)	7
(256)	L	7L	(=263)	LZ
C	C	LZC	(=264)	C
(256)	L	CL	(=265)	LZ
M	M	LZM	(=266)	M
(258)	W	MW	(=267)	WL
Z	Z	WLZ	(=268)	Z
A	A	ZA	(=269)	A
P	P	AP	(=270)	P

Illustration 214: *LZW encoding*

LZW encoding and its variants have become so important for the compression of digital signals and data that they are to be explained here pictorially in detail. Without knowing the message that is to be compressed compression is carried out efficiently and loss-free. The decompression algorithm automatically recognises the code, creates a new identical lexicon and uses this to reconstruct the source signal. If you look at the illustration long enough it is often possible to understand the process intuitively.

Decoding also starts with the standard lexicon described. It again contains the entries from 0 to 255. In our example "L" is again the longest known "chain of signs". It is therefore given and stored in the variable P (as in "prefix"). The next entry is also a known sign "Z". This is first stored as the variable C.

Now the content of P and C is bracketed and the result "LZ" included in the lexicon under the index number (256). The whole thing proceeds in this way and the lexicon and the original input sign sequence are created by means of the decoding algorithm.

Try encoding and then decoding a different input sign chain according to the LZW principle. Only then will you notice that the input sign chain must have a special form for compression to be carried out quickly and efficiently ("the rain in Spain falls mainly on the plain").

Source encoding of audio signals

A/D and D/A conversion was already dealt with in Chapter 7 (section "Quantisation") and above all in Chapter 9. Illustration 174 shows the principle of an A/D converter/encoder, which outputs "measurement" serially as a sequence of 5 bit strings of numbers. From a technical point of view at present 8 bit to 24 bit A/D and D/A converters are usual.

This process is generally referred to as PCM (pulse code modulation). Every measurement is allocated – as in ASCII encoding – a code of the same length independent of the frequency of occurrence. According to Illustration 210 a transmission rate of 172 kbyte/s (roughly 620 Mbytes per hour) results for an audio-stereo-signal.

The compression strategies described up to now ought to be applicable to audio-signals. These are, after all, available as a sequence of bit patterns. Usually processes are used in A/D conversion which themselves have a "compression effect". In addition, with this process the share of analog circuitry is reduced and is replaced by DSP (digital signal processing).

Delta encoding or delta modulation

So-called *screencam videos* are very popular in order for example to record the installation or use of a program in the form of a screen video. This is also used frequently on the CD accompanying this manuscript. Most of the time the cursor with which menu items are clicked on to simply moves around the screen.

It would not make sense here to save the entire content of the screen 25 times a second.

> Note:
> in mathematics and in the technical and scientific field the Greek capital letter Δ is used to describe a difference or change. Thus, Δt is the time difference between two points t_1 and t_2 .
>
> The notation Δ-modulation is more usual than delta modulation.

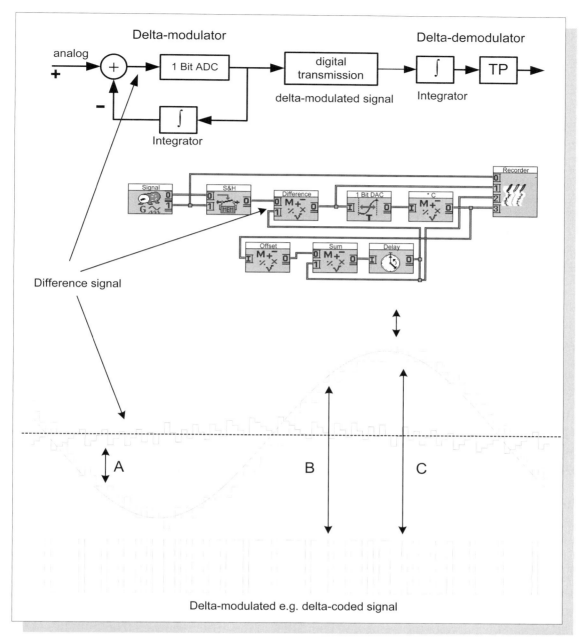

Illustration 215: **Schematic block diagram, DASYLab circuit and signals in Δ-encoding**

*In this form of encoding or modulation only the **change** in the audio signal is transmitted. In a way the original signal is differentiated digitally, i.e. the gradient is measured. The way the circuit functions is explained in detail in the text.*

The (digital) differentiation must be reversed in the receiver by (digital) integration. You remember: integration reverses differentiation and vice versa (see Illustrations 123 and 124).

In the case of screencam video it is therefore sufficient to register *changes* in the picture – i.e. the cursor movement and change of picture. For this only a fraction of storage capacity is needed compared to saving all the pictures.

This Δ-encoding can be used efficiently if the signal only changes slightly between two measurements or within the time segment $\Delta t = T_S = 1/f_S$ (f_S is the sampling frequency).

In this case the Δ-coded signal has a smaller amplitude than the original signal. In other words: the probability for measurements around zero rises. Far fewer measurements will be far away from zero.

These are favourable conditions for Huffmann encoding as described above. If the original signal does not change or rise linearly the Δ-coded signal has a sequence of the same bit patterns. They occur most frequently. A typical prodedure is therefore after (lossy) Δ-encoding to use the Huffmann or RLE encoding as loss-free entropy encoding.

In Illustration 215 the schematic block diagram of the Δ-encoder (often also called Δ-modulator), the transmission route and the Δ-decoder are shown. Δ-encoder consists of a feedback loop. The 1 bit ADC (analog digital converter) has only two states: + and –. If the difference signal at the input of the 1 bit ADC is greater than zero at the output, for instance +1, otherwise –1 (corresponding to low and high). In the final analysis the 1 bit ADC is simply a special comparator.

In Illustration 215 the "zero decision line" can be seen around which the difference signal fluctuates. It can be clearly recognised (see arrows A, B and C) that: as soon as the difference signal is above the zero line, the Δ-coded signal is "high"; otherwise it is low.

What is the – here sinusoidal – input signal compared with? In this connection you should recapitulate what an integrator does (see Ilustrations 123 and 125). If there is a positive signal segment at the imput of the integrator – here on the right – the integrator moves upwards, otherwise downwards. This can be clearly seen in Illustrations 215 and 216. Where the Δ-coded signal has the value +1 steps rise; in the case of –1 they lead downwards.

In the centre of Illustration 215 the realisation of the schematic block diagram using DASY*Lab* is to be seen. The "digital integrator" is a feedback analog adder. In the last output value +1 is added or –1 subtracted. This produces the step-like curve. In the case of DASY*Lab* and similar programs a delay is part of all the feedback circuits. This makes sure that the causal principle is adhered to. The reaction of a process at the output can only be fed back to the input with a delay (first the cause, then the result). Before the analog adder two arithmetic modules (creating a multiplicative and additive constant) makes certain that the curve of the input signal and that of the difference signal are roughly in the same area. The real 1 bit ADC is the trigger or comparator. The subsequent component serves the correct signal setting of the Δ-encoder.

As the output signal has only two states, a kind of *pulse length modulation* takes place. The quantisation error which occurs in this connection can be reduced almost at will by oversampling. The smaller the interval between two sampling values the smaller is also the difference.

A characteristic of Δ-modulation or Δ-encoding is oversampling. The factor n is important by which the sampling frequency lies above the limit defined by the Sampling Principle. As nowadays in the case of these Δ-procedures sampling frequencies of 1 MHz and more are used, n may have a value of around 25. As a result of oversampling the noise is distributed over a greater frequency range (Illustration 220).

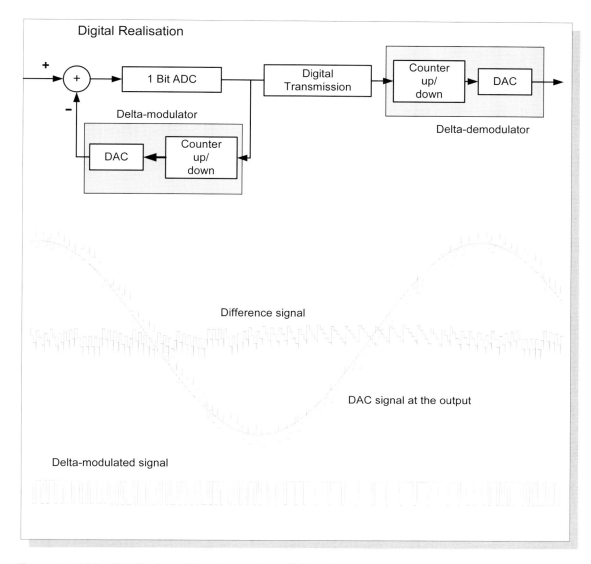

*Illustration 216: **Realisation of the Δ-encoder and signal curve in the case of higher sampling rate***

In this representation of the digital schematic block diagram it can be clearly seen where analog technology is still used: at the transmitter input, the receiver output and – this always applies – on the transmission route. The time and value discrete digital integrator is realised by means of a forward-backward counter followed by a DAC (digital analog converter).

Note for the following illustration the technical identity of the negative feedback branch and the receiver circuit. Both are Δ–encoders or Δ–modulators. The reference signal fed into the difference modul corresponds exactly to the time and value discrete (here sinusoidal) input signal recoved in the receiver. The difference between the two, i.e. the difference signal is equivalent to "quantisation noise".

In the signal curve shown at the bottom a higher sampling rate was selected. The Δ-encoding can be clearly seen. The maximum and minimum values of the deltamodulated signal lie where the upward or downward gradient are greatest. The floating averaging would reproduce the gradient curve of the original signal, i.e. the differentiated input signal.

The Δ–encoded signal at the bottom is roughly equivalent to a pulse length modulated signal (see Illustration 177). Unlike the continuous time pulse length modulated signal there here we have a kind of pulse length modulated time discrete signal.

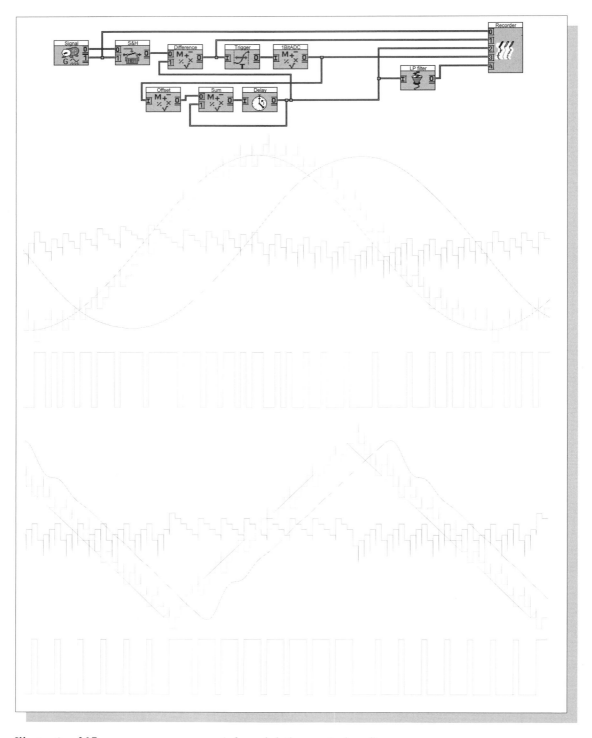

Illustration 217: ***Δ-demodulation or Δ–decoding***

Here you see in addition to the signals present in the Δ-modulator the final analog signal recoverd in the receiver. As described in the text of Illustration 216 the output signal of the digital integrator and the time and value discrete signal recoverd in the receiver must be identical. For this reason there is no receiver in the DASYLab circuit. The latter signal must only be passed to an analog lowpass.

Note: the triangular input signal ought to look like the recoverd (filtered) signal. It ought to have been band-limited (Sampling Principle).

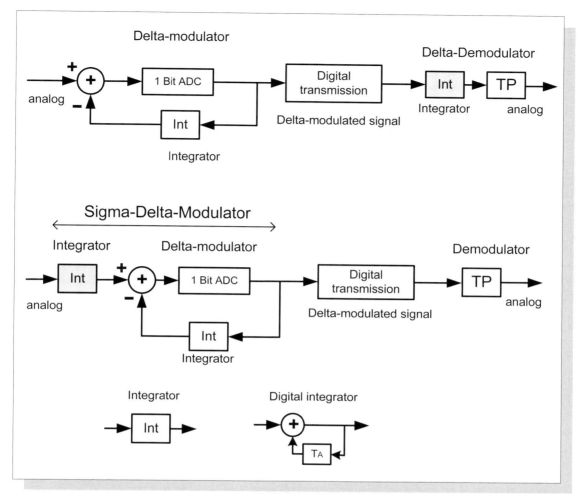

Illustration 218: **From the delta modulator to the sigma-delta modulator**

At the top the block diagram of the transmission channel of a delta-modulator and demodulator is shown. Below this only the integrator of the demodulator was placed at the beginning of the transmission channel. This can hardly have any effect on the output signal as the sequence of linear processes – such as integration – can be changed around (see Chapter 7 from Illustration 123).

As interference may also occur on the transmission route this results in advantages which will be explained in more detail in the text.

The two integrators can be replaced by a single integrator after the differentiator (see Exercise 10 in Chapter 7). Thus the Σ–Δ–modulator/encoder is complete. The demodulator now consists only of the analog lowpass. This functions like a "floating averager" and recovers the analog input signal from the digital sigma-delta modulated signal.

At the bottom the integrator is realised by means of a feedback adder i.e. the instantaneous output value is added (with a delay) to the input value following. This circuit variant is also used in the DASYLab simulation in Illustration 218.

Sigma-delta modulation or encoding (Σ–Δ–M)

A disadvantage of the Δ-modulator is that the bit errors which occur in transmission lead to an "offset" i.e. to a offset in the signal received. So-called sigma-delta modulation (Σ–Δ–modulation) in which significant improvements can be achieved by a skilful changing round of components is a solution here.

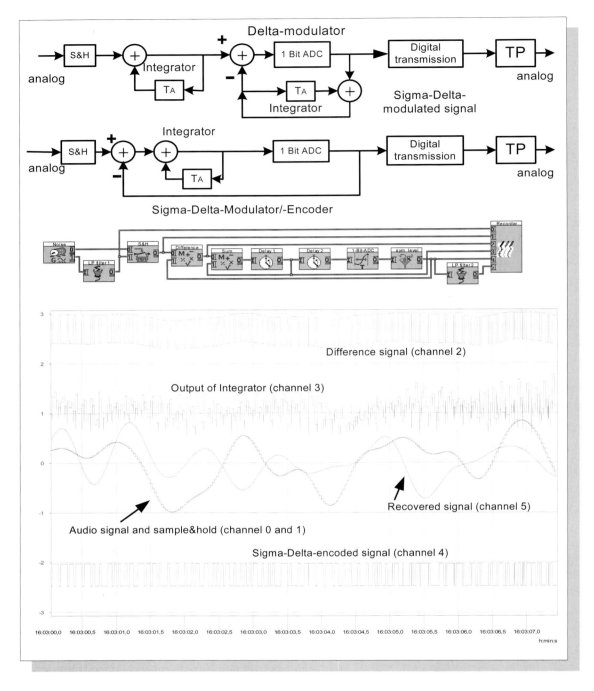

Illustration 219: **From delta modulator to sigma-delta modulator**

The form of integrator used here results from the DASYLab circuit in Illustrations 217 and 218: an adder which adds the last output value of the 1 bit ADC to the next input value. If an integrator of this kind is connected additionally before the delta-modulator it can be done without in the receiver. The circuit below in which the the two integrators (in actual fact, the adder, for which the Greek letter Σ (sigma) is used) are replaced by a single integrator before the 1 bit ADC is then equivalent. This circuit is therefore called a Σ–Δ–modulator/encoder.

By means of this process of re-arrangement from the receiver to the transmitter numerous other advantages are achieved which are described in more detail in the text. As a result of these advantages the Σ–Δ–modulator/converter has become a kind of standard ADC for high-quality A/D conversion (audio, measuring technology).

The schematic circuit diagram of the Δ-encoder and Δ-decoder with the two integrators can be simplified. Hardly anything is changed in the output signal at the receiver if the integrator changes its positiom from the end of the transmission channel to the input. Linear processes can be changed round in sequence, as described in Chapter 7.

Why should this be done? In this way the two integrators in the Δ-encoder can be replaced by a single integrator directly in front of the 1 bit ADC (see Exercise 10 in Chapter 7). We now have the so-called sigma-delta modulator/encoder with a demodulator/decoder as shown in Illustration 219. There the integrator is shown as a feedback adder. The capital sigma (Σ) is the symbol for "sum total" in the technical and scientifc field. The circuit variant is therefore known as a Σ–Δ–modulator/encoder.

The demodulator only consists of an analog LP filter. This recovers the analog input signal from the digital Σ–Δ–modulated signal. It functions like a floating averager.

Noise shaping and decimation filter

The higher the oversampling the smaller the quantisation error. This is shown in both Illustration 216 and 220. By the n-fold oversampling the quantisation noise distributes itself over an n-fold higher frequency range as the horizontal bar in Illustration 220 shows.

A further reduction in this level of interference is possible by means of the Σ–Δ–modulator. This shaping effect known as "noise shaping" arises as a result of the re-grouping of the integrator at the beginning of the 1 bit A/D converter. This gives preference to the slow changes in the signal as the integrator is a kind of averager.

The quantisation error increases linearly with the frequency and has therefore a highpass characteristic. By this means only the level of interference is present after the retrieval of the signal in the demodulator which corresponds to the (pink) triangular area.

The sampling frequency f_S of the audio signal should be selected as high as possible in this process, but the transmission rate is usually fixed and may be much lower than the sampling rate. This "adjustment" is brought about by so-called "decimation filters" (see Illustration 220). They divide up the bit sequence of the 1 bit A/D converter into blocks with an odd-numbered bit number (e.g. 7). They then check whether more "0" or "1" symbols occur within the block. The symbol which occurs more frequently is then indicated at the output of the decimation filter. Of course, this is also a kind of averaging with a lowpass characteristic.

Exploiting psycho-acoustic effects (MPEG)

The frequency domain so far does not seem to play a part in the encoding or compressing of audio signals. However, hearing is entirely in the frequency domain, that is we hear only sinusoidal signals of different frequencies. A really intelligent method of compression ought to include the frequency domain.

Audio-signals such as language and music do not contain redundant features, unlike text where the letter "e" occurs much more freequently than "y". As a result it is difficult to allocate a shorter code to a sound or tone.

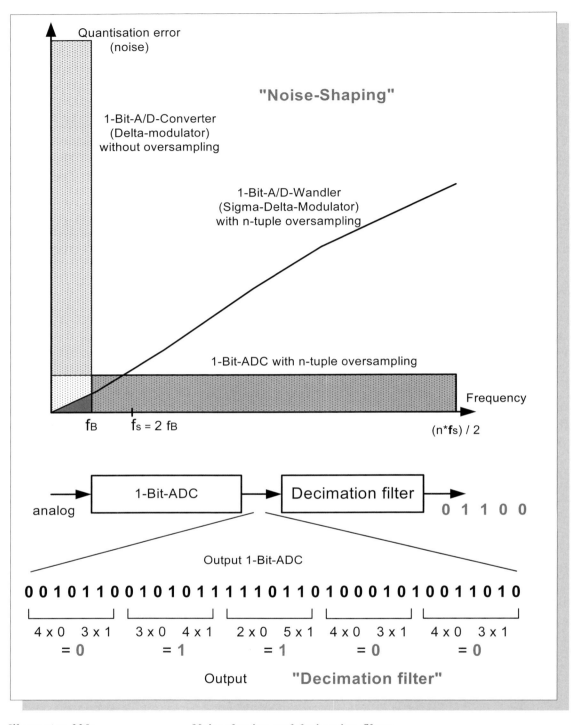

Illustration 220: **Noise shaping and decimation filters**

The upper half of the illustration shows the distribution of quantisation noise in different cases over the frequency domain. The frequency f_B designates the bandwidth of the original and recovered signal, f_S the minimum sampling frequency of the source signal. In the case of Σ–Δ–modulation only the quantization noise designated by the triangle is effective.

The decimation filter "decimates" the number of bit/s at the output of the D/A converter by an odd-number factor. Within a block it is only checked which of the two symbols 0 or 1 occur most frequently. This then appears at the output. If there are more 1s the signal rises within the block instantaneously "on average". If there are more 0s it falls.

However, acoustic-physical phenomena which have something to with the "inaccuracy" of our hearing organs and the brain can be exploited:

> *In the case of audio-signals at least those parts of the signal can be omitted in the encoder which are imperceptible to the human ear as a result of its limited resolution capacity in the interplay of the time and frequency domain and amplitude. This is referred to as the **irrelevance reduction** of signals or data, i.e superfluous information can be omitted without any loss of quality.*

A compression process which achieves this must in consequence take into account the psycho-acoustic properties of our hearing organs in the perception of audio-signals. As far as the recognition of irrelevant information is concerned this includes:

- The frequency response and the so-called threshold of audibility and auditory sensation area (see Illustration 221) and

- The masking effects which describe the inaccuracy of our hearing organs (see Illustration 222).

The auditory sensation area in Illustration 221 shows the frequency-dependent perception of amplitude. This also applies to the so-called threshold of audibility. According to this our hearing is at its most sensitive around 4 kHz. This is determined by measuring the point from what amplitude a frequency appears to be audible for a large number of different frequencies.

> Note: the volume L is a logarithmic measurement which is given in decibels (db). If the amplitude uncreases by 20 decibels the amplitude has increased by the factor 10 compared with the reference quantity, at 40 decibels by the factor 100 etc.

Usually the frequency scale is selected logarithmically in acoustics. Logarithmic calculation is exponential calculation. As a result the intervals on the frequency axis between $0.1 = 10^{-1}$ and $1 = 10^0$, $10 = 10^1$, $100 = 10^2$, $1000 = 10^3$ etc are equal.

Illustration 222 describes the so-called the masking effect. Imagine you are in a club. Loud music blares from huge speakers. This is very hard work for the sense of hearing as sound levels of 110 decibels and more are reached. Because of the extreme loudness of the music it is almost impossible to talk unless you shout. In acoustics this is referred to as masking. In order to remove this effect the sound level of speech has to be raised to such an extent that it is no longer masked by the interference signal (in this case music).

In principle this describes the property of the ear which cannot hear weak tones in the frequency context of a strong tone. This masking has greater bandwidth the louder the tone in question.

Conclusion: weak tones in the immediate vicinity of loud tones do not need to be transmitted as they cannot be heard anyway.

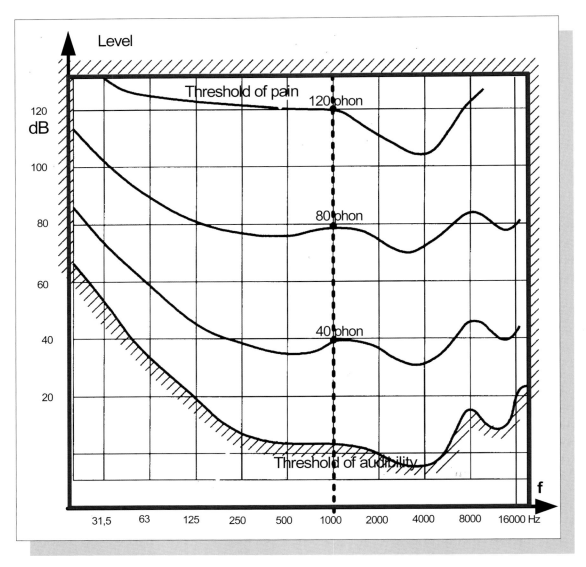

Illustration 221: **Auditory sensation area with audibility threshold**

The so-called auditory sensation area is indicated by hatching. Note the logarithmic scaling on both axes. An increase of 20 decibels implies a tenfold increase in the signal level. In acoustics it is both useful and appropriate to use a logarithmic measure for the frequency axis: here the frequency doubles from marking to marking.

What is important in connection with encoding/compression is the frequency dependency of the sensitivity of our hearing organs. It is greatest around 4 kHz.

Illustration 223 looks at masking from a different angle. The more inaccurate the quantization in the A/D conversion the more unpleasantly loud the quantisation noise. In the immediate vicinity of loud tones "coarser" quantisation – i.e. with fewer bits – is possible than outside masking areas. Thus, considerably more quantisation noise could occur within masking areas than outside.

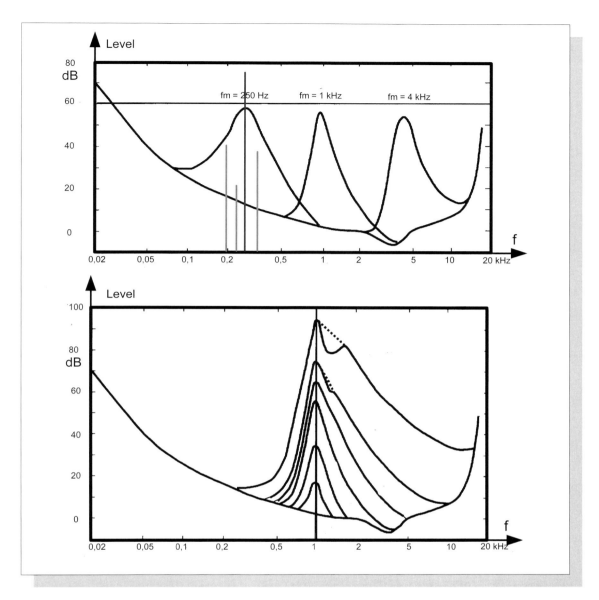

Illustration 222: ***Masking and threshold of audibility***

Top diagram: A loud 250 Hz tone causes a masking area or masks an area so that the quieter tones in the vicinity cannot be heard.

Bottom diagram: the height and width of the masking area increases considerably with loudness; at the tolerance limit of 100 decibels it ranges from roughly 200 to 20,000 Hz.

In the case of the state of the art MPEG audio encoding (MPEG = moving pictures expert group; it is responsible for compression processes for digital audio and video; its standards are recognised worldwide) for the two reasons given the frequency band of the audio-signal is divided up into 32 frequency bands of equal size. Each of these frequency bands contains a narrowband filtered part of the original audio signal. For each of these bands the masking properties are exploited. Weak tones (frequencies) are eliminated if loud ones are present, and at the same time the coarseness of the quantisation is adapted to the masking.

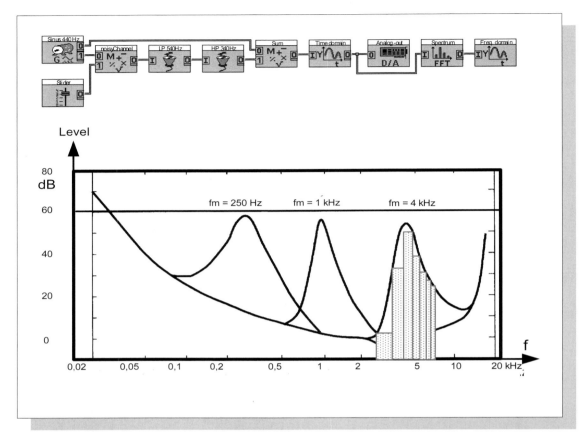

Illustration 223: **Masking thresholds and quantization noise**

The masking areas are frequency and amplitude-dependent. In the case of high frequences the masking areas are much wider than in the case of low ones. This is concealed here by the logarithmic frequency scale.

In the masking area of a loud 4 kHz tone 7 of the 32 equally wide frequency bands are entered. It can now be seen how strong in this area the quantisation noise can be without being perceived acoustically.

At the bottom a straightforward experimental circuit is shown. A narrowband noise signal – equivalent to the quantization noise within the masking area - is added to a loud 440Hz tone by means of a manual regulator. Only from a certtain level can the narrowband noise signal be heard alongside the sinusoidal tone. However, it is clearly visible on the screen in the time and frequency domain before this.

As Illustration 224 shows a MPEG encoder works on a psycho-acoustic model which is taken into account in Illustrations 221 – 223 In order to be able to code/compress optimally the optimal psycho-acoustic model at any given moment must be used. This depends on the audio-signal. The output signal and the signals of the 32 channels are evaluated to this end. As a result this model provides the maximum acceptable quantization for any sub-band, taking account of the masking effects.

By means of the subsequent bit stream formatting the bit patterns of the quantized sampling values of all 32 channels and additional data (for the reconstruction of the audio-signal in the decoder) are formatted to form a bit stream and made largely immune to interference by means of an error protection code (optional).

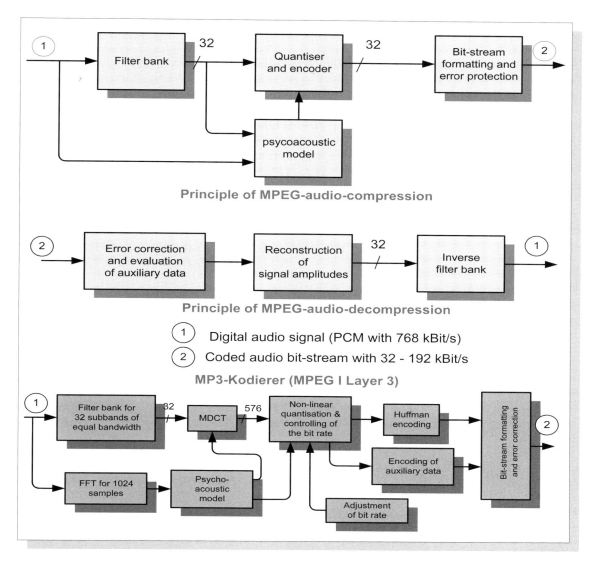

Illustration 224: **Data reduction according to MPEG**

The top half shows the schematic block diagram of the MPEG encoder and decoder. The PCM audio-signal is divided up into 32 subbands of equal width by the filter bank. The quantization and encoding is controlled by the "psycho-acoustic model"; this takes account of the characteristics of the given audio-signal and works out the masking within the 32 subbands. Quantisation is accordingly coarse or fine. After that the 32 channels are formatted to form a single bit stream and provided with error-protection (see in the following the section "Error protection encoding").

At the bottom details of the at present most effective audio-compression according to MP3 are shown. The FFT at the input shows the psycho-acoustic model in which frequency bands strong or weak masking characteristics occur. To go to extremes, the 32 frequency bands are subdivided into a total of 576 (!) subbands by means of an MDCT (modified discrete cosine transformation, a simplified FFT). These bands are so narrow that there are problems with the Uncertainty Principle: the build-up time and dying out time and the pulse response become large as a result and at the same time the resolution of the encoder in the time domain becomes too small. Therefore, depending on the signal it is possible to switch over from the time to the frequency domain resolution.

All this effort serves the purpose of making the data rate of the MP3 signal as low as possible. This approaches the limits of physics and an enormous amount of calculation for the encoder is accepted. The decoding process requires much less calculation. All in all this example shows that the calculation-capacity of processors of multimedia PCs cannot be large enough.

The compression rates are determined above all by the setting of the quantisation thresholds in each partial band. These are determined by the instantaneous masking characteristics of the 32 channels, which determine which tones (frequencies) must be transmitted. A digital PCM audio signal of CD quality and for mono requires 768 kbit/s. By means of the MPEG audio encoding the signal can be compressed to under 100 kbit/s.

The MP3 process (or more precisely: MPEG 1 Layer 3) is particularly popular for audio compression. By means of numerous technical refinements – see Illustration 224 – the compression rate can be increased to less than 10% of the PCM audio signal. Hence, at least 10 normal audio CDs go on to one MP3 CD.

Encoding and physics

Encoding is described here as one of the most important possibilities of modern communications technology. It is striking that with the exception of the compression of audio data – the fundamental physical phenomena of the first chapters – FOURIER, Uncertainty and Symmetry Principle – seem to have little to do with this. They were scarcely used in this chapter.

This has to do with the historical development of communications theory. All the physically based information theories associated with names such as Hartley, Gabor, Clavier, and Küpfmüller were overshadowed by the grandiose success of the information theory of Claude Shannon (1948) which was based purely on mathematics (statistics and the calculus of probability). This theory made modern applications of signal processing such as satellite and mobile telephony possible.

It is still difficult to get a clear grip on the concept of information from a physical point of view. There is still a yawning gap between Shannon's theory and physics, to which all technology is subject.

Shannon's theory will be examined in more detail at the end of the next chapter.

Exercises on Chapter 11

Exercise 1

Explain using concrete examples the enormous appetite of multimedia applications for transmission capacity. What importance could ADSL have as a medium-term solution.

Exercise 2

The term "encoding" does not exist in analog technology. Give a meaningful definition of this term.

Exercise 3

ASCII encoding is a long-established standard. Summarize why it is regarded as obsolete and what qualities a new standard shoulod have.

Exercise 4

Audio and video signals are lossily compressed, but program and text files are not. Try to show the boundary between the two kinds of compression. Why is compression not always loss-free?

Exercise 5

Check using the example of Illustration 213 whether this Huffmann code is an "optimum" code, by producing a different Huffmann tree (e.g. take "0" instead of "00" for A)

Exercise 6

Carry out the LZW encoding for "the rain in Spain falls mainly on the plain".

Exercise 7

Explain the virtues of delta and sigma-delta encoding. Under what conditions does this encoding make sense with audio signals compared with the traditional PCM process? Why are 1 bit-converters gaining more and more ground?

Exercise 8

Describe the structure and purpose of a "decimation filter" at the output of a sigma-delta-encoder.

Exercise 9

Explain the psycho-acoustic effects which are exploited in the MPEG encoding of audio signals (irrelevance reduction).

Exercise 10

In MPEG audio compression the input signal is distributed over 32 equal frequency bands. What is this intended to achieve?

Exercise 11

The masking effects of neighbouring frequencies and frequency domain has already been explained. According to the Symmetry Principle there ought to be masking effects in the time domain (which is, in fact, the case. What are they like?).

Exercise 12

Audio and video MPEG encoding and decoding at the PC can be carried out by software or hardware. What requirements are made of the computer in each case?

Chapter 12

Digital transmission technology II: channel encoding

If a device transmits a continuous stream of bit patterns it should arrive without distortion at the receiver. This is also true of the retrieval of data from a storage medium. The yardstick par excellence for the quality of transmission or storage is the so-called bit error probability.

Error protection encoding for the reduction of bit error probability

In what way can a signal be transmitted as safely as possible via a noisy or distorted channel? The *theory of error-correction encoding* attempts to provide answers to this question. Research into this problem has many direct effects on Communications and computer technology.

While this is clear as far as communications technology is concerned it is perhaps not so clear for computer technology as such. But, think for example of the storage and compression of data. Storage can be understood as a kind of temporary data transmission. The transmission of data is equivalent to writing in a storage medium and receiving data is equivalent to reading. In between time passes in which the storage medium may be scratched or changed in some other way. Error protection encoding can also be advantageous in the storage of data.

In retrospect there have always been transmission errors in the past which could be hardly avoided with the means available. In the case of a telephone call over a noisy and crackling line you can still usually understand the gist of what the person you are talking to says even if you can only hear half of it. And if you can't undestand you can ask him to repeat. Obviously, language is highly redundant and this appears to be the true source of error correction.

For the storing of a longer straightforward text file it would scarcely be necessary to take precautions. If some bytes are erroneous they only make themselves felt as a kind of typing error. It is possible usually to correct the errors from the context as a result of redundancy.

The situation is very different with a packed zip file. The compression is achieved by eliminating redundant data and a single erroneous sign could make the the entire file unusable. Think for instance of the packed *.exe file of a program.

As already mentioned the strategy of modern transmission technology is as follows (see Illustration 210):

- In *source encoding* data are compressed via the elimination of redundancies.

- In *entropy encoding* the attempt is made independent of the type of source signal to optimise the code of the source encoder with respect to compactness.

- In *channel encoding* redundancy in the form of additional bits (check data) is added systematically according to a particular plan in order to recognise and eliminate errors more effectively.

Note:
the development of professional storage media represents a special problem because the error accuracy had to be increased at least one thousandfold compared with (error protection coded) audio CDs. On an audio-CD player a sector makes the data available for 1/75 seconds of the piece of music. If one of these sectors or blocks are defective the previous sector could simply be taken again without the listener noticing.

The data from professional storage media must be supplied practically error-free to the computer. In practice error rates of one error byte per 10^{12} (!) data bytes are accepted for this type of storage medium. For 2000 CD-Roms only 1 error is expected! This is only possible with the given materials and processes with a much more effective process for error protection encoding than with audio.

Let us now take up again the idea of *repeating the signal* in order to be absolutely sure about the transmission. Instead of "1" it is better to transmit "11" and instead of "0", "00". In the case of a high level of noise a mistake or two can occur, for instance, "01" is received. Thus the signal must have been received with errors (error recognition) But what was transmitted: "00"or "11"? Was the first symbol or the second distorted?

In order to increase the reliability of transmission even more redundancy is selected: "000" and "111". Then if "101" were received the original value "111" would all things being equal be more likely than "000". By a kind of "majority voting system" the *string of signs* received ("*vector*") can be decoded and the correct "1" reinstated. Therefore this code does not only recognise errors it also eliminates them (error correction).

Distance

The elimination of errors is possible in the case of "101" because this sequence is closer to "111" than "000". The concept of distance is the key to error recognition and correction.

> *To construct a code which recognises and corrects errors means adding exactly the amount of redundancy so that the code words which are part of the stock of symbols "lie as far as possible apart" without making the string of signs unnecessarily long.*

At this point a formal definition of a code is useful:

> *A code of the length n (e.g. n= 5) is a subset of all possible strings of signs or "vectors" which can be formed from the symbols (e.g. letters of an alphabet). Any two code words from this subset differ in at least one of the n positions of the string of signs. In digital signal processing d = 2, that is we always use binary codes which consist of strings of signs made up of "0" and "1".*

A source with four different symbols is given (for example control of a forklift "to the right", "to the left", "down" and "up"). They are coded in binary form:

> **00** for "right"
> **01** for "down"
> **10** for " left"
> **11** for "up"

Now a parity bit is added for error recognition: If the number of the "1" is even, 0 is added, otherwise 1:

> **000** for "right"
> **011** for "down"
> **101** for "left"
> **110** for "up"

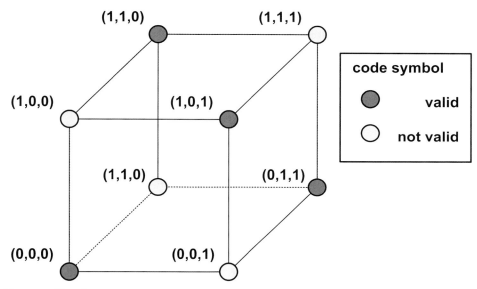

Generalization:
A binary code word of the length n can be represented as a corner of a die in the "n-dimensional room". Then 2^n different corners of the die result.
For n = 3 we have 2^3 = 8 corners in the picture. The distance between the neighboring valid code symbols is larger than between valid and not valid.

Illustration 225:   Binary codes from a geometrical point of view

A binary code word of the length n can be seen from a mathematical point of view as the corner point of an n-dimensional cube even if this cannot be conceived in a spatial sense for n > 3. This way of looking at things makes concepts such as Hamming distance or the importance of the minimal Hamming-distance of a code, the Hamming sphere (it represents the environment around a certain code word point) and the Hamming limit (it indicates how many check bits are necessary to be able to correct a certain number of errors reliably) more easily imaginable, at least in the three-dimensional field.

R.W.Hamming discovered this encoding process in 1950. Alongside Claude Shannon, the true founder of information theory, he is one of the outstanding personalities of this discipline.

The bigger the (minimum) number of places by which code words differ the more reliably errors can be recognised and corrected. This number indicates the minimal distance.

Hamming codes and Hamming distance

In our example a code of length 3 with two data bits and one check bit was selected. With one check bit it is only possible to differentiate $2^1 = 2$ states. If more are selected, say k check bits, 2^k states can be represented. If a binary code word of length n is present, it should be possible in error correction to tell at which of the n places the error occurred or whether no error has occurred at all. Different states n + 1 are considered.

The k check bits must at least be able to represent the n different places at which an error occurred or show that no error has occurred. Thus a useful condition for error correction in the form of an inequality results:

$$2^k \geq n + 1$$

For k = 3 and n = 7 this inequality becomes an equation; thus conditions for these values are optimal. Illustration 226 shows these relationships clearly.

The k check bits form a binary number. In addition, in this type of encoding called after R.W.Hamming the place where the error occurs should be indicated by this binary number. If no error has occurred the check bits should all have the value 0.

For reasons of speed one would like to have a short code, i.e. n should be as small as possible. The number M of the code words should for reasons of efficiency be as big as possible. This is also true of the minimum distance D. Of course, these aims are contradictory. Encoding theory tries to find an optimal compromise between n, M and D. An (n,M,D) code already shows in these values how good it is.

If a code word is distorted on the transmission path then the number of errors is exactly the Hamming distance between the received and the original code word. If the error is to be recognised the (erroneous) code word received must not be allocated to a different code word. This is why a minimal distance D which is as large as possible is so important:

> *If fewer than D errors occur at least an error can be detected.*
>
> *If fewer than D/2 errors occur the code word received is closer to the original code word than to all the others.*

After a code has been used to transmit a message the code word has to be decoded in the receiver. Our intuition suggests that we should look for the code word which is closest to the "vector" received. As long as the transmission path provides 50% reliability for every bit transmitted, the best bet is the code word which differs least from the vector received, i.e. the code word with the smallest Hamming distance. This strategy is called the *maximum likelihood method*.

The **Hamming** code is a binary error patching code that is able to patch 1-bit-errors in one 4-bit data word. This code requires three error control bits which must be computed before stored. The data bits are a1, a2, a3 and a4, the error control bits c1, c2 and c3. Therefore a **7-bit code** results :
Example:

a1	a2	a3	a4	c1	c2	c3
1	0	0	1	?	?	?

The error control bits in the transmitter are to be computed as follows:

$c1 = a1 + a2 + a3$ $c1 = 1 + 0 + 0$
$c2 = a2 + a3 + a4$ $c2 = 0 + 0 + 1$
$c3 = a1 + a2 + a4$ $c3 = 1 + 0 + 1$

Now we applicate the "modulo 2-sum": It corresponds exactly to the exclusive-or-function (EXOR).

$0 + 0 = 1$
$0 + 1 = 1$
$1 + 1 = 0$

This results in $c1 = 1$, $c2 = 1$ und $c3 = 0$

$1 = 1 + 0 + 0$
$1 = 0 + 0 + 1$
$0 = 1 + 0 + 1$

a1	a2	a3	a4	c1	c2	c3
1	0	0	1	1	1	0

Now a data error on transmission channel

a1	a2	a3	a4	c1	c2	c3
1	0	0	0	1	1	0

$1 = 1 + 0 + 0$
$1 = 0 + 0 + 0$
$0 = 1 + 0 + 0$

In the receiver a **parity test** is now carried out as follows::

$e1 = (a1 + a2 + a3) + c1$
$e2 = (a2 + a3 + a4) + c2$
$e3 = (a1 + a2 + a4) + c3$

e1, e2 and e3 are new check bits which must always be 0 in case of perfect transfer

$e1 = (1 + 0 + 0) + 1 = 0$
$e2 = (0 + 0 + 0) + 0 = 1$
$e3 = (1 + 0 + 0) + 0 = 1$

The 2nd and 3rd equation is wrong. Both equations contain a2 and a4. However, since a2 would also make the 1st equation wrong, a4 must be wrong !

An error control bit was falsified on the transmission channel

a1	a2	a3	a4	c1	c2	c3
1	0	0	1	1	0	0

$1 = 1 + 0 + 0$
$0 = 0 + 0 + 1$
$0 = 1 + 0 + 1$

$e1 = (1 + 0 + 0) + 1 = 0$
$e2 = (0 + 0 + 1) + 0 = 1$
$e3 = (1 + 0 + 1) + 0 = 0$

Only the 2nd equation is wrong. Since c2 occurs only in this and no other equation, c2 must be wrong !

Illustration 226:   ***Codes which discover errors and codes which correct them***

Individual bit errors can be discovered and corrected by the Hamming code with 100% certainty. If several bits are transmitted or read defectively this code is no use. There are however codes which correct errors and make it possible to correct more than one bit, on the basis, of course, of additional check bits or redundancy. Practically all these codes that correct errors essentially follow the example described: "upstream" information is calculated according to a preset pattern and the data is stored. This information is used to recalculate the bits after being read. The check bit pattern makes it possible to locate the site of the error and recognise the original correct value.

With reference to the case described above with D/2 errors error correction can now be precisely defined:

> *A code with a minimal distance from D can correct up to (D-1)/2 errors reliably.*

Comparing the vector received with every possible code word in order to find the "closest" code word is theoretically the most promising method but in the case of lengthy codes it requires far too much calculation. For this reason a large part of encoding theory is concerned to find codes which can be decoded efficiently and therefore rapidly.

All this might seem rather theoretical and not very practical. The contrary is true. The following text is taken from a tribute to R.W. Hamming underlining the importance of his life's work:

"In everyday life we measure a distance by counting the metres which are required to get from one place to another.

In the digital world as a sequence of zeros and ones, that is "bits", the Hamming distance is the number of bits which must be changed in order to get from one bit sequence to another. This distance was introduced by R.W. Hamming at the beginning of the 50s and has been used since extensively by information technology and informatics. Thus it is possible in the typical task of "artificial intelligence" of recognising a face to measure the similarity of two faces by means of their Hamming distance.

The first application for which Hamming introduced this measurement of distance was connected with protection against errors in the transmission, storage or processing of bit sequences. The greater the number of transmission errors the greater also the Hamming distance transmitted and defectively received bit sequences. Hamming therefore used only bit sequences for transmission which among each other have a more than twice as large a Hamming distance than can arise through errors of transmission. Thus a mistake can be recognised and even corrected. Several bit sequences with a large Hamming distance form a "Hamming code".

This work which which was carried out at the same time as Shannon's information theory forms the foundations of the general theory of error-correction systems and codes. Their importance lies above all in the fact that systems of any complexity with a definable level of reliability can be constructed from less reliable subsystems. The effort required can be calculated precisely.

The mathematically highly sophisticated theory of error-correction encoding has an indispensable application in every CD-player and mobile phone and in every large-scale computer and worldwide information networks. It is only by means of such codes that the unavoidable errors in transmission, storage and processing of information can be corrected".

Convolutional encoding

In error protection encoding two types of encoding are distinguished: block and convolutional encoding. So far only codes of a fixed block length have been dealt with. Here the information is transmitted blockwise.

In convolutional encoding, on the other hand, the input data – as with digital filters – are "spread" over several output data in a convolution-like process. Illustration 211 shows the structure of a straightforward convolutional encoder. The input signal – e.g. a long lasting digitalised audio signal – in the form of a bit pattern is fed bit by bit into a two-step shift register. At the same time the input signal is linked via an "addition" (EXOR) to the bits of the shift register at the outputs. The shift register can take on 4 different states (00, 01, 10, and 11). In order that these states (the next bit appears here as a righthand bit in the shift register) should correspond pictorially to the states in the shift register the input of the convolutional encoder is registered on the right and the outputs on the left.

The two outputs receive the same clock pulse frequency as the input signal, the redundancy of the total output signal is therefore 50% greater than that of the input signal, the prerequisite for error protection encoding.

We should now be on the look out for possibilities of visualising the signal flow at the output in its dependency on the input signal and the states of the shift register. Two methods have proved their worth:

- *State diagram:*

    The states diagram shown in Illustration 228 describes completely the "system of rules" of the convolutional encoder represented there. The four different states of the shift register (state circles) are in the four circles. An "0" or a "1" may lie at the input of the convolutional encoder which is why two arrows point away from each state circle.

    Let us assume an initial state "00". If a „1" is connected with the input, "11" appears at the outputs (all the operations of the shift register result in 1 + 0 = 1). The next state is then "01". 1/11 is entered at the arrow, the lefthand 1 is the input signal and the righthand 11 is the output signal.

    If an "0" is connected to the input the next state is "00". This is why 0/00 is entered at the arrow.

- *Network diagram:*

    Usually called a *trellis diagram* in the literature. The sequence in time is now an additional factor. The four possible states of the shift register are arranged vertically. Every additional bit at the input implies a further step to the right.

    Let us begin at the top left with the state "00". A thick line means a "0" at the input and a thin line an "1". From each point of the trellis reached a thick and a thin line – corresponding to "0" and "1" – go to a different state. The input and output signals are registered by the thin and thick lines.

While the state diagram forms the "set of rules" for the convolutional encoder the encoding of a specific input signal bit pattern can be followed closely by means of the trellis diagram.

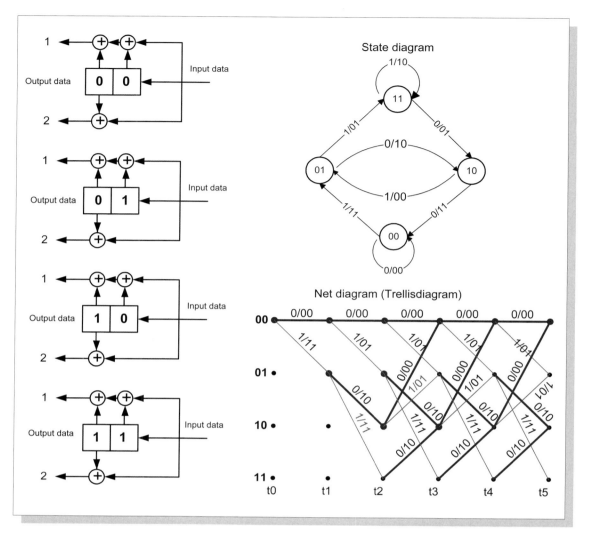

Illustration 227: ***Example of a convolutional encoder, a state diagram and a trellis diagram***

On the left there is the same convolutional encoder with its total of 4 "inner" states of the shift register). The "+" links carry out a "modula-2" or EXOR-operation out (see Illustration 226).

The state diagram represents the "set of rules" of the convolutional encoder in a very straightforward way. In the four circles there are the different states of the shift register (state circles). An "0" or a "1" may lie at the input of the convolutional encoder; for this reason two arrows lead away from each circle. Two arrows "land" accordingly on each state. It is possible for the arrow to begin and end on the same state. Moving from one state to another is only possible in a quite specific manner; for instance, a change from "11" to the state "01" is not possible in one step.

The trellis diagram shows the possible change from state to state in the time sequence, starting here from the state "00". The states of the shift register are arranged vertically. Starting from the state "00" only two other states ("00" and "01") are possible, depending on whether there is a "1" at the input (thin red line) or a "0" (thick blue line). From each point in the trellis reached in principle two paths are possible, an "0" path and a "1" path.

The input signal is on each path and after the diagonal line there is the output signal (e.g. 0/10). Among the possible paths listed here there is the path for a specific bit pattern (see Illustration 226) and also the most likely path for the Viterbi decoding of a signal or bit sequence which has been changed on the transmission path or in the storage medium. (see Illustration 229).

Chapter 12: Digital transmission technology II: channel encoding

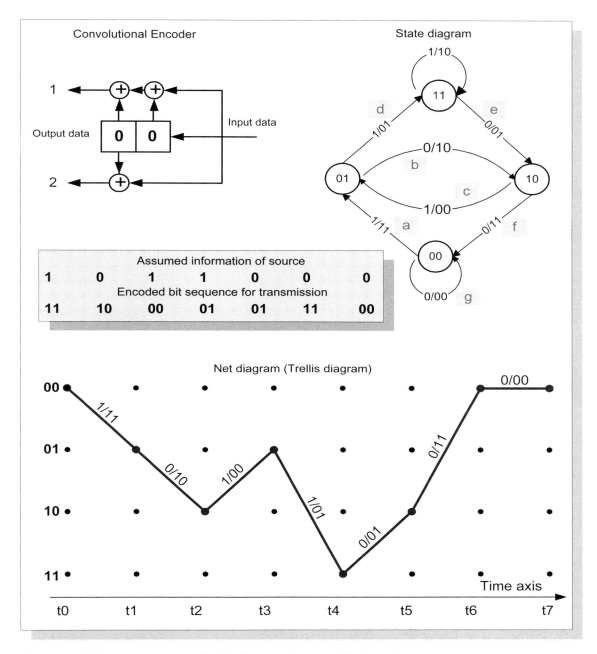

Illustration 228: *Trellis diagram for a specific input bit sequence*

The state diagram applies for all kinds of input bit pattern sequences and not just for a specific one. On the other hand the trellis diagram can describe the course of the state sequence and the output signals for a specific input bit sequence.

The above example leads to a specific coded bit sequence at the output. The trellis diagram shows the path for this signal. In order that you can follow the course, changes of state and the output signals the path in the state diagram is marked with the letters a,b,c....

In decoding the following problem awaits us: how can this path be reconstructed if the bit sequence of the output signal was distorted at one or several points on the transmission path or by the storage medium? Viterbi decoding provides the answer (see Illustration 229)

Example: the bit pattern at the input consists of the sequence:

1 0 1 1 0 0 0

The coded bit sequence which can be checked by means of the state diagram results at the outputs:

11 10 00 01 01 11 00

Illustration 228 shows the relevant trellis diagram.

Viterbi decoding

What happens in decoding if the above coded bit sequence is distorted on the transmission path and is decoded in the receiver. How can the errors be recognised and corrected?

Illustration 229 shows this for the sequence received:

1<u>0</u>   10   <u>1</u>0   01   01   11   00

The underlined values represent the errors on the transmission path or the storage medium. Now the trellis diagram is developed step by step in the decoder by means of the state diagram which contains the whole set of rules for the convolutional encoder.

The decoding process which is called Viterbi-decoding after its inventor tries several paths and tries to find the most likely path or *the most likely coded bit sequence* (maximum likelihood method).

Now for decoding:

- Let the decoder be in the state "00" and receive the bit sequence "1<u>0</u>". A glance at the state diagram shows: the encoder cannot have generated this bit sequence because starting from the state "00" there are only two alternatives:

    - Sending "00" and retention of the state "00". However, the decoder already "knows" that in this case only 1 bit was correctly transmitted. As "sum of the correct bits" the "1" is registered at the trellis point (top).

    - Sending "11" and transition to the state "01" (in the trellis diagram the diagonal line from top left). Here too only one correct bit was received, that is a "1" at the trellis point.

- The decoder again receives the bit sequence "10"

    - Starting from the state "00" the same procedure is repeated once more. If "00" had been transmitted the states "00" and "01" would be possible; in both cases only 1 bit was correctly transmitted, however. The sum of the correct bits has thus risen to "2" (out of altogether 4 bits) and this has been registered at the trellis points.

Chapter 12: Digital transmission technology II: channel encoding

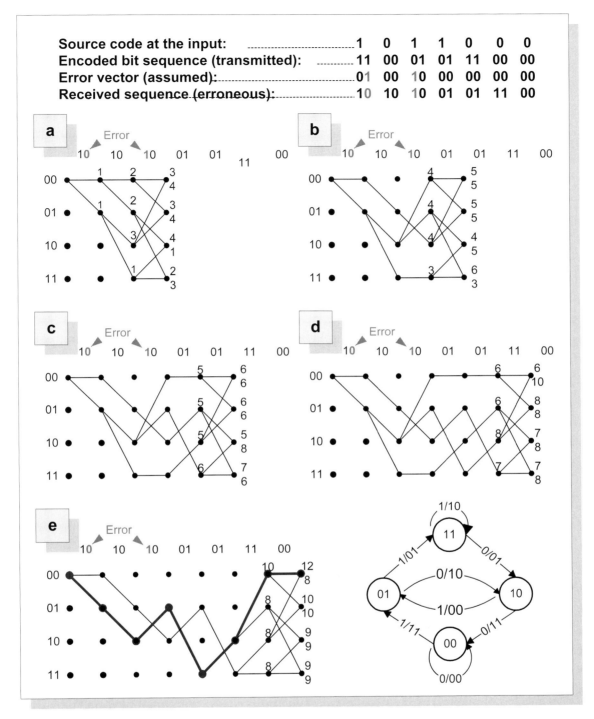

Illustration 229: **Viterbi decoding**

Viterbi decoding is carried out to ensure error-free reception of the bit pattern sequence. The state diagram shows that the bit pattern in this sequence cannot have been generated by the convolutional encoder, i.e. it recognises immediately that a bit must be wrong in this case. However, as it is not clear which of the two bits must be wrong, both possibilities are investigated. The paths diverge. The "sum of the correct bits" is registered at the trellis points.

Two different paths may intersect at a trellis point. The path with the smaller "sum of correct bits" is deleted. Thus gradually the path with the greatest likelihood emerges. It is in fact identical with the path for the correct sequence of bits in Illustration 228. The details are described more closely in the text.

- Starting from the state "01" the reception of a "01" leads to the state "11". But then both bits would be wrong compared with "10" and the sum of the correct bits remains "1". The alternative reception of a "10" with the transition to the state "10" gives two correct bits, that is on this path so far 3 correct bits. After two steps this path is the most likely..

- The third bit sequence "10" is now evaluated by means of the state diagram and the sum of the correct bits registered. Two transitions end in each state point. In state "00" for example we find two transitions with different sums of the correct bits (3 and 4). Viterbi decoding is based on deleting the transition with the smaller number of correct bits as this is the less likely path through the trellis diagram. If the sum of the bits in the state point is equal in number no path is deleted and both alternatives are pursued further.

- By evaluating the other bit sequences received "01", "01", and "11" and constant deletion of the least likely transitions a "most likely" path in the trellis diagram is formed.

- The last received bit sequence "00" leads to a decision. The most likely path of the trellis diagram for the Viterbi decoder is identical to the correct path of the decoder in the trellis diagram in Illustration 228.

Errors in the sequence received have thus been corrected. By comparing the sum of the correct bits registered on the right the number of errors which have occurred can be estimated (14 – 12 = 2).

Hard and soft decision

After encoding the signals in the area of the transmitter, modulation follows to make transmission possible. The signal is demodulated in the receiver. As a result of interference of the most various kinds during transmission the "0" and "1" states cannot be as clearly differentiated as in the transmitter.

Apart from interference by noise or other signals the properties of the transmission media influence the form of signal. In the case of wireless transmission this is for example the multiple reflection of the transmitted signal by a variety of obstacles which may produce echo effects or extinguish the signal.

Dispersive characteristics of cable (see Illustration 106) are apparent too. The speed of transmission in a cable is always frequency-dependent. As a result sinusoidal oscillations of different frequencies propagate at different speeds which implies a change in the phase spectrum. This involves a change in the form of the signal.

The signals or bit sequences received are distorted and noisy. If a hard decision is being used the possibility for the decoder of finding the most likely path in the trellis diagram is reduced considerably if not completely eliminated. More subtle possibilities of making decisions should be offered (soft decision).

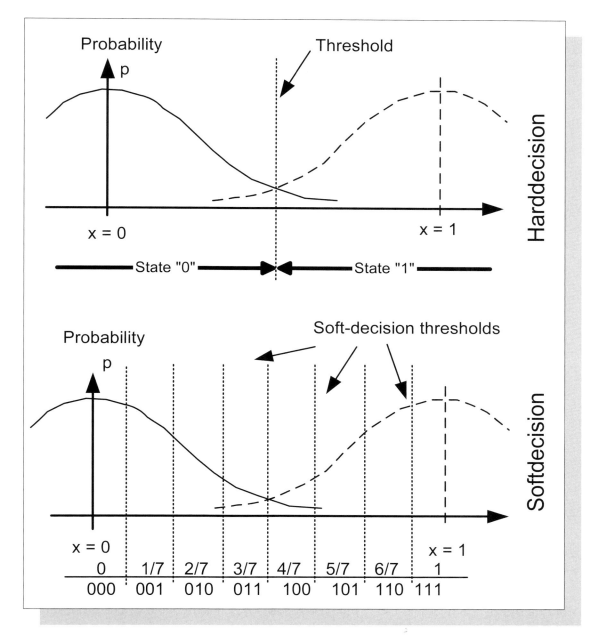

Illustration 230: **Hard and soft decision**

The curves represented here can be interpreted as follows: the signal for the state "1" (or "0") usually arrives in good condition but the degree of deviation is sometimes so great that it is sometimes mistaken for a different state and interpreted wrongly (the areas of deviation overlap in the middle). As, however, some bit patterns contradict the "set of rules" of the convolutional encoder the decision on "0" or "1" should be taken in conjunction with the Viterbi decoder.

It is the task of the demodulator (see Illustration 210) to provide support to the decoder. The demodulator passes a 3 bit value to the decoder which ranges from "000" (considerable likelihood for the reception of "0") to "111" (considerable likelihood of the reception of 1); it selects the value depending on the grade allocated to the signal. The simplest allocation of steps is shown in the diagram (bottom). This has proved its worth in many cases.

In the case of soft decision not only integer values are possible for the "sum of the correct bits". This leads to a much more precise assessment of the likelihood of a particular path through the trellis diagram. The typical encoding gain lies in the region of 2dB

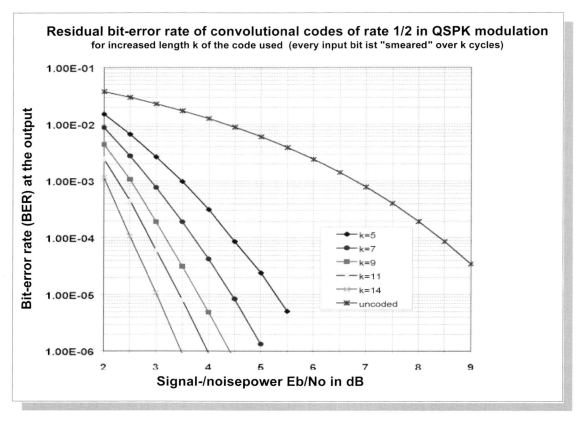

Illustration 231: *The efficiency of convolutional codes*

The efficiency of error correction increases as can be expected with the length of influence of the encoder used. It is determined by the storage steps of the shift register. The larger these are the longer the input bit influences its predecessors and successors. The input information is "spread" more widely and therefore protected more effectively against individual bit errors.

K = 5 already implies considerable error protection and a much lower bit error rate (BER) compared with non-coded signals. However, the amount of calculation increases steeply for the Viterbi decoder with increasing influence length.

For an optimal decision strategy the probablility of receiving a distorted signal should be known (see Illustration 230). In practice a choice of 8 decision levels has proved to be appropriate.

Channel capacity

The information theory has a definite date of birth marked by the famous "encoding theorems" of Claude Shannon from the year 1948. Before anyone could have foreseen the present development of digital signal processing he recognised the fundamental problems and provided brilliant, definitive solutions at the same time. It took decades for his work to be understood properly and put to practical use. His contributions are among the most important scientific insights of the last century, a fact which is, however, recognised by very few people. From a sociological, technological and scientific point of view his work overshadows everything else. Read the theses at the end of Chapter 1 again if you do not believe this. For this reason it is appropriate to look at his fundamental ideas with reference to the information channel.

Chapter 12: Digital transmission technology II: channel encoding

The selection of symbols chosen by the information source in the generation of the information represents a source of uncertainty for the receiver in the sense of unpredictability. This is only removed when the receiver recognises the message.

> *Thus the aim and the result of communication processes is the resolution of uncertainty.*

In order to measure the amount of information a yardstick is necessary which increases with the extent of the freedom of decision of the source. At the same time the uncertainty of the recipient increases as to what message the source will produce and transmit. This yardstick for the quantity of information is called entropy H. It is measured in bits. The information flow R is the number of bits/s.

How can the transmission capacity of a distorted or imperfect channel be described? The information source should be selected in such a way that the information flow of of a given channel is as great as possible. Shannon calls this maximum information flow "channel capacity" C. Shannon's main thesis on the distorted transmission channel is:

> *A (discrete) channel has a channel capacity C and a (discrete) source with the entropy H is connected with it. If H is smaller than C there is a encoding system by means of which the information from the source can be transmitted via the channel with a minimal error frequency (bit error likelihood).*

This result was noted by other scientists and engineers with astonishment. Information can be transmitted via distorted channels with the desired level of reliability!?.

If the probability of transmission errors increases, that is, errors occur more frequently, the channel capacity C as defined by Shannon becomes smaller. It decreases the more frequently errors occur. Thus the information flow must be reduced to such an extent that it is smaller or at most equal to the channel capacity C.

How can this be achieved? Shannon "only" proved the existence of this cut off value – he didn't say how to achieve this. Modern encoding processes are approaching this ideal more and more. This can be seen in the development to date of modem technology. If 20 years ago the data rate which could be achieved via a 3 kHz telephone channel was 2.4 kbit/s today it is more than 56 kbit/s!

It now seems clear why source encoding (including entropy encoding) as a method of compression should be dealt with quite separately from (error protection) channel encoding. In source encoding as much redundancy as possible should be removed from the signal in order to get as close as possible to the channel capacity C of the undisturbed signal. If the channel is affected by interference exactly the amount of redundancy can be added to get as close as possible to the (smaller) channel capacity of the disturbed channel.

Exercises on Chapter 12

Exercise 1

List the types of files which it is especially necessary to protect against defective transmission. In what context does the concept of redundancy occur?

Exercise 2

Define the concept of the (Hamming) distance of a code and describe its importance for error recognition and correction. What is the significance of the inequality $2^k = n + 1$ (k check bits; n is the length of the code word).

Exercise 3

Explain the maximum likelihood method for the recognition and correction of transmission errors.

Exercise 4

Create an example (as in Illustration 226) and determine the correction and check bits via "modula 2 addition".

Exercise 5

Describe the principle of convolutional encoding as compared with block encoding.

Exercise 6

Explain in connection with convolutional encoding the representation of the signal flow as a state diagram or a trellis diagram.

Exercise 7

Describe the strategy of the Viterbi decoding of a defective bit pattern reception sequence in the trellis diagram by means of the state diagram

Exercise 8

Soft decision is usually advantageous compared with hard decision. Explain.

Exercise 9

Formulate and interpret Shannon's main thesis in encoding theory.

Chapter 13

Digital Transmission Techniques III: Modulation

Digital signals at the output of an encoder are a sequence of numbers in the form of a bit stream. In order to be able to transmit such bit patterns via a physical medium (cable or space) they have to be converted into a modulated, continuous-time signal, which is, in the final analysis, analog.

Bit patterns as a particular type of signal seem to consist of a random sequence of rectangular impulses. These contain steps in quick succession and therefore their bandwidth is very large. This is why prior to transmission the bandwidth needs to be restricted by means of filters. According to the Uncertainty Principle this leads, however, to an extension in time of each bit-pulse whereby neighbouring bits may overlap (ISI Intersymbol Interference).

A rectangular pulse always has a Si-like spectrum. This is also shown in Illustration 232. A "black box" provides an arbitrary sequence of binary pulses. The spectrum has a Si-shaped curve. The task is, by means of filters, to try and restrict the bandwidth in an optimal way so that the receiver can still reconstruct the bit pattern.

> *Note:* eye diagrams
>
> The eye diagram in Illustration 232 provides information on the quality of a digital signal. For this purpose the horizontal deflection of the screen is synchronized with the clock of the (random) bit pattern. A more detailed, comparative analysis of the diagrams leads to the following results:

- The curves of the "eyes" are the result of the curves of the pulse transitions.

- The horizontal lines passing through at the top and bottom are a measure of the existence of rectangular pulses.

- If the openings of the eye diagram disappear the distortion of the bit pattern makes it almost impossible to reconstruct the original transmission signal in the receiver.

- Eye diagrams make it possible to estimate the quality of ISI and of so- called phase jitters (irregular phase fluctuations caused by instable oscillators) and overlaid noise.

ISI depends to a great extent on the type of filter used and above all on its edge steepness. This is shown in Illustration 232 for four different types of filters where the cutoff frequency of the filter is identical with the so-called Nyquist-frequency (the minimum frequency with which a band-limited signal is to be sampled according to the sensing principle: $f_N = 2 * f_{max}$ (see Illustration 190). Illustration 232 shows two types of filters where the eye diagrams are closed on the right hand side. Butterworth and Chebycheff filters have a greater edge steepness than Bessel filters (with the same cutoff frequency and "quality"), but they have a pronounced non-linear phase curve. This means that the sinusoidal oscillations at the output of the filter are in a different phase position towards each other than at the input of the filter. That is why the pulse form becomes blurred and neighbouring bits overlap each other. It should hardly be possible to reconstruct the original bit pattern in the receiver in these two cases.

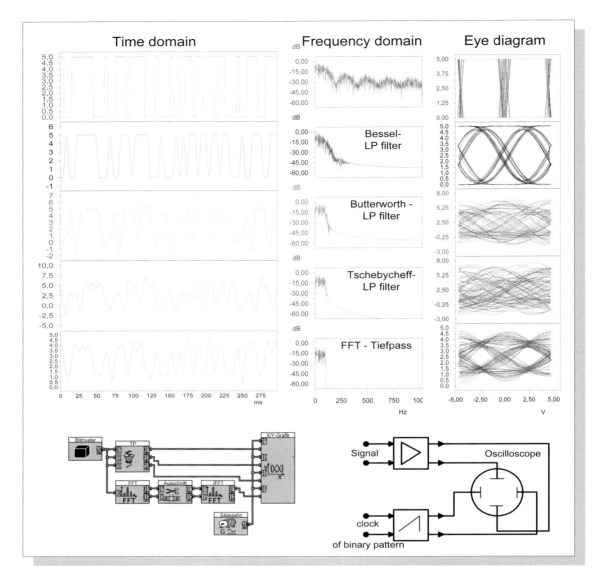

*Illustration 232: **Lowpass filtering of the bit stream to restrict the transmission bandwidth***

At the top left you can see the bit patterns which are generated by the encoder, in the centre the relevant Si-shaped spectrum. The frequency of the basic pulse rate is in this case 200 Hz. Because of its large bandwidth the signal cannot be modulated. It is important to find an optimum filter which allows the original bit pattern to be reconstructed in the receiver and which at the same time reduces the bandwidth to a minimum range. A sampling frequency of 200 Hz means that at best a signal of 100 Hz can be filtered out. This is attempted using four different types of filters (with the same cutoff frequency of 100 Hz and with the same – 10th – order).

At the top you can see the experiment using a Bessel filter, below that with a Butterworth filter, followed by an experiment with a Chebycheff lowpass filter (see Illustration 131) and finally, at the bottom, using an FFT filter (see Illustration 196 ff). The Bessel filter has a very pronounced edge, but an almost linear phase response. Butterworth and Chebycheff filters both have a relatively steep edge, but a non-linear phase response. The "ideal" FFT filter, by contrast, has an extremely negative edge and complete phase linearity.

The eye diagram is a standard method in measuring technology to find out whether a bit pattern is reconstructible. The larger the free area of the eye, the better. According to this method both the Bessel and the FFT filter meet these requirements. The non-linearity of the phase response obviously leads to an extreme distortion of the pulse forms of the other two filters. This can also be seen in the time domain.

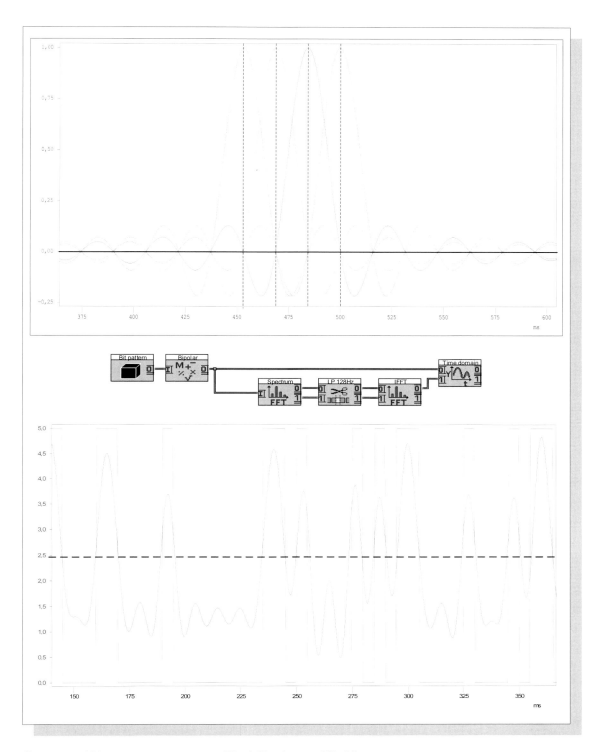

Illustration 233: ***Ideal filtering and limiting case***

In Illustration 232, bottom, the limiting case (minimal bandwidth and complete reconstruction) is reached. At a cutoff frequency of 100 Hz and an ideal – i.e. rectangular - bandwidth limitation the pulse response is Si-shaped. In this special case, however, the Si-functions overlap as is shown in the illustration, top. The aggregate signal of the individual pulse responses is free of interference at the sampling instant because the Si-functions of all neighbouring pulses have a zero crossing at that moment. The illustration at the bottom shows that in this case the "0" and "1" values of the bit pattern are definitely reconstructible at a discrimination threshold of 50%.

The most interesting case is the FFT filter. You would really expect the worst eye diagram because of the "ideal" filtering with an almost rectangular edge. The pulse response and the transient of this filter take, according to the Uncertainty Principle, an extremely long time and they resemble a Si-function. Therefore all these Si-functions overlap considerably. So, why is the eye diagram relatively open?

The explanation is given in Illustration 233. With the Nyquist frequency the aggregate signal of the individual pulse responses is free of interference at the sampling instant because the Si-functions of all neighbouring pulses have a zero crossing at that moment. The illustration at the bottom shows how well the original bit pattern might be reconstructed. However, there are no analog FFT filters with these characteristics.

Keying of discrete states

Classical modulation procedures only have a limited frequency range at their disposal and the same is true of digital modulation methods. These also use, directly and indirectly, sinusoidal oscillations as carriers. The basic difference between analog and digital modulation methods is in the keying of discrete states.

A sinusoidal oscillation can at best be modulated with regard to amplitude, phase and frequency. In other words, digital modulation procedures change amplitude or phase (or a combination of the two) in steps or discretely in one or several (adjacent) sinusoidal oscillations.

Amplitude Shift Keying (2-ASK)

Illustration 234 shows the keying of a sinusoidal carrier oscillation by means of a unipolar bit pattern. The keying in this case is the multiplication of the two signals and the information is – just as with AM – in the envelope. The modulated signal takes on two states (ASK: Amplitude Shift Keying).

ASK in this form is hardly ever used nowadays. If the clock pulse frequency and the frequency of the carrier do not have an integer relationship there are sporadic steps which widen the bandwidth. Switching off the transmitter would give the receiver the impression of a permanent transmission of zero states. Besides, synchronization between transmitter and receiver does not seem to be entirely unproblematical.

Phase Shift Keying (2-PSK)

For the above reasons it seems to make more sense to use a bipolar bit pattern for modulation which is symmetrical to zero (NRZ-Signal: No return to zero). In this procedure the sinusoidal carrier is alternately multiplied by a positive and a negative value, e.g. +1 and −1.

This, however, means that the amplitude does not change, but the procedure is a phase shift keying between 0 and π rad (PSK: Phase Shift Keying). Phase shift keying can also be interpreted as amplitude shift keying where the amplitude is not switched on or off but where it is inverted.

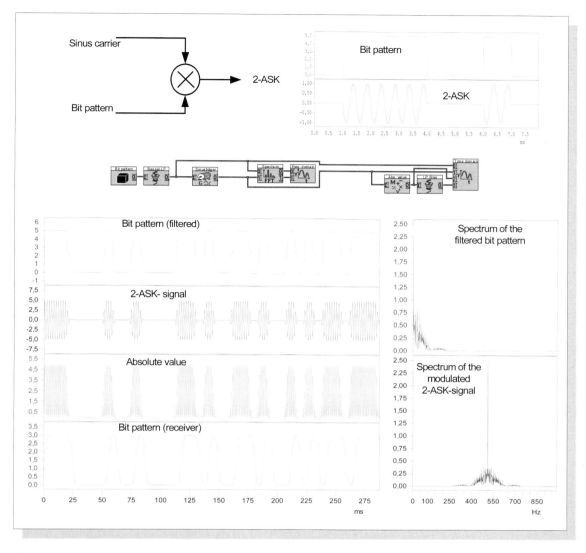

Illustration 234: ***Amplitude Shift Keying 2-ASK***

Illustration top: In principle, 2-ASK does not differ from classical amplitude modulation AM with regard to circuits. Instead of a continuous signal, a discrete bit pattern is modulated which only chooses between two states.

Illustration bottom: this example is much more realistic. At first the bit pattern is band limited using a suitable (Bessel) filter. The signal does not have any steps and cannot change faster than the highest frequency it contains. The modulated or "keyed" signal has a continuous curve. 2-ASK is a double-side-band AM with carrier (see bottom right).

Here demodulation is conducted in a conventional way using rectification (or formation of absolute value) prior to Bessel lowpass filtering. Thanks to the phase linearity of the filter the pulse form – the bit pattern – is hardly distorted. The original bit pattern can easily be reconstructed. For this purpose the basic pulse rate of the binary pattern needs to be extracted from the received signal. The components which are needed are a PLL (Phase Locked Loop, see Illustration 171), a comparator (with discriminator potential) and a Sample & Hold circuit.

A considerable amount of transmission energy is used by the carrier which does not contain any information. This means that the distortion liability of the information transmitted is relatively high. The following Illustration shows that phase shift keying 2-PSK is a preferable method. It is, however, hardly used anymore because quadrature phase shift keying is a much more efficient method which guarantees a much higher transmission rate.

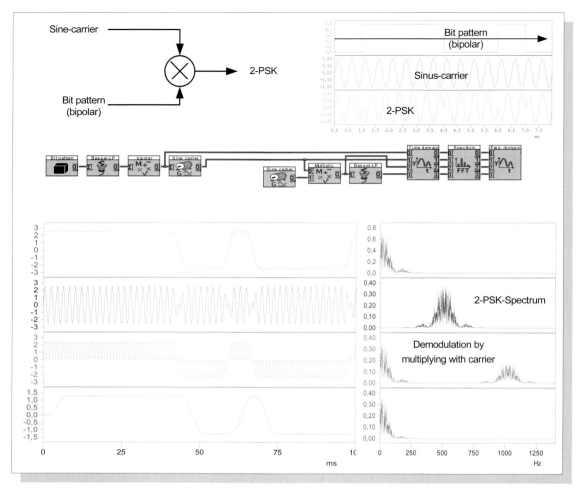

Illustration 235: **Phase Shift Keying 2-PSK**

Phase shift keying is the result of the multiplication of a bi-polar signal with the sine carrier. This automatically leads – as can clearly be seen above –t o phase shift keying. The Illustration in the centre shows a modulation and demodulation circuit for phase shift keying with DASYLab. An arbitrary sequence of bit patterns is band-limited by a lowpass, transformed into a bipolar signal and multiplied by the carrier. The pre-filtered signal clearly shows the phase steps. In contrast to 2-ASK no carrier is recognizable in the spectrum, the total transmission energy is used by that part of the signal which carries the information.

Demodulation is carried out by multiplying the incoming signal by the sinusoidal carrier which can be reconstructed using a PLL or a Costa's loop. Aggregate and differential signals are – as with AM – shown in the spectrum. The source signal is reconstructed by lowpass filtering.

Frequency Shift Keying 2-FSK

2-FSK allocates two different frequencies to the two states of the bit pattern. Illustration 219 shows signals, a block diagram and the DASY*Lab*-system (2-FSK with two different states).

Additionally, the input bit pattern is inverted. This branching means that only one signal at a time is "high", whereas the other one is always "low" at that moment. Both bit patterns are then filtered and finally multiplied by the frequencies f_1 and f_2. Thus only one frequency at a time is – in line with the bit pattern – switched on. The sum is the 2-FSK signal.

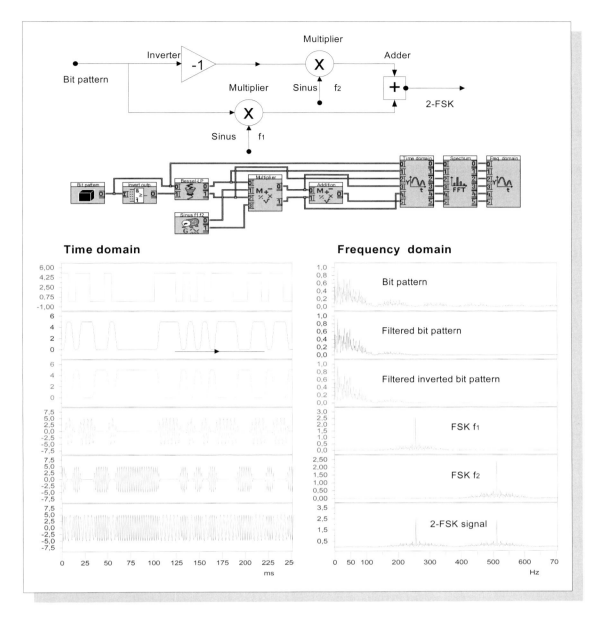

Illustration 236: **Frequency Shift Keying (2-FSK)**

The DASYLab circuit additionally inverts the incoming bit pattern. Both signals – the inverted and the original bit pattern – are then lowpass filtered. Only one of the two signals is different from zero. Multiplication of each signal by the frequencies f_1 or f_2 results in two ASK signals of different frequencies, only one of which is present at a certain time. The sum of both signals is the 2-FSK signal.

The frequency domain (bottom) shows that the two frequencies need to be relatively wide apart from each other so that the two signals do not overlap. This method is therefore not applied where a particularly efficient transmission on a band-limited channel is aimed at.

Signal space

Discrete states of sinusoidal oscillations are allocated to the discrete states of the bit pattern with regard to amplitude, phase and frequency. We will see that combinations of all three possibilities are generally used with the particularly efficient digital modulation method. This results in a kind of APFSK (amplitude-phase-frequency-shift keying).

The GAUSSian numerical plane which was introduced in Chapter 5 (see Illustrations 90 ff.) provides a clear presentation of these possible states of digitally modulated signals. In contrast to the symmetry of frequency presented in that chapter (any sinusoidal oscillation consists of two frequencies +f and −f), it is in this case sufficient to use just one vector +f.

In the signal space the various different discrete states of the carrier oscillation are represented as vectors with regard to amplitude and phase. It is, however, common practice to represent only the corners of the vectors. In addition, a cosine-shaped carrier oscillation is usually chosen as carrier or reference carrier with 2-ASK and 2-PSK modulations so that the ends are on the horizontal axis on the GAUSSian numerical plane.

In Illustration 237 the ends of 2-ASK and 2-PSK are on one line, i.e. from a mathematical point of view in a *one*-dimensional space. So why is the representation on the GAUSSian plane (*two*-dimensional) so important with regard to the signal space? Consider the following aspects:

- Precisely one area can be determined in which the (discrete) state is to be located in a disturbed channel to be clearly identifiable in the receiver.

    - This shows the advantage of 2-PSK where the ends have twice the distance compared with 2-ASK (provided that the amplitude of the carrier frequency is identical).

    - In a disturbed channel the ends will not always be in the same position as in an undisturbed ideal state. If, for instance, the channel is noisy the end is randomly distributed within a given range with each procedure (see Illustration 237).

- The one-dimensional space could easily be enlarged to become a two-dimensional space, if not only the phase angles 0 and 180 degrees, but other phase angles and amplitudes were permitted, too.

    - The more different (discrete) states of a carrier oscillation (constant frequency) are permitted the better the bandwidth utilization of the transmission should be.

    - On the other hand: the shorter the distance between the different ends of the signal space, the greater the interference proneness of the signal. This could perhaps be avoided by linking the channel encoding and channel modulation.

- The most important question now is to what extent the bandwidth of the signal changes when the number of discrete states (amplitude and phase) of the carrier oscillation increase.

Chapter 13: Digital Transmission Techniques III: Modulation

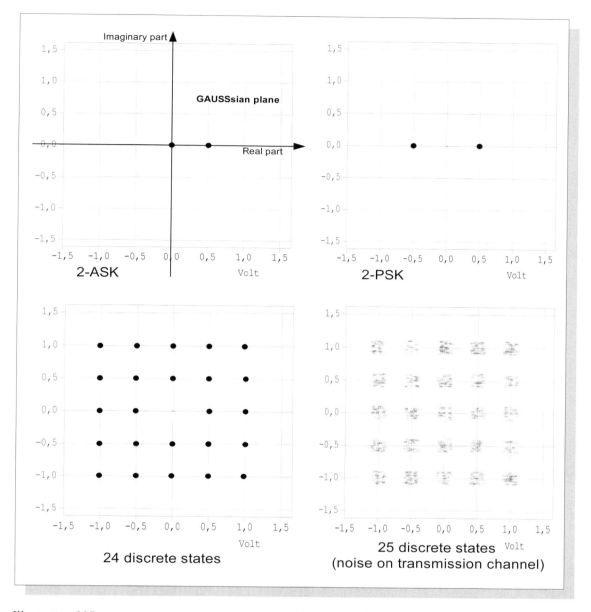

Illustration 237: **Discrete signal states and signal space**

The GAUSSian numerical plane is represented as a two dimensional signal space for carrier oscillations where amplitude and phase are in a discrete state. 2-ASK and 2-PSK are positioned on the horizontal axis. It is obvious that the number of possible discrete signal states should be increased to form a two dimensional signal space.

This is possible – as shown in Illustration 91 ff – by splitting up the phase shifted sinusoidal oscillations into a sine and cosine part. Quadrature phase shifting (QPSK) is the simplest example - described in Illustration 238 and the following text.

- Another point which needs to be clarified is the use of several adjacent carrier frequencies (frequency multiplex) each of which has a certain number of discrete states. Different frequencies theoretically require different GAUSSian planes. If you follow this train of thought the signal space is really *three* dimensional with the frequency as the third dimension.

- Is it preferable to work with just one carrier frequency and many different discrete states, or with many different carrier frequencies and only a few discrete states?

- But there are even more situations imaginable. For instance a mobile telephony network where all participants use the same frequency range at the same time with channels being separated by means of a special encoding. In such a case all other channels would make themselves felt as additional interference of the channel used.

This is an interrelationship of entirely different discrete states (amplitude, phase, frequency) of a signal with regard to the optimum utilization of the limited bandwidth of a transmission medium. Encoding is added as the *fourth* "dimension". This means that Claude Shannon's information theory needs to be taken into account with regard to this problem and there remain only two ways to gain a deeper insight:

- Well-aimed experiments with DASY*Lab*, always bearing in mind that in signal physics as well as in other fields of science the motto always is: the results of proven experiments that can be explained by physics are scientific truth. In technology nothing is possible which would contradict the laws of nature.

- Interpretation of the fundamental statements of Shannon's information theory. Which of the results of this theory are of relevance for our questions? To what extent do they contribute to solving our problem?

Quadrature Phase Shift Keying – QPSK

QPSK is the first step into a two-dimensional signal space. The aim is to use a trick to transmit twice the mount of data (per time unit) using the same bandwidth.

For this purpose the input bit sequence is converted into two bit sequences of half the pulse frequency (see Illustration 238, top right). Two successive serial bits of the frequency f_{Bit} are transformed into a "parallel dibit" with the frequency f_{Dibit}. In this case f_{Dibit} is only half as large as f_{Bit}.

This task is assumed by the so-called demultiplexer (multiplexers combine several channels to form one channel, demultiplexers reverse this process, i.e. they split one receiver channel into several output channels).

One component of this dibit is to denote the horizontal, the other one the vertical part in the state of the signal space. This means that $2^2 = 4$ different states are possible.

For a successful outcome (see Illustration 90 ff) a real part – which is related to cosine-shaped oscillations – and an imaginary part – which is related to sinusoidal oscillations – are needed. Both parts are phase-shifted by just 90 degrees.

Chapter 13: Digital Transmission Techniques III: Modulation   Page 381

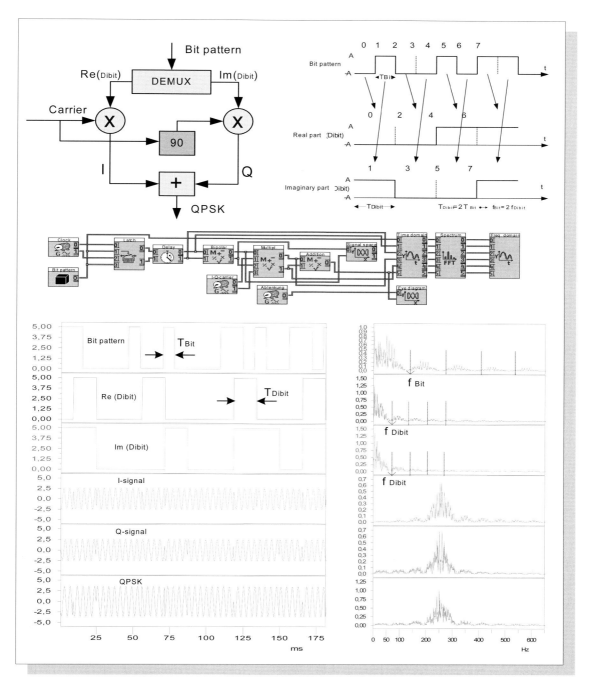

Illustration 238:   *Quadrature Phase Shift Keying (QPSK)*

The text gives a detailed description of how a QPSK-circuit functions. In order to be able to present all signals in a realistic way an elaborate DASYLab-circuit had to be designed in the first place. The delay circuit has the function of achieving an exact synchronization of the two dibit channels. The bit sequences were not filtered so that the representation is clearer.

The frequency domain proves that twice the amount of data per time unit are transmitted by the QPSK-signal compared with the original bit pattern and compared with 2-ASK and 2-PSK. The Illustration clearly shows that the zero position distance of the Si-shaped spectrum with the QPSK-signal is just half as large as with the above input bit pattern.

This points to how to further increase the amount of data transmitted per time unit: instead of four discrete states possibly 16 (4*4), 64 (8*8) or even 256 (16*16).

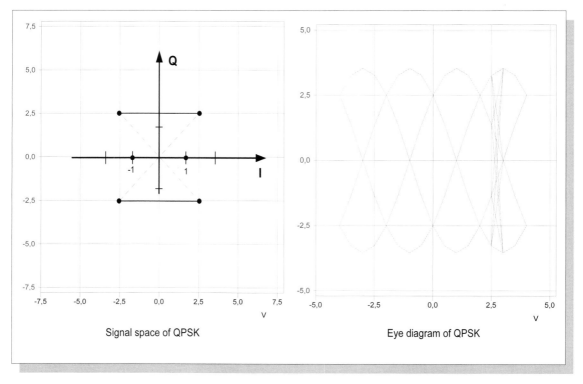

Illustration 239: **Signal space and eye diagram for QPSK**

In the signal space on the left-hand side you can see four discrete states represented as points. The connecting lines between them show possible transitions from one state to another.

In this particular eye diagram the phase shifting of 90 degrees between the four different sinusoidal oscillations – which correspond to the four points of the signal space – can clearly be seen.

This explains the structure of the QPSK-circuit. Real and imaginary parts now need to be added. Please keep in mind that the sum of a pure cosine and a pure sinusoidal oscillation always results in a sinusoidal oscillation shifted by a particular phase angle.

> *By adding two sinusoidal oscillations which are phase –shifted by 90 degrees to each other (sine and cosine) sinusoidal oscillations of any phase shift can be generated provided the relationship of the amplitudes towards each other can be arbitrarily selected.*

> *This means that any point on the Gaussian numerical plane can be reached if the two relevant amplitudes are selected.*

As dibit channels have only two states, A and –A (bipolar), there are four different phase angles. Illustration 239 shows the four states and the possible transitions from one state to the next one as connecting lines between the points.

The eye diagram not only shows the four sinusoidal oscillations shifted by 90 degrees towards each other (corresponding to the four points in the signal space), but also the 45 degree angle of the diagonal.

Digital Quadrature Amplitude Modulation (QAM)

The utilization of bandwidth has been doubled thanks to QPSK and this points the way to a possible further drastic increase in utilization: instead of four discrete states possibly 16, i.e. 4 * 4 states arranged in a grid, 64 (=8*8) or even 256 (=16*16).

How can such grids be generated in the GAUSSian numerical plane? This is demonstrated in Illustration 240. The schematic circuit diagram shows a *mapper* which generates two signals with four different amplitude levels each from the incoming serial bit stream. The black box of the DASY*Lab* circuit reveals that this mapper has a rather complex structure. First, a serial parallel transducer or demultiplexer generates a 4-channel sequence of bit patterns from the serial bit stream. Its pulse frequency is lower than that of the serial bit stream by the factor 4.

Accordingly, this "4-bit-signal" can take on $2^4 = 16$ different states instantaneously. Thus the relevant signal space needs to consist of 16 different discrete states. The mapper now allocates a 4-step signal to each of the 4-bit-patterns on two outputs, one for the I-signal, the other one for the Q-signal. The 16-QAM-signal with a total of 4*4=16 signal states arranged in a grid is created by adding the I- and the Q- components.

> *From a mathematical point of view one can say that the mapper depicts a 4-bit-pattern on 16 points in the signal space.*

The most important result is in the frequency domain. The bandwidth decreases by the factor 4 as is demonstrated by comparing the zero distance of the Si-shaped spectra with the serial bit stream and the 16-QAM-signal.

Illustration 241 shows the signal space, an eye diagram and a detail of a signal to be able to check the regularities:

- The signal space shows 10 (there are a total of 16) different signal states which were taken up within a short period of time. In addition, it shows which transitions to other states took place.

- The eye diagram is very complex due to the numerous possible signal states. The diagrams shown here are different from conventional eye diagrams in signalling technology in that a periodical triangular oscillation with a rising and a falling ramp is used instead of a periodic sawtooth with a rising ramp. In contrast to an oscilloscope, the backshift of the sawtooth cannot be suppressed. This is why the symmetry of the eye diagrams is slightly different.

- The detail of the signal permits a better examination of the relation between I-signal, Q-signal and 16-QAM-signal. The vertical line shows very clearly that the sum of a sine and cosine can result in a sinusoidal signal with a different phase position and amplitude.

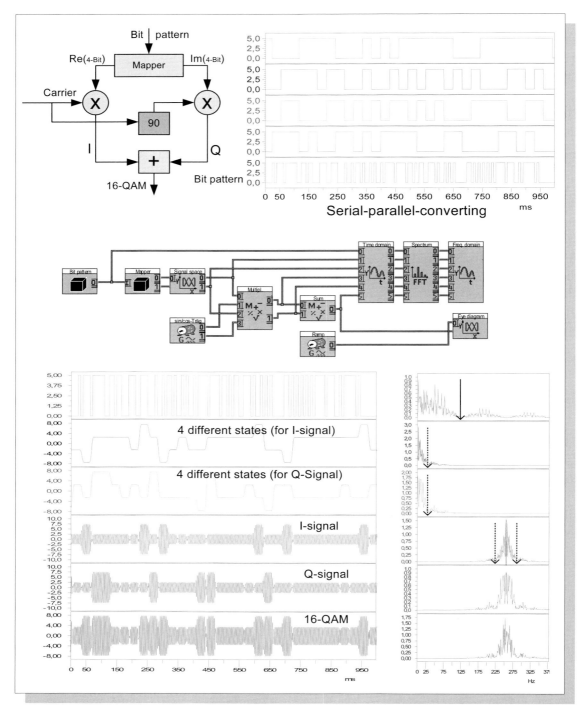

Illustration 240: **16-QAM: serial parallel conversion, mapper, signal formation**

The block diagram top left shows the basic set up which has been modelled as realistically as possible using DASYLab. The so-called mapper, which depicts the incoming bit stream on 16 grid points in the signal space, has been represented as a black box in order to make the representation not too confusing. This black box contains a demultiplexer with 4 output channels. A signal with a width of 4 bits is generated from a serial bit stream by means of serial parallel conversion.

Instantaneously, a signal with the width of 4 bits has $2^4=16$ possible signal states which are to be found in the signal space. The mapper uses a mathematical rule ("illustration") to generate 2 bipolar 4-step signals for X- and Y-deflection in the signal space from this signal with a width of 4 bits.

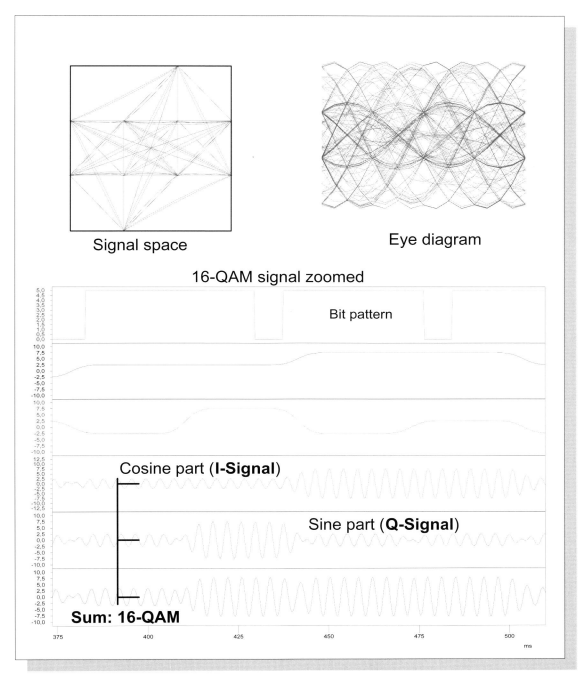

Illustration 241: **16-QAM: signal space, eye diagram and signal detail**

The signal space shows 10 (there is a total of 16) signal states which were taken up within a short period of time. In addition, it shows which transitions to other signal states took place.

The eye diagram is highly complex due to the numerous signal states. An interpretation is only possible in the context of the other illustrations.

The signal detail permits a better examination of the relation between I-signal, Q-signal and 16-QAM signal. The vertical line shows very clearly that the sum of a sine and a cosine result in a sinusoidal signal with a different phase position and amplitude.

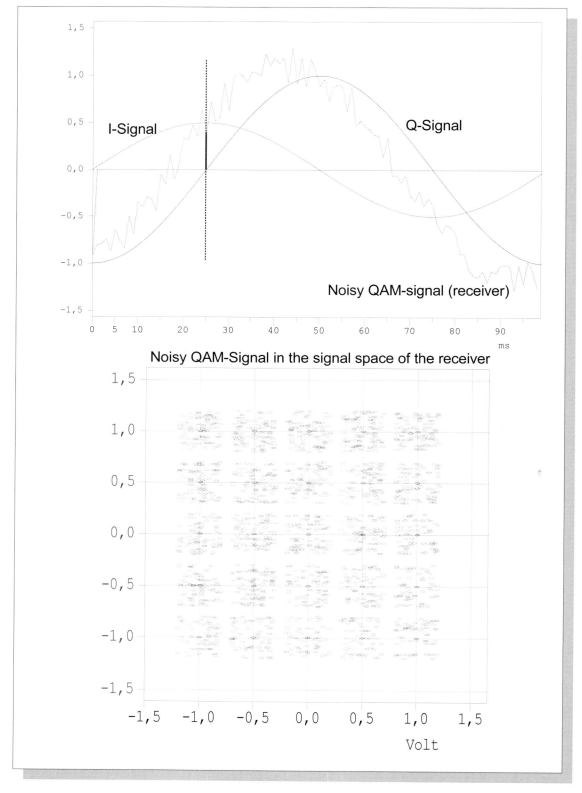

Illustration 242: **Reliability of reception of a QAM-signal**

The simulation shows the detail of a noisy 25-QAM received signal. The signal space shows free corridors between the 25 ranges. This means that the signal can be completely reconstructed.

Multiple Access

You have already been familiarized with the first lesson of Shannon's information theory in connection with the statements on source encoding and channel encoding. They can be summed up as follows:

> *Lesson 1:*
> Never cut off the information of a signal over-hastily before all decisions regarding further signal processing have been taken.

> *Lesson 2:*
> Source encoding (which compresses the signal by removing redundancies) should be completely separated from channel encoding (which makes transmission via noisy channels more reliable by targeted addition of redundancies).

Scientists are still dealing with Lesson 2. There is still complex research in this field with the aim of reaching the limits of the transmission rate – channel capacity C – which was defined by Shannon's theory.

But there is a third lesson which is increasing in importance and which has not yet been fully understood and the consequences of which still remain to be fully assessed. And although the theory was published more than 50 years ago, experts are still at loggerheads. This lesson has enormous economic implications and investments amounting to hundreds of billions of dollars depend upon it. On an abstract level this theory can be put as follows:

> *Lesson 3:*
> *If a channel is disturbed by interference, protective measures can be optimized even if the interference comes in its worst form, namely as white (wideband) "GAUSSian" noise!*

Before I start describing the latest modulation techniques I should like to explain what this lesson implies with regard to concrete, practical use.

The most lucrative business with an extremely promising future is to supply a technology for worldwide communication which is as perfect, efficient and economical as possible. Any decision for or against a particular transmission technology can either mean victory or defeat. I am talking about Multiple Access to a transmission medium.

At first glance, the traditional technologies of frequency-division multiplex and time-division multiplex appear to be relatively unproblematical. Frequency-division multiplex has already been presented in-depth in Chapter 8 and presented in some of the Illustrations (e.g. Illustration 155). This technology is still used in shielded cables and in radio relay, broadcasting and television.

> *Frequency-division multiplex means that all channels are transmitted frequency-staggered and **simultaneously**.*

You all know the drawbacks of this method. As long as the transmission medium is a (shielded) cable, an optical waveguide or a radio relay link the problems are more or less under control. FM-reception in a car, however, shows that due to multiple reflections of the transmission signal at obstacles such as traffic lights, interferences can appear which may extinguish the overall signal locally. When the driver moves on a few yards the signal often reappears. The transmission signal causes interference to itself when at the place of reception the phase shift between the carrier of the direct transmission signal and the reflected transmission signal is 180 degrees or π rad.

For reasons of symmetry – frequency and time are interchangeable – a time-division multiplex method should also be possible:

> *Time-division multiplex means that all the channels are transmitted staggered in time and in the **same** frequency range.*

Time-division multiple access (TDMA) was the first fully digital transmission technology. Its plan is shown in Illustration 243 This method is also highly interference-prone (e.g. caused by multiple reflections in mobile telephony) unless special protective measures are taken.

These interference problems are of course greatest with mobile, wireless, cellular communication (mobile telephony). In the worst case, the following multiple effects can occur:

- Multiple interference when many users access the system simultaneously.

- Multiple interference caused by multipath reception when obstacles lead to multiple reflections.

- Multiple interference caused by several adjacent cells which transmit and receive in the same frequency band (mobile telephony network).

The transmission channel has so far been regarded as "one-dimensional", but here the influence of space and the influence of the movement between receiver and transmitter are taken into consideration.

Let us return to Shannon's 3rd Lesson. It says that interferences of all kinds – e.g. those caused by multiple access – can be controlled if they behave like white (Gaussian) noise. Possible control strategies can partly be deduced from Shannon's law on channel capacity C:

$$C = W \log_2(1+S/N)$$

The letters stand for: C := channel capacity in bits/s   W := bandwidth of the signal

S := signal level   N := noise level

For those who find logarithms too complicated, there is another way of presenting the above law (logarithmic functions are exponential functions):

$$1+S/N = 2^{C/W} \quad \text{or} \quad S/N = 2^{C/W} - 1$$

Example:
let us assume that the noise level N equals the signal level S, i.e. the signal almost disappears in the noise.

In this case the following is true: $S/N = 1$ or $1 = 2^{C/W} - 1$ or $2^1 = 2^{C/W}$ and therefore $C = W$.

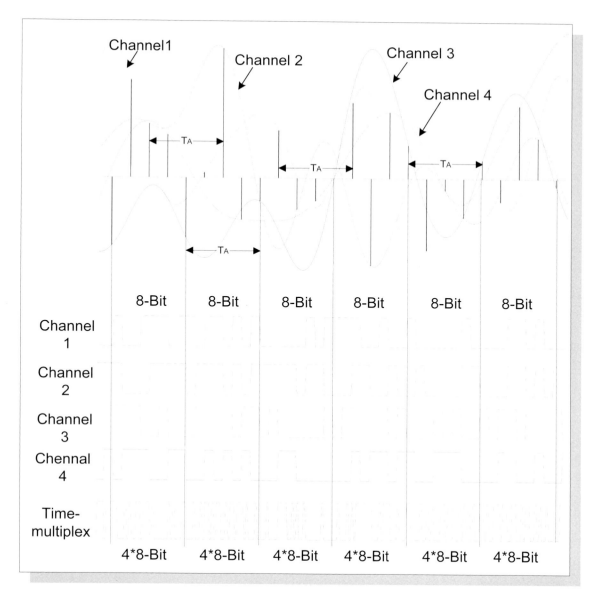

Illustration 243: **Time-division multiplex process**

In this case the digitalization of analog signals comprises four phases which are represented by means of four analog signals: scanning, quantization, encoding, and time-division multiplexing.

The four (parallel) channels are all scanned at the same scanning rate (e.g. $f_A = 8$ kHz or $T_A = 125$ μs). This, however, is done phase-shifted by 90 degrees each or time-shifted by $T_A/4$. This results in a total scanning rate of $4*8$ kHz $= 32$ kHz for the four channels.

Quantization and encoding are executed in one stage. The result is available in the form of four digital – i.e. time and value discrete – signals which in this case have a width of 8 bits. The basic pulse rate of each individual digital signal is 32 kHz.

Time-division multiplexing is a parallel-serial conversion in which the transmission rate of 32 kbit/s is quadrupled and increases to 128 kbit/s.

The receiver conducts this procedure in reverse order: first, demultiplexing, i.e. serial-parallel conversion and then D/A conversion.

In telecommunications at least 30 different channels are combined using time-division multiplexing (PCM 30). Similar to frequency-division multiplexing, several small groups can be combined to form larger units (e.g. PCM 120 etc).

Result: the theoretical limit for the transmission rate C equals the bandwidth provided signal and noise level are identical; e.g. at W = 1 MHz the result is C = 1 Mbit/s.

This results in the following possibilities:

- In principle, it is possible to make the signal disappear completely in the noise. This has already attracted the attention of the armed forces. A signal which disappears in the noise can hardly be tapped or located. The precondition, however, is a "pseudo random encoding" which means that the signal not only has to be broadband but also needs to appear as random as possible.

- If all the signals are broadband and pseudo random in connection with multiple access all users can access the same frequency range at the same time – provided the correct procedure is used.

This is where, alongside the time and frequency domains, *encoding* comes in. Hardly anybody knows that it was cryptography, i.e. codes which cannot be decoded by unauthorized persons, which led Claude Shannon to his insights. An undecipherable code or a coded signal does not show any regularities or "tendency towards preservation" which could be utilized. From a physical point of view it is merely a kind of noise.

> *Shannon's 3rd Lesson shows ways of overcoming the problem of multiple access and reciprocal interference by a combination of signal bandwidth, duration of time and encoding.*

Illustration 244 shows the new dimension of coding (code-multiplexing or code division multiple access CDMA) compared with time division multiple access TDMA and frequency division multiple access FDMA in a very graphic way.

> *CDMA is today regarded as the most intelligent, efficient and comprehensive solution for largely interference-free access in all its variations.*

> *CDMA facilitates simultaneous multiple access in the same frequency range. The individual connections are marked by means of coding.*

Modern digital transmission systems, and probably all future systems, will utilize all three "dimensions" using CDMA. I would like to add that there is not just one possible CDMA procedure, just as there is no such thing as just one random sequence.

> *CDMA can incorporate FDMA and TDMA at the same time. Several different combinations are imaginable. Channel encoding can also be incorporated in an optimum way (coded modulation).*

Discrete Multitone

Discrete Multitone (DMT) facilitates the transmission of a large amount of data in critical transmission media. The frequency band is divided into numerous equidistant sub-bands in the shape of a comb. Each sub-band has a carrier which can be modulated individually using QPSK or multilevel QAM.

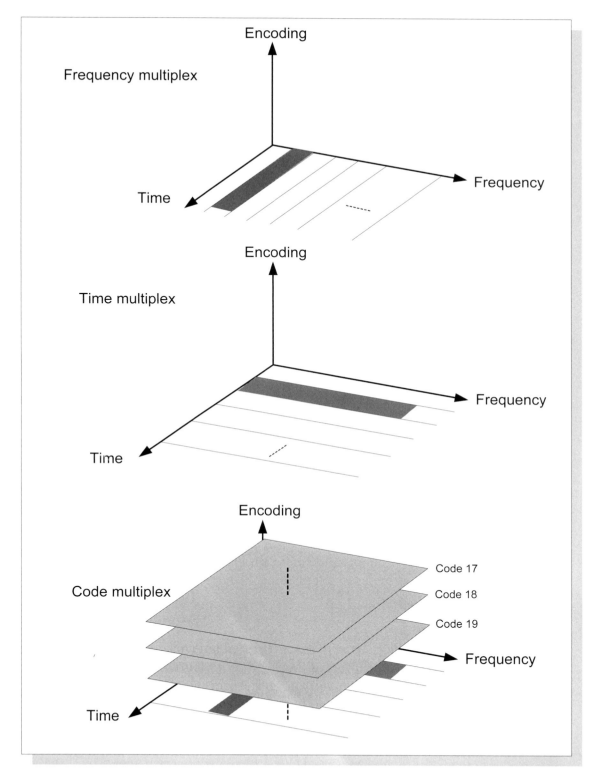

*Illustration 244:   **Frequency-, Time-, and Code Division Multiple Access (FDMA, TDMA and CDMA)***

Time-Division Multiple Access: all channels are time-graded and transmitted in the same frequency band.

Frequency-Division Multiple Access: all channels are frequency-graded and transmitted simultaneously.

Code-Division Multiple Access: CDMA facilitates multiple access simultaneously and in the same frequency band. Channel spacing is done using the code.

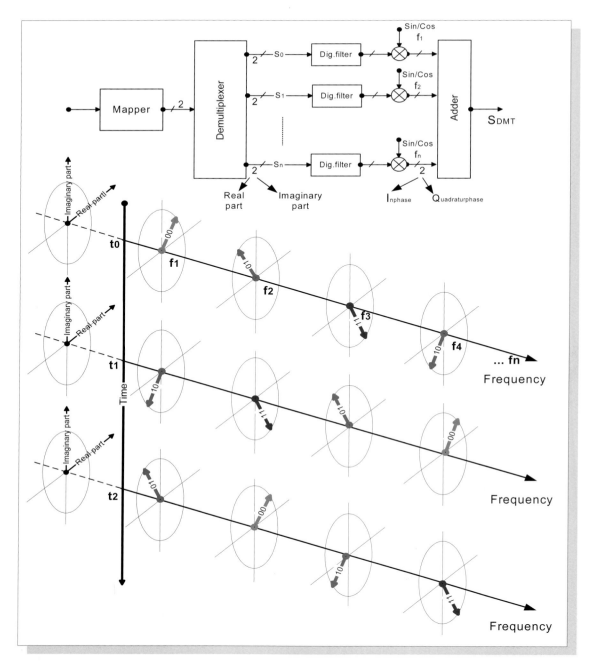

*Illustration 245: **4-PSK-DMT as an example of a discrete multiple carrier system***

The block diagram at the top shows the mapper (which in this case is very straightforward) which "parallelizes" two successive bits of the incoming bit stream (real and imaginary part). It connects the two bits with twice the symbolic length with two circuits (real and imaginary part). The demultiplexer then distributes these bits to, for example, 1000 channels with two circuits each, which additionally increases the symbolic length by the factor 1000. These bits need to be buffered by the demultiplexer and output synchronously so that they are time-synchronous. Then the real part is multiplied by a cosine and the imaginary part by sinusoidal oscillation of the frequency f_k ($0 < k < 1000$) per channel. All signals are added in the adder; s_{DMT} is composed of 1000 frequencies of the same amplitude which can take on just four phase positions: 45, 135, 225 and 315 degrees (4-PSK).

The bottom of the Illustration shows the instantaneous phase position of four of these 1000 carrier frequencies at three successive points in time. The vertical axis is the time axis, the horizontal axis is the frequency axis.

Note:
Critical transmission media are linearly distorting channels, i.e. the medium causes a change in the amplitude- and phase spectrum of the signal. Physical reasons are frequency-dependent absorption, dispersion (sinusoidal oscillations of different frequencies propagate at different speeds), and multipath propagation caused by reflection at obstacles. The latter indirectly has the same effects.

DMT procedures are used in:

- Speech band modem technology (300 – 3400 Hz)

- Fast digital transmission via metallic wire pairs. Examples are HDSL (High Speed Digital Subscriber Line) and ADSL (Asymmetric Digital Subscriber Line).

- Radio transmission via channels with multipath propagation (mobile communication or mobile telephony) including DAB (Digital Audio Broadcasting).

DMT transmits multilevel PSK and QAM using many equidistant carrier frequencies. This is shown in Illustration 245. It presents a DMT system for a particularly straightforward case (4-PSK). A very simple mapper distributes the bit stream on two channels (serial parallel conversion). One channel represents the real part, the other the imaginary part. The effect is that the symbolic length (bit duration) doubles.

In principle, this is what a demultiplexer does. It conducts a serial parallel conversion by splitting the two input channels into, for example, 1000 parallel channels. Thus the symbolic duration is additionally increased by the factor 1000. A precondition is that the 2 times 1000 output signals are synchronous. Therefore the demultiplexer needs to buffer the 2 times 1000 channels and output them simultaneously. In addition, the output signals need to be bipolar.

After pre-filtering to reduce the bandwidth of the rectangular sequences each real-part bit pattern is multiplied with a cosine-shaped carrier, and the imaginary bit pattern with a sinusoidal carrier of a given frequency f_k ($0 < k < 1001$). The situation is quite simple because with PSK the amplitudes of all carrier oscillations are always the same.

Cosine and sine-part of one carrier oscillation are added – depending on the 2-bit pattern – to form a carrier oscillation with the phase shifts 45, 135, 225 and 315 degrees. The output signal s_{DMT} of the adder contains a total of 1000 carrier frequencies the phases of which change in line with the rate of symbolic duration between these four values. This is shown in Illustration 245, bottom.

This process needs to be reversed in the receiver by "projecting" the received signal into the (three-dimensional) signal space. Illustration 242 shows that certain ranges surrounding the signal points of the undistorted signal should not be exceeded. But errors can still be identified by means of an effective channel encoding so long as they do not occur too often.

The complexity of a DMT system appears hardly imaginable. 1000 carrier frequencies and more, multipliers, pre-filters, a highly complex demultiplexer etc. We tend, however, to think in terms of *analog* systems. If such an analog system had been built in pre-digital days, it would have filled several rooms. The magic word, however, is DSP (Digital Signal Processing) which makes such a complex system possible. The precondition for such a system is above all computing power because the DMT system is largely a *program*, i.e. it is virtual. Real-time processing of such a large amount of signals in the DMT system make extremely high demands on the processors. Recently chips have been produced which take up the bit stream at the input and issue the DMT-baseband signal directly at the output. This means that they contain all the processes which have been described above, including channel encoding.

The baseband can be shifted into any frequency range by means of one single high-frequency carrier. This is done using analog technology.

Which alternatives to broadband DMT systems are there along the lines of Shannon's 3rd Lesson? Would it not be easier if for instance the input bit stream was directly modulated by one single carrier via QAM?

Experts have long been at loggerheads, but now they seem to have come to an agreement. CAP (Carrierless Amplitude/Phase) Modulation is a special multi-level form of QAM where the many different states of the signal space are allocated to one single carrier oscillation by means of discrete amplitude and phase shifting. The 16-QAM in Illustration 240 also shows the principle behind CAP.

The differences between DMT and CAP point to the subtle differences with regard to Shannon's 3rd Lesson:

- DMT operates more in the frequency domain, CAP more in the time domain. The QAM/CAP technology operates with a relatively high symbol rate. Each symbol has a brief duration and has therefore a wider frequency band. DMT has a much longer symbol duration and a large number of narrower frequency bands.

- DMT is therefore much less prone to multipath reception in connection with mobile telephony or digital broadcasting DAB. As long as the numerous reflections caused by obstacles arrive during the duration of the symbol, the transmission signal can be reconstructed.

- DMT can be adapted much more flexibly to the physical properties of the channel (e.g. cable). In frequency ranges with high damping and/or interferences the carrier frequencies in these ranges can be PSK-modulated at a low level. As a result the distance between the points in the signal space becomes greater which means that noise immunity increases and the likelihood of bit errors decreases. Thus DMT can be much better adapted to the interference-dependent channel capacity.

- Until recently the QAM/CAP concept had the advantage of being the more straight-forward principle and could thus be realized more easily from a technical point of view. This is no longer the case. On the contrary: DMT has become the accepted standard and chips by numerous manufacturers are now available. CAP, by contrast, is not standardized and there are no chips available because DMT has learnt Shannon's 3rd Lesson in a more intelligent way.

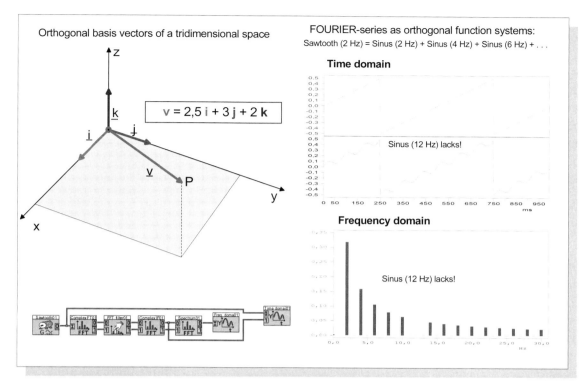

Illustration 246: ***Orthogonality***

*How can a point P in space be reached? Quite easily by following these instructions: go 2.5 standard steps (of length 1) in x-direction (2.5**i**) plus 3 standard steps in y-direction (3**j**) plus 2 standard steps in z-direction (2**k**). Thus the vector leading to P can be defined as v = 2.5**i** + 3**j** + 2**k**. Vectors are underlined and in bold print. They represent numeric values which indicate a direction.*

***i**, **j**, and **k** are linearly independent because they are in a vertical, i.e. orthogonal position to each other. This means: Point P could never be reached without one of the three standard vectors **i**, **j** and **k**. The equation **v** = 2.5**i** + 3**j** + 2 (**i**+ **j**) would be wrong. **i**, **j** and **k** are indispensable if you want to set up a three-dimensional space, i.e. if you want to be able to reach any point in this space.*

Orthogonal function systems*: a periodic sawtooth of 2 Hz contains, as we know, all integer multiples of the frequencies of 2Hz, i.e. sinusoidal oscillations of 2,4,6, ... Hz up to "infinitely high" frequencies.*

This infinite number of discrete frequencies form – in the above mentioned sense – a "vector space with an infinite number of dimensions". The reason is: if only one of these frequencies is missing – in the Illustration at the top it is the sinusoidal oscillation of 12 Hz – the sawtooth oscillation cannot be reconstructed as a sinusoidal oscillation. In that sense each of these innumerable oscillations is indispensable and their total constitutes the minimum number of oscillations required to form the sawtooth. In mathematical terms this means: all these sinusoidal oscillations have an "orthogona"l relationship towards each other.

Orthogonal Frequency division Multiplex (OFDM)

The new European digital broadcasting standard DAB operates using a special DMT procedure: OFDM (Orthogonal Frequency Division Multiplex). "Orthogonal" means "to be in a vertical position to..." and is a basic term in mathematics, in particular in vector space mathematics. Illustration 246 shows the connection between orthogonality and modulation technology.

The carrier frequencies used in OFDM are always integer multiples of a basic frequency.

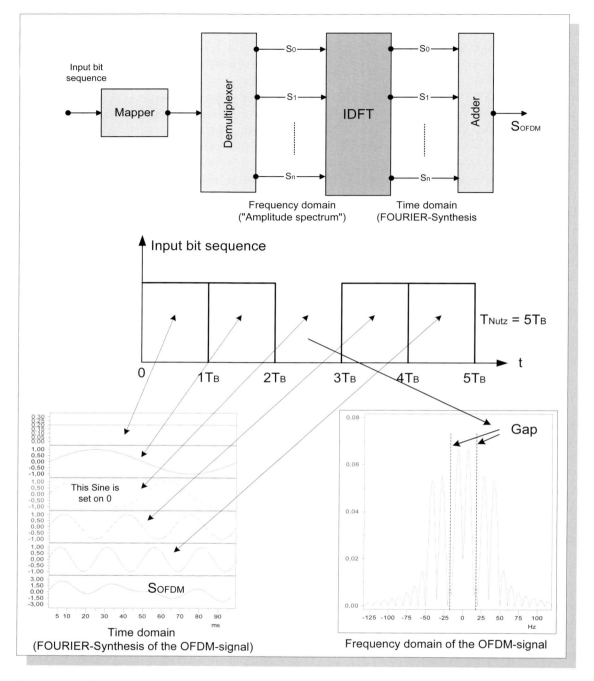

Illustration 247: **Block diagram and simplified representation of OFDM**

In Illustration 245 there are a number of sinusoidal oscillations at the input of the adder. If they are integer multiples of a basic frequency their sum can be defined as a FOURIER-synthesis. The FOURIER-synthesis is, however, the result of an Inverted FOURIER-Transformation IFT (see Illustration 88) or in this case of an IDFT (Inverted Discrete FOURIER-Transformation). As a consequence, the bit patterns at the output of the demultiplexer can be defined as a discrete spectrum (real and imaginary part) which is transformed via IDFT by FOURIER-synthesis into a DMT-signal in the time domain.

Instead of the many multipliers and oscillators in Illustration 245, an IDFT-block is used in OFDM-systems. Each bit at the numerous inputs of the IDFT-block has a discrete frequency with a discrete amplitude and phase position at the output.

To be precise, each input/output consists of 2 lines (real and imaginary part).

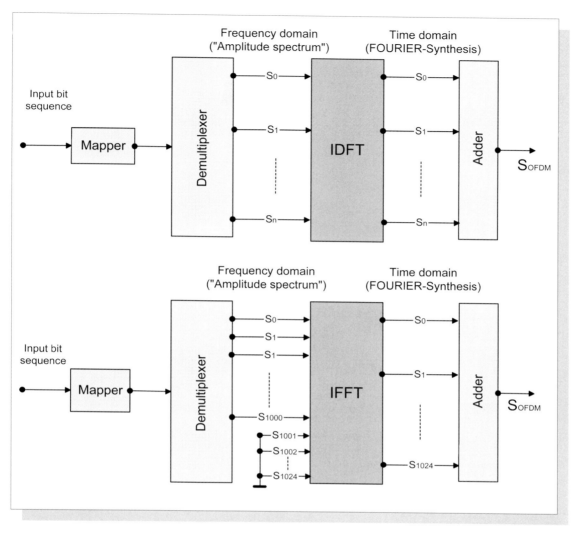

Illustration 248: **IDFT and IFFT**

The Discrete FOURIER-Transformation DFT and the Inverse Discrete FOURIER-Transformation IDFT facilitate a variable block length, i.e. they can be adapted to the number of outputs of the demultiplexer. Both algorithms are, however, extremely compute-bound when the number of outputs is high.

FFT, just like IFFT, is a DFT which has been optimized with regard to speed. They utilize the symmetry principle in the algorithm. FFT and IFFT always require a block length which is a power of 2, e.g. $2^{10} = 1024$.

In this case a 1024-IFFT takes considerably longer than a 1000-IDFT. Therefore it is possible to assume that the 24 input values missing equal zero. The information on all input bit patterns is maintained.

This has surprising consequences and unexpected advantages. Please take another close look at Illustration 245, top. At the output of the demultiplexer there is an instantaneous wide bit code. Each of the bits contained in that code is then multiplied by a sinusoidal or cosine-shaped carrier and therefore there are e.g. 1000 (equidistant) sinusoidal oscillations with a variable phase position at the input of the adder, the total of which is the DMT–signal in the time domain at the output of the adder. This might have something to do with the FOURIER-synthesis, i.e. with signal formation in the time domain by adding relevant sinusoidal oscillations.

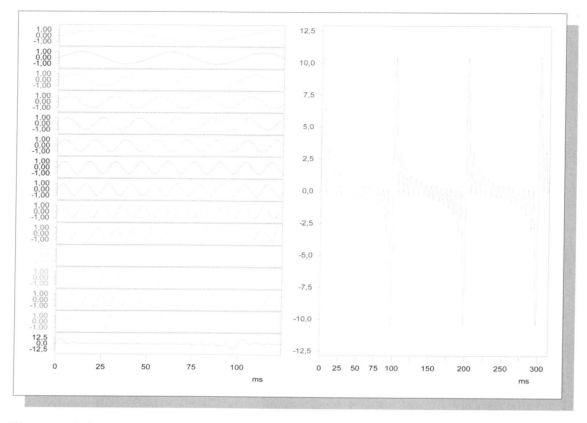

Illustration 249: **OFMD-signal with zero phases (ASK-OFDM)**

The Illustration shows why there should not be a pure ASK-OFDM because it would mean that all carrier oscillations would have the same phase. This would have serious consequences at the output of the adder. The sum of all carrier oscillations might have high local peaks ("needle pulses"). This in turn would mean that the transmission appliances – e.g. amplifiers - would not be able to cope.

If the whole AFDM-signal is to resemble white noise in the frequency domain, pseudo-random phase key shifting (as an image of the relevant bit pattern) of carrier oscillations of the same amplitude is the suitable option.

But which sinusoidal oscillations are suitable? As we know, only those sinusoidal oscillations which are integer multiples of the basic frequency can be considered in connection with (periodic) signals as a digital signal of a certain block length is defined as periodic (see Chapter 9 "Digitalization). A FOURIER-synthesis, however, is obtained by moving from the frequency into the time domain, i.e. via an Inverted FOURIER-Transformation IFT, or in this case via an IDFT (Inverted Discrete FOURIER-Transformation).

Thus the bit code at the output of the demultiplexer can be interpreted as a frequency spectrum and the enormous number of multipliers and oscillators can be replaced by an IDFT block. Thanks to the orthogonality of the carrier frequencies the procedure using IDFT is simplified.

Illustration 247 demonstrates in a simplified way that the symbol duration with OFDM equals the period length of the basic oscillation. This means that an integer multiple of the period length of all carrier oscillations fits into each OFDM-symbol.

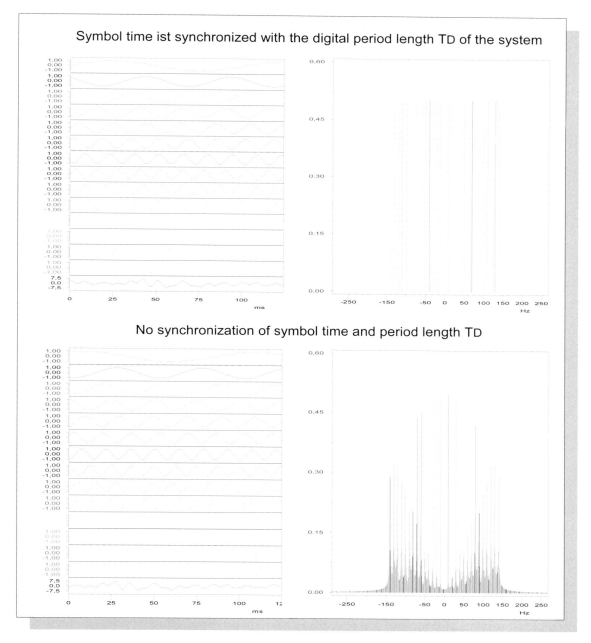

Illustration 250: **OFDM and digital period length T_D**

The symbol length is regarded as a time reference in connection with OFDM. In addition, its synchronism with the digital period length T_D of the signal processing overall system is of particular importance. Only then is the spectrum – the sum of all carrier oscillations – represented or reconstructed correctly. This Illustration shows a random selection of the four possible phases of the individual carriers.

The bottom half of the Illustration shows the consequences of non-synchronism of symbol length and the digital period length T_D of the overall system as simulated by DASYLab. This is why the block length (in this case 2048) does not quite correspond with the scanning rate (in this case 2040).

Symbol length and digital period length T_D of the overall system thus need to have an integer relation towards each other.

A special problem with OFDM-systems is therefore the exact time synchronization of all the transmitters with regard to interference and synchronized reception of adjacent transmitters.

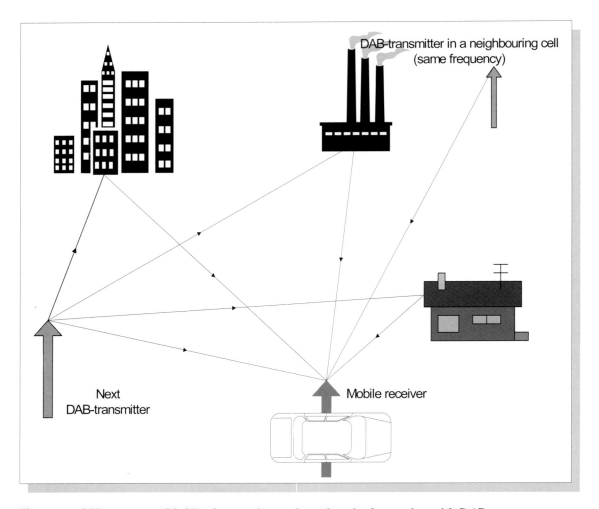

Illustration 251: ***Multipath reception and synchronized reception with DAB***

Coded OFDM (COFDM) and Digital Audio Broadcasting (DAB)

Digital Audio Broadcasting DAB is a joint development of several European countries. Characteristic of DAB are its different transmission modes facilitating adaptation to audio, text, images and mobile TV according to demand. For that reason DAB is not only an improvement over its predecessor, FM broadcasting, but part of a completely new transmission system for digital data – with a strong emphasis on radio and TV.

1536 sinusoidal carriers are transmitted at a distance of 1 kHz (mode I). For the duration of each symbol length T_S of 1.246 ms each carrier can take on four different discrete phase states (4-DPSK). This is the equivalent of 2 bits ($2^2 = 4$). Within a DAB time frame of 76 symbols (94.7 ms) $76 * 2 * 1536 = 233{,}472$ bits can be transmitted. The result is 2.4 Mbit, i.e. the transmission rate with DAB is 2.4 Mbit/s.

Illustration 251 shows multipath and digitalized reception in mobile telephony. The resulting problems such as interferences (fading) etc could not be solved by means of analog methods. DAB has largely been able to solve these problems using a unique combination of protective measures. Therefore DAB is referred to as COFDM (Coded Orthogonal Frequency Division Multiplex).

Chapter 13: Digital Transmission Techniques III: Modulation

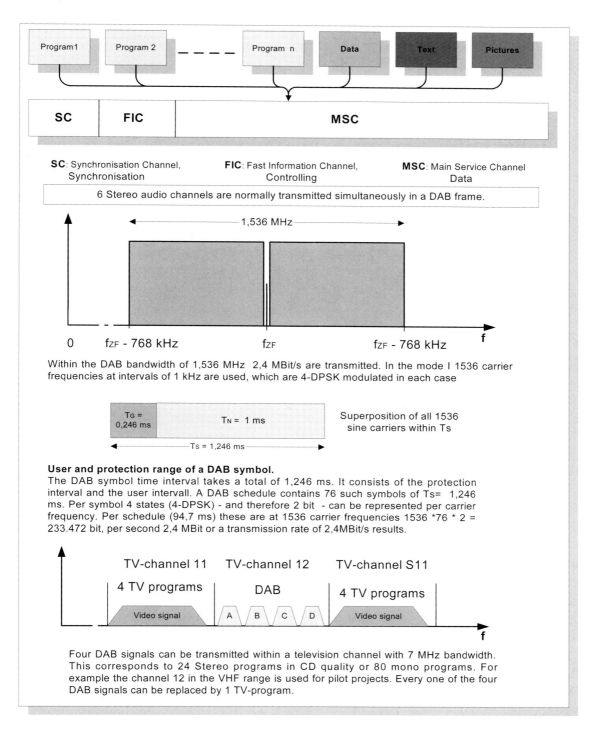

Illustration 252: ***DAB-frame, frequency- and time structure***

The synchronization component SC contains a zero symbol which enables the receiver to conduct a coarse synchronization of the frame and the symbol structure. The phase reference symbol which is also contained in the synchronization part serves to conduct a precise synchronization. The control component FIC transmits data to control and decodes the DAB multiplex signal. As these data require error-free transmission they have been specially coded to avoid errors.

The signal received is only evaluated during the time that the interval is used T_N. The protective interval with the length T_G reduces (linear) distortions caused by multipath and synchronized reception. The data component MSC contains the user data.

Audio signals are coded as a protective measure against errors (convolutional encoding and Viterbi decoding). Another effective protective measure is interleaving.Time-interleaving facilitates the restructuring of bits of a program that originally succeeded each other using a fixed pattern to interleave them in such a way that they are wide apart in time. Frequency interleaving means that the data of different programs are also restructured and allocated to the 1536 carrier frequencies in such a way that the data of a program are wide apart in the spectrum.

> *The audio signals of a DAB signal are literally "ripped apart" in the time and frequency domains. The DAB signal and its neighbours therefore appear almost like wideband white noise.*
>
> *If a narrow frequency band is disturbed by fading only a tiny proportion of the signals is generally corrupted and can be recognized and corrected in the decoding process.*

Global System for Mobile Communications (GSM)

GSM is also a joint European development, in this case for mobile telephony. In the D-network GSM uses the 890 – 915 MHz frequency range for the uplink and the 925 – 960 MHz range for the downlink. In the E-network the uplink is in the 1760.2 – 1775 MHz range and the downlink in the 1855.2 – 1870 MHz range. Each of these bands is divided into channels with a width of 200 kHz. Owing to uplink and downlink two frequencies or channels always correspond to each other (duplex). This means that GSM also uses FDMA (see Illustration 253).

The objectives of GSM are protection against misuse and tapping (cryptology), high user capacity, high utilization of bandwidth, transitions to fixed-line telephony, optimizing telephone services, high data quality and low bit error probability, and integrated data- and supplementary services.

GSM uses TDMA to separate users of a (local) cell. 8 users of a base station share one carrier frequency.

The combination of TDMA and FDMA results in 400 transmission paths in the D-network with 50 duplex channels of 200 kHz each and 8 users per channel. The bit rate of a channel is 271 kBit/s at a bandwidth of 200 kHz. Again, the 4-DPSK process is used for modulation. The bit rate per call is 13 kBit/s. The remaining transmission capacity is used for data protection. Convolutional encoding and interleaving is used for this purpose.

Asymmetric Digital Subscriber Line (ADSL)

ADSL (Asymmetric Digital Subscriber Line) procedures have already been described at the beginning of Chapter 11. It is a special variation of DMT for cable sections. By optimizing all the procedures of encoding and modulation described so far it is possible to transmit, alongside ISDN, up to 8 Mbit/s additionally via a regular 0.6 mm Cu-wire for telephony.

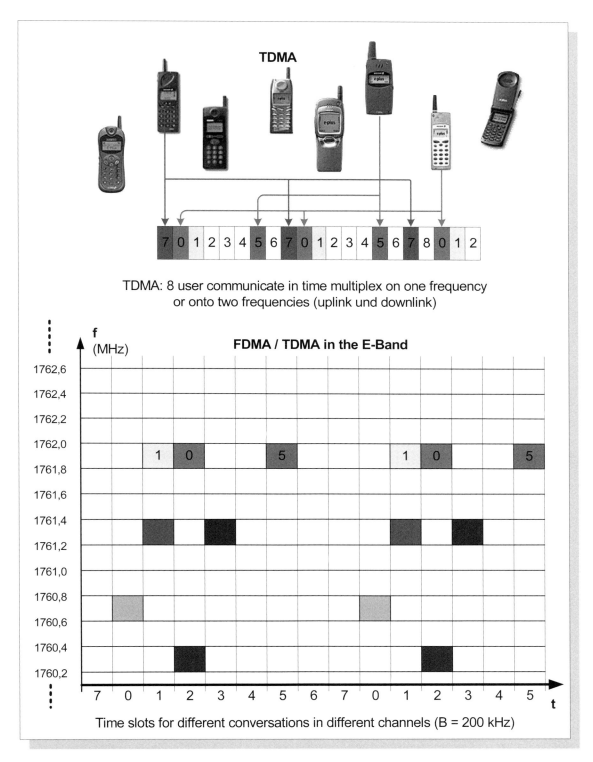

Illustration 253: Frequency- and time-multiplexing in the GSM-procedure

Modern ADSL-systems operate above the ISDN frequency band and utilize a frequency range of approximately up to 1.1 MHz. A total of 256 carrier frequencies with a distance of 4 kHz (for downstream) are used. Each carrier can as a rule take on up to 32 states in the signal space.

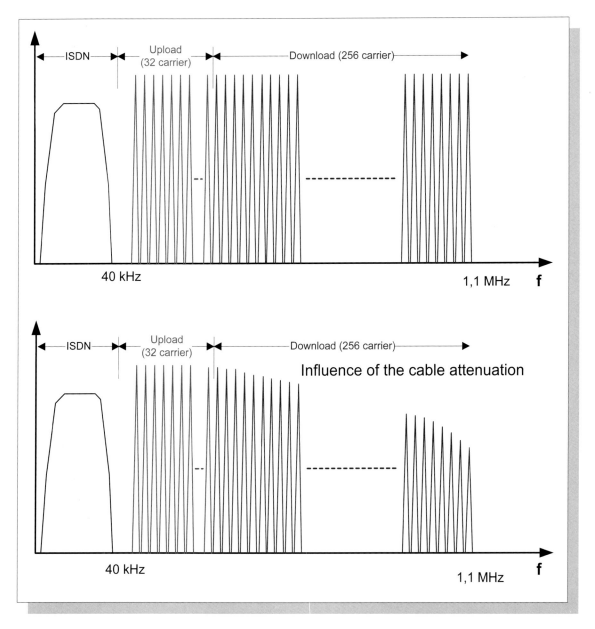

Illustration 254: ***ADSL as a special – asymmetrical – DMT-procedure***

The system is, however, adapted to the properties of the transmission path by automatically measuring the signal-noise distance. With regular Cu-wires transmission loss at around 1 MHz increases drastically. Therefore the possible states of the carriers in that range are reduced so that only 4-QAM instead of e.g. 32-QAM is made use of. If there are strong narrowband interferences in that range – e.g. caused by broadcasting – the carriers are "switched off".

> *Transmission loss increases proportionally to the length of a cable. Therefore the transmission rate in ADSL is proportional to the length of the line. Manufacturer information on ADSL usually refers to lengths of 2 to 3 km.*

As ADSL has been conceived for faster internet communication, the downstream is at up to 8 Mbit/s considerably larger than the upstream (up to 768 kBit/s). Generally speaking, a much greater amount of data is transmitted from the internet to the user than vice versa.

Modulation and demodulation is carried out as OFDM. i.e. an IDFT or IFFT is used in the transmitter to transform the discrete amplitude levels at the outputs of the demultiplexer into the time domain. (see Illustrations 240 and 247). This procedure is reversed in the receiver by an FFT. 256 ($=2^8$) carriers for the downstream and 32 ($=2^5$) for the upstream are a clear sign that IFFT and FFT are used. A 512-point IFFT needs to be used as the complex signal in the transmitter consists of 256 real parts and 256 imaginary parts each.

Telekom, a German telephony provider, currently offers T-DSL, a variation with a much lower rate. For this reason it is not necessary for each digital switching station involved to install internet connections with an extremely high transmission rate.

A symmetrical HDSL-procedure with up to 50 Mbit/s is offered for short connections.

Spread Spectrum

CDMA (Coded Division Multiple Access) is often identified with the spread spectrum procedure. With Shannon's Lessons in mind, this may be the most interesting transmission method. The "natural" frequency band of a signal is deliberately spread so that the band transmitted is much wider. What is the purpose of such a procedure?

Let me first explain how to transform a narrowband signal into a broadband signal. The simplest method would be to use shorter pulses because in a limiting case a needle pulse generates an infinitely wide spectrum. Such a procedure would be comparable with time multiplexing (e.g. in GSM, see Illustration 243) where 8 users share one signal which leads to an increase of the data rate by the factor 8. Each of the users would only have an eighth of the time at their disposal. On the other hand, the use of short pulses causes technical problems in most cases because the transmitter needs to change the transmission power quickly.

Spreading narrowband sequences by means of pulse sequences which behave like white noise (pseudo noise (pn)-sequences) seem to be a more feasible option. These sequences, however, need to behave in such a way that they are identifiable as noise since everything needs to be reversed in the receiver. This pseudo noise is generated by means of feedback shift registers and is repeated after a certain period of time. In contrast to white noise, pseudo noise has a fixed pulse rate.

Any error-protection encoding results in a spreading of the frequency band because a longer code is used. Therefore it seems plausible that spreading should be carried out in the form of encoding which makes a signal much less susceptible to errors.

The principle of direct sequence spread spectrum is quite straightforward. In principle it is the multiplication of a low-rate bipolar bit sequence by a high-rate bipolar pn bit sequence (see Illustration 256), otherwise as with AM.

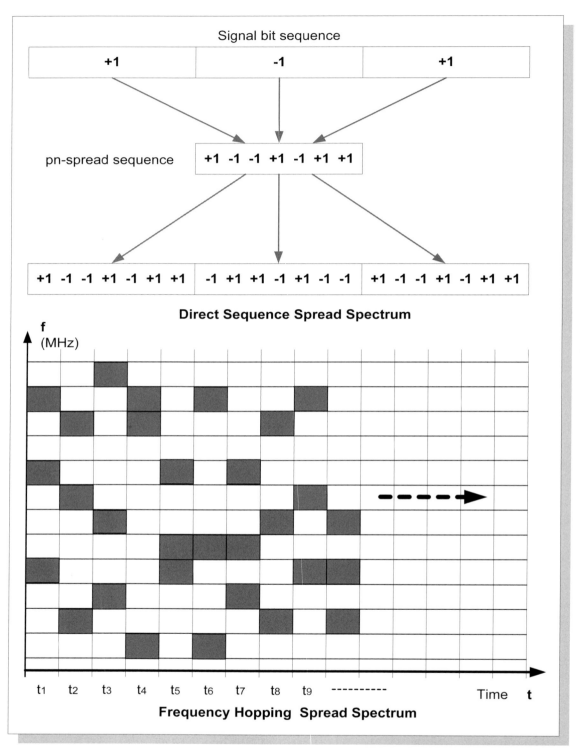

Illustration 255: ***Direct Sequence Spread Spectrum and Frequency Hopping***

The principle of the Direct Sequence procedure seems to offer the easiest method. The spread signal can be generated by simple logical operations.

It is probably not very easy to produce a set of frequency synthesizers for "frequency hopping" that can be tuned very fast. "Time hopping" as a third method should be easier.

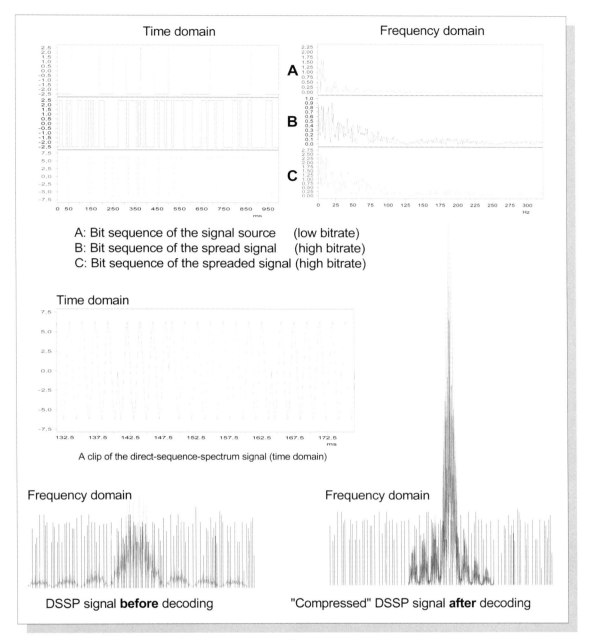

Illustration 256: ***Direct Sequence Spread Spectrum DSSS***

Top: the spread sequence clearly widens the spectrum. The bandwidth is identical with the spread sequence. It should theoretically recur periodically with the pulse of the low-rate bit sequence. This is, however, of no consequence for the spectral bandwidth.

Centre: A carrier is used to modulate the spread bit sequence for transmission. The Illustration shows a small detail in the time domain of the modulated spread-spectrum-signal. In this case it is obviously a PSK-modulated signal.

Bottom: If the receiver knows the code the received signal can be reconstructed in a compressed form. This compressed signal stands out above the signal and contains the information on the original low-rate bit sequence (see top).

In the case of spread-spectrum signals the bandwidth – unlike all the digital transmission processes described so far – is much greater than the information rate.

The widening of the spectrum increases interference immunity (also) towards narrowband sources of interference. This can be advantageous if someone tries intentionally to jam the signal.

Another possibility is the Frequency Hopping Spread Spectrum. The frequency hopping is controlled by a pn sequence. This system consists mainly of a code generator and a rapidly controllable frequency synthesizer which is controlled by the code generator. Unlike the FSK (see Illustration 236) the number of possible frequencies is not just two; it may be very large.

For reasons of symmetry time hopping spread frequency must also exist. Here the pulse hops randomly backwards and forwards between possible transmission times as though there were a time multiplex system for just one signal.

If the signal is spread frequency-wise in the transmitter it is compressed correspondingly in the receiver. As the energy of the signal is not changed the amplitude curve must be increased by the spread factor. The signal protrudes from the noise. This is shown clearly in Illustration 256.

The signal can only be compressed frequency wise in the receiver if the code is known. As the code is different for each signal many signals can use the same frequency range so that the disadvantage of the frequency band spread is more or less compensated for. The signal is retrieved by means of a correlation (see Illustrations 77 and 78). By this means the similarity or preservation tendency of the original signal is "filtered out".

Spread spectrum can be used above all where very weak signals which have disappeared into noise have to be recovered. This procedure is deployed in GPS (global positioning system), for instance, or in wireless microphones. Spread spectrum might be the solution for a broadband, wireless communication network in the maximum frequency range.

Exercises on Chapter 13

Exercise 1

(d) Describe the possibilities of limiting a digital signal – i.e. a bit pattern – in frequency before modulation and transmission.

(e) How can overlapping by adjacent pulses (interference) be limited?

(f) How is it possible to determine qualitatively using measuring techniques whether the information of a received signal can be reconstructed?

(g) Why are many filter types not suitable for filtering "rectangular" bit sequences?

Exercise 2

Summarize the general differences between classical and digital modulation processes.

Exercise 3

Compare amplitude, phase and frequency key-shifting.

(a) What advantages and disadvantages do you see?

(b) What is the easiest way to transform an ASK into a PSK?

Exercise 4

Describe the idea behind the (two-dimensional) signal space in the GAUSSian plane.

Exercise 5

Describe using the four phase-key shifting QPSK how practically every point in the signal space can be reached.

Exercise 6

In digital quadrature amplitude modulation QAM any number of points arranged in grid-form can be reached in the signal space using a mapper.

(a) Describe its fundamental structure

(b) How many signal states (points in the signal space) are needed for a 4 bit signal at the input?

(c) How can this be realized in terms of circuits in mapper?

(d) How many discrete amplitude levels are required by the real and imaginary part of a signal?

Exercise 7

From the representation of the QAM reception signal it can be determined exactly whether the source signal can be reconstructed.

(a) How can this be seen

(b) On what principle does a QAM receiver work?

Exercise 8

Summarize the three "lessons" of Shannon's information theory.

Exercise 9

PCM 30 system uses the time multiplex process TDMA. 30 subscribers can use the same transmission channel. How large is the bit rate of the PCM 30 system with an NF bandwidth of 4 kHz and using 8 bit A/D converters?

Exercise 10

(a) DMT processes are at present at the forefront of modern digital transmission technology.

(b) What advantages are expected from DMT in spite of its complex structure?

(c) What transmission media seem particularly suitable for DMT processes? Give reasons.

(d) What is the special feature of OFDM

(e) By means of what digital signal processes can effort be reduced considerably?

(f) To what extent is a DMT system virtual, i.e. a program?

(g) What "standard chips" are required above all for digital signal processing DSP?

(h) What applications work on the basis of DMT?

Exercise 11

Describe the structure of an OFDM transmitter and receiver

Exercise 12

What is behind the frequency band spread and CDMA?

Bibliography

It would be possible at this point to provide a comprehensive bibliography full of well-known textbooks. However, only very few of these books were really used as this "Learning System " uses a new approach to deal with the subject "Signals – Processing – Systems".

I do not know of any textbook that mentions and evaluates the *Uncertainty Principle* **UP** and the *Symmetry Principle* **SP** alongside the *FOURIER Principle* **FP** as the fundamental basis.

The *Uncertainty Principle* is mentioned briefly in a number of books but the *Symmetry Principle* is not mentioned at all except for the observation that mathematics also produces negative frequencies.

Many interesting notes and sources are taken from the internet. It is therefore planned in the next edition to provide a list of footnotes on the text in which interesting sources on the internet can be directly retrieved via the PC from the electronic document.

I found the book by Ulrich Reimers (Technical University of Brauschweig) very useful in dealing with digital transmission technology. I have taken many ideas and examples from this book and have modified, re-designed and simulated them.

As further reading in this discipline I would recommend the following book in which the methods of explanation are not purely mathematical but also pictorial and content-related:

- *Allan V. Oppenheim, Alan S. Willsky: Signals and Systems; Prentice Hall, New Jersey, 1997; ISBN 0-13-814757-4*

On the problems of digital transmission technology I would recommend especially:

- *Ulrich Reimers: Digital Video Broadcasting; Springer Verlag Berlin/ Heidelberg/New York 1998; ISBN 3 – 540 –60946-6*

An older textbook helped me to understand better signal processing in relation to its physical basis in oscillations:

- *R.B.Randall: Frequency analysis: Brüel & Kjaer, Glostrup, Denmark 1987; IBSN 87-87355-07-8*

The following is a textbook on digital signal processing DSP written in relatively straight-forward English:

- *Craig Marven and Gillian Ewers: A simple approach to Digital Signal Processing; Texas Instruments 1993; ISBN 0-904047-00-8*

For those who would like to have information on up-to-date specialist literature I would recommend the various online bookshops. You quickly find what you are looking for by entering keywords and there are summaries of books and reviews by other readers.

The internet is by far the cheapest specialist bookshop particularly in the case of articles and essays and pointers to other literature.

Additional advanced publications from Springer Verlag:

Bremaud, P.: *Mathematical Principles of Signal Processing*
2002. Approx. 280 pp. 47 figs. Hardcover
ISBN 0-387-95338-8

Written for: Graduate students, applied mathematicians

Contents: **Fourier Analysis in L1**: Fourier Transforms of Stable Signals. Fourier Series of Locally Stable Periodic Signals. Pointwise Convergence of Fourier Series. **Signal Processing**: Filtering. Sampling. Digital Signal Processing. **Subband Coding**: Fourier Analysis in L2. Hilbert Spaces. Complete Orthonormal Systems. Fourier Transforms of Finite Energy Signals. Fourier Series of Finite Power Periodic Signals. **Wavelet Analysis**: The Windowed Fourier Transform. The Wavelet Transform. Wavelet Orthonormal Expansions. Construction of a MRA. Smooth Multiresolution Analysis.

Lacroix, A. : *Fundamentals of Digital Signal Processing*
2003. XII, 250 pp. Hardcover
ISBN 3-540-41340-5

Written for: Graduate students and scientists

Contents: Discrete Time Signals and the Z-Transform. Properties of Discrete Time Systems and Analysis. Design of Discrete Time Systems

Gibson, J.D. : *The Mobile Communications Handbook*
2nd ed. 1999. XII, 600 pp. Hardcover
ISBN 3-540-64836-4

Written for: Practitioners, planners, managers, researchers, and students in the mobile communications field; electrical engineers; computer science professionals; professionals and specialists in the telecommunications industry and the microwave industry

Contents: **Basic Principles**: Complex Envelope Representations for Modulated Signals. Sampling. Pulse Code Modulation. Baseband Signaling and Pulse Shaping. Equalization. Line Coding. Echo Cancellation. Pseudonoise Sequences. Optimum Receivers. Forward Error Correction Coding. Spread Spectrum. Diversity Techniques. Digital Communication System Performance. Standards Setting Bodies. **Wireless**: Overview. Modulation Methods. Access Methods. Fading Channels. Statistical Distributions of the Fading Signal. Space-Time Processing. Location Strategies for Personal Communication Services. Analysis of IS-41 C Authentication Protocols for PCs. Cell Design Principles. Microcellular Radio Communications. Fixed and Dynamic Channel Assignment. Radio Location Techniques. Power Control. Enhancements in Second Generation Systems. Pan- European Cellular Standard. IS-54 North American Cellular Standard. British Cordless Telephone Standard. RACE Programs. Half-Rate Standards. Wireless Video Standards. Fixed and Dynamic Channel Assignment. Wireless LANS. Wireless Data. Wireless ATM. Wireless ATM 2. Third Generation Systems.